CONFLICTS OF INTEREST IN SCIENCE

How Corporate-Funded Academic Research Can Threaten Public Health

SHELDON KRIMSKY

FOREWORD BY NANCY OLIVIERI

HOT BOOKS

First paperback edition, 2022

Hot Books may be purchased in bulk at special discounts for sales promotion, corporate gifts, fund-raising, or educational purposes. Special editions can also be created to specifications. For details, contact the Special Sales Department, Skyhorse Publishing, 307 West 36th Street, 11th Floor, New York, NY 10018 or info@skyhorsepublishing.com.

Hot Books® and Skyhorse Publishing® are registered trademarks of Skyhorse Publishing, Inc.®, a Delaware corporation.

Visit our website at www.skyhorsepublishing.com.

10 9 8 7 6 5 4 3 2 1

Library of Congress Cataloging-in-Publication Data has been applied for.

Cover design by Kai Texel
Cover photo credit: Getty Images

Hardcover ISBN: 978-1-5107-3652-8
Paperback ISBN: 978-1-5107-6952-6
Ebook ISBN: 978-1-5107-3653-5

Printed in the United States of America

ACKNOWLEDGMENTS

My collaborators over thirty years have been an inspiration and provided special knowledge that made our research possible. I want to give special thanks to: the late James Ennis, sociologist at Tufts; Les Rothenberg, bioethicist, formerly at University of California at Los Angeles; to my collaborator Lisa Cosgrove, psychologist U. Mass Boston, who conceptualized the studies of conflicts of interest in the DSM and clinical guidelines for psychopharmacological drugs; Harold J. Bursztajn, psychiatrist, Harvard Medical School; Carey Gillam, journalist; and Tim Schwab, journalist, formerly with Food & Water Watch. My appreciation also goes to Anna Sangree, who helped edit and format the manuscript in its early stages.

CONTENTS

2007–2017

FOREWORD

BY NANCY OLIVIERI

Conflicts of Interest in Science: How Corporate-Funded Academic Research Can Threaten Public Health, by Professor Sheldon Krimsky, is a collection of twenty-two essays—some originally written for scholarly journals, others published as book chapters; the earliest published in 1985, the latest in 2017. It is both remarkable and remarkably depressing that Professor Krimsky's central theme—that the integrity of scientific research has been severely compromised by corporate funding—remains as relevant in 2021 as it was when he began to explore the issue thirty-five years ago.

"Conflicts of interest" provide the thread tying these essays together. Professor Krimsky has long demonstrated that Big Pharma's role in funding scientific research is pervasive. Because scientists cannot conduct their research without proper funding, and because governments have steadily retreated from supporting clinical research including drug research, scientists who want to develop and retain their careers find themselves obliged to study issues of interest to their pharmaceutical sponsors.

Here is where the problems start. The obligation of researchers is to pursue, honestly and rigorously, advancement of scientific knowledge. The commitment of Big Pharma is to maximize shareholder profit. This clash of fundamental values creates conflicts of interest. Pharma decides what problems will be explored: unsurprisingly, research which will produce costly "blockbuster drugs" and devices. With little or no profit to be had from non-pharmacologic approaches (think: diet, physical activity, reducing environmental toxins), scientists interested in evaluating these questions, but unwilling or unable to win corporate sponsorship, may find themselves frozen out of the (many and large) government grants that are now tied to corporate funding. Lose-lose.

In parallel, Pharma-funded researchers whose findings show their sponsors' products to be unsafe or ineffective have a big incentive to "disappear" the critical findings or to give them a positive spin: the "funding effect," upon which Professor Krimsky has expanded in many of these essays. In practice, this means that published research findings constitute only a small slice of actual research findings—and a highly unrepresentative slice, at that. Many negative findings simply do not see the light of day.

This massively pro-corporate bias in academic research and publishing, described and analyzed throughout this volume by Professor Krimsky, means that published biomedical research is no longer trustworthy. This matters. It matters a lot, because when clinicians go to "the literature" in an effort to discover which treatments are most appropriate for their patients, what they find is pseudo-evidence masquerading as scientific research.

My own experience as a clinical researcher at the University of Toronto amply confirms both Professor Krimsky's excoriating analysis and his dire predictions. The University of Toronto is Canada's largest and wealthiest research university. Despite this, or perhaps because of it, the University's recent history has been distinguished by a series of explosively publicized scandals. Conflicts of interest have been at the heart of every one.

At the historical juncture when Professor Krimsky begins writing on these issues, in 1985, I was a junior physician who, the previous year, had read the prescient analysis of the threats posed to research integrity by conflicts of interest by Dr. Arnold Relman, then-editor of the *New England Journal of Medicine*. Yet it was only later, through bitter personal experience, that I came to understand the morally salient role of the involvement of corporate funding in research—in my case, research potentially affecting hundreds of thousands of patients.

In 1989, I began clinical trials of an experimental drug to treat children with an inherited blood disease. My initially publicly funded trials were later supplemented by pharmaceutical company money, to facilitate the process of taking the drug to market. Later, when I raised safety concerns, the company CEO threatened me with "all legal remedies" if I were to publish my concerns or inform patients. After I ignored these threats and proceeded both to inform my patients and publish my concerns, he sued me for "defaming" "his" drug. Thus began a 25-year saga of litigation, smears, dismissals, and all manner of harassment: Canada's most publicized, certainly its most enduring, biomedical research scandal.[1]

Sheldon Krimsky has described the punitive treatment of researchers who dare to blow the whistle against corporate interests as "industry denigration." He has shown how such denigration is a by-product of the blurred boundaries between academic research and corporate profit-seeking. I can confirm that it isn't altogether pleasant to watch one's career implode and one's character be assassinated by the wealthiest players in the wealthiest sector of our economy. But while it's not entirely surprising to discover that corporations protect their bottom-line profitability with ruthless ferocity, it is the complicity of our public institutions—our leading research universities and hospitals—that comes as a shock, though it was predicted by Professor Krimsky in much of his early work.

My academic institutions provided little effective support in the struggle to protect research integrity and patient safety. Worse, over more than two decades, both institutions took harassing actions against my colleagues and me. Perhaps coincidentally, when the conflict commenced, the university was secretly in negotiation with the same Pharma CEO, for what would have been the largest donation in Canadian university history. The importance of corporate funding appeared to trump issues of both patient safety and research integrity. The earliest failures of accountability snowballed. The conflict was never resolved, and tragic outcomes later emerged for many patients,[2] and families.[3]

---

1 David Healy, "Repetition Compulsion to the Death or Beyond?" March 29, 2020. https://davidhealy.org/repetition-compulsion-to-the-death-or-beyond/.

2 Arthur Schafer, "Institutional conflict of interest: attempting to crack the deferiprone mystery," Journal of Medical Ethics, January 16, 2020. https://jme.bmj.com/content/early/2020/01/16/medethics-2019-105498.

3 Nancy Olivieri, "Decades after Motherisk wrongful conviction, we've cleared a woman but fixed nothing," Toronto Star, May 16, 2021. https://www.thestar.com/opinion/contributors

My experience was by no means unique. By the late 1990s, as the middle section of this essay collection was launched, researchers at other universities were enduring firings, harassment and lawsuits very similar to mine. In each case, the university was beholden to corporate funding. The stories of Dr. David Kern at Brown University,[4] Dr. Betty Dong at UCSF (*Chapter 12*),[5] and of Dr. Aubrey Blumsohn at the University of Sheffield,[6] and more recent exposures of the perils for clinicians who disclose adverse events in Pharma-funded drug trials,[7] make clear that, as Professor Krimsky has observed (*Chapter 16*), "for many institutions, some gifts may come at too high a cost."

The University of Toronto appears to have been a particularly slow learner in recognizing the moral dangers posed by corporate funding. A scant few years after its very public failures described above, its Faculty of Medicine became embroiled in another conflicts of interest controversy. That new scandal centered on the hiring, and subsequent firing, of Dr. David Healy, an internationally prominent psychiatrist. Healy showed himself to be "not a good fit" for the university's department of psychiatry by raising concerns about the safety of certain antidepressant drugs. The termination of his appointment was predictable: to read Krimsky (*Chapters 10, 13, 18, 19, 21, and others*) is to understand how the field of psychiatry, possibly more than any other in medicine, has been damagingly shaped by conflicts of interests.

/2021/05/16/haunted-by-injustices-decades-after-motherisk-wrongful-conviction-weve-cleared-a-woman-but-weve-fixed-nothing.html?rf.

4 "Right, Wrong, and Occupational Health: Lessons Learned," International Journal of Occupational and Environmental Health, Vol. 4:1, 33-34, 1998, https://www.tandfonline.com/doi/abs/10.1179/oeh.1998.4.1.33.

5 Science News Staff, "Quashed Study Sees Light of Day," Science, April 15, 1997. https://www.sciencemag.org/news/1997/04/quashed-study-sees-light-day.

6 Government Accountability Project, "Dr. Aubrey Blumsohn," May 8, 2020. https://whistleblower.org/whistleblower-profiles/dr-aubrey-blumsohn/.

7 Charles L. Bennett, Shamia Hoque, Nancy Olivieri, et al., "Consequences to patients, clinicians, and manufacturers when very serious adverse drug reactions are identified (1997–2019): A qualitative analysis from the Southern Network on Adverse Reactions (SONAR)," January 2021. https://www.sciencedirect.com/science/article/pii/S2589537020304375#.

Throughout this volume, Professor Krimsky also addresses the critical issue of disclosure of researchers' conflicts of interest, including his exposures of the conflicts of interest of individuals represented on speakers' bureaus and expert panels, and the widespread failures of medical journals to implement clear policies of disclosure. Many scholars have recommended disclosure as a way "to manage" researchers' conflicts of interest but, as Professor Krimsky highlights, merely acknowledging such conflicts has little beneficial effect. If we are serious about protecting the integrity of biomedical research as clarified here, such conflicts need to be eliminated, not merely acknowledged.

Professor Krimsky is a charter member of that small group of bioethicists, which includes Professors Carl Elliott and Arthur Schafer, who have rejected the reformist agenda of "managing" conflicts of interest, "adapting to them," or "learning to live with them." Professor Krimsky tells the story straight: that what academic research needs is not a "balanced" relationship, but a clean break from corporate funding; that "those who produce knowledge, and those with a financial interest in that knowledge, should be kept separate and distinct."

To demonstrate the inadequacy of disclosure, Professor Krimsky invites us to consider a judge with financial interests in for-profit prisons. Would you, would anyone charged with a crime, want to be tried and sentenced by a judge who stands to profit financially from a guilty verdict, even if he disclosed this interest? The question answers itself.

Amidst the cacophony of voices urging "balance" in the relationship between academia and Pharma, that of Professor Sheldon Krimsky is an almost lone and revolutionary voice. And as Orwell observed, "In a time of deceit, telling the truth is a revolutionary act." This volume of collected essays should serve as a bible for the revolution still to come.

NANCY OLIVIERI, MD, MA, FRCP(C), is a professor of pediatrics, medicine, and public health sciences at the University of Toronto, Canada, and senior scientist at Toronto General Hospital Research Institute (TGHRI) Cancer Clinical Research Unit (CCRU), Princess Margaret Cancer Centre. The author of over 200 scientific papers, reviews, and book chapters, she was elected to The American Society for Clinical Investigation in 1996.

INTRODUCTION

The three cornerstones of any democratic society are the right of every adult citizen to vote in free and open elections, an independent free press, and a scientific establishment dedicated to the pursuit of credible, verifiable knowledge unbiased by political, economic, or personal interests. Without these conditions in place a society cannot guarantee that its citizens will have access to reliable information that will enable them to make informed political, social, and medical decisions. This book addresses the third cornerstone, also referred to as "scientific integrity."

Threats against free science have appeared episodically for over 500 years from Papal authority, dictatorships, and unfettered corporate influence. The term "conflicts of interest," which the framers of the U.S. Constitution had recognized as a cause for concern among government elected officials, was not applied to scientists until the last quarter of the twentieth century. The reasons that professional societies, journals, and agencies of government began to acknowledge conflicts of interest in scientific research have been widely discussed in various books and journal articles.[1]

In 1979 the editors of *Nature*, one of the world's preeminent science journals, invited Nobel Laureate David Baltimore, then a professor of biology at MIT, and me to a several hour discussion on the new collaborations between industry and biology. They recorded our responses and wrote them up in a two-page spread in the journal.[2] Our discussion was prefaced by a series of questions: "As industrial corporations become more involved in developing new biological techniques, where does this leave the scientist? How will university biology departments maintain their integrity and autonomy? How will individual scientists react to growing corporate demands?" The ensuing discussion can be crystallized by the following interchange.

> **Krimsky:** Our academic institutions and academic scientists need to be free from corporate influence . . . through direct funding by corporations to departments and investigators, by scientists serving as consultants to or on advisory boards of industrial firms, or by academic scientists playing a double role by setting up their own small venture-capital firms . . . Just as war-related research compromised a generation of scientists, we must anticipate a demise in scientific integrity when corporate funds have an undue influence over scientific research.

> **Baltimore:** There is nothing wrong with universities serving as a source of technical expertise for newly developing industries or even for mature industries. And that kind of symbiosis between industry and academic departments has been very good for both over history—take the chemical industry. The positive side of industrial involvement in the academic world should not be discarded out of fear that university-based scientists will lose their autonomy owing to industrial pressure. We need to maintain a balance between industrial influences and influences from outside the corporate world.

The papers contained in this volume are the result of three decades of research addressing the questions posed in this dialogue. What impact would the increased commercialization of basic, applied, and medical science have on the integrity of the scientific community and the trust in the knowledge they produce?

From my undergraduate and graduate education in physics and my doctoral training in the philosophy of science, I have always been interested in the methodology of the sciences and in the production of knowledge. The methods deployed by scientists are primarily an internal affair, because science is largely a meritocracy. Achievement determines one's status in the scientific community. The best scientists float to the top of the pecking order by virtue of their contributions to the discipline. They define and are the gatekeepers of the methods embraced by their discipline. The decision about what is good science is not arrived at by a popular vote. For example, the National Research Council, of the National Academies of Sciences, set the standards for risk assessment that have been adopted by the Environmental Protection Agency and the Food and Drug Administration.[3]

Leaders of various scientific disciplines are invited to serve on prestigious panels such as those created by the National Academies. These panels set the standards for their discipline and help inform policy recommendations for government bodies. A recent report by one of these panels criticized the methods used in animal studies, popular in the European Union, to test the safety of genetically modified crops. They concluded that statistical significance was not sufficient to show an effect. It also had to have biological significance and had to be accompanied by a validated mechanism of action.[4] This recently-established standard disqualifies many animal studies that have shown that genetically engineered crops may be harmful.

Scientific institutions are a different matter. They interact with society and adopt rules that are accountable to public bodies, such as in the responsible conduct of research. Scientists must conform to federal laws of informed consent in clinical trials or in maintaining a safe workplace.

When students are introduced to science in college, it is presented as an objective body of truths. The textbooks contain the canons of the discipline. At that stage, the focus is on the laws, principles, and theories or generalized empirical findings—sometimes referred to as "facts." No attention is given to how knowledge is produced. There is scarcely room in the curriculum for the study of contested ideas, alternative explanations, and bias in research. Other courses of study in the social studies of science—sometimes referred to as Science, Technology Studies (STS), address contested knowledge and competing explanations.

Besides being a meritocracy, science is also a conservative institution. New ideas do not replace established ways of thinking easily. A single paper rarely changes the scientific consensus. As an example, stomach ulcers, also known as peptic ulcers, were once believed to be caused by stress and treated as a chronic disease. Physicians prescribed antacids or bismuth salts not as a cure, but merely to address the symptoms. It took over a decade for Barry J. Marshall, an Australian physician, to convince the medical community that bacteria can survive in the gastric juices of the stomach and that peptic ulcers were caused by the colonization of *Helicobacter pylori* and could be cured. He infected himself, contracted ulcers, and then cured himself with antibiotics to gain the attention of the medical community.[5]

There are also ample cases where corporate funding of science is designed to dispute reputable toxicology studies of substances such as lead, asbestos and tobacco. In one notable case, authors describe how the tobacco industry influenced the science and the scientists in Germany by distorting or concealing associations between smoking and disease.[6] An example of a distinguished researcher who has been the target of industry denigration for his findings on a chemical's toxicity is Frederick vom Saal, an endocrinologist at the University of Missouri. Vom Saal has been studying bisphenol A (a chemical widely used in plastics) for over twenty years. From his many animal studies, he has concluded that human exposure to the substance is dangerous. The chemical industry has continued to dispute vom Saal's findings and fund scientists to show that the substance is safe in its current uses.

The deeper one delves into the scientific literature the more one becomes aware that science is driven by contested viewpoints until a consensus is reached on some area of a field, and the number of naysayers begins to diminish. The distinguished late Columbia University sociologist Robert Merton referred to one of the central norms of scientific inquiry as "organized skepticism." Before a scientific paper is approved for publication, it must cross the threshold of acceptance by skeptical reviewers, who will critically examine its methods, data and the conclusions drawn from the data.

Even under the best peer review process reviewers can always overlook something. And, of course, no one watches the scientist at the laboratory bench collecting data. But a thorough and well-written scientific paper has detailed information that gives another scientist the opportunity to replicate the study. Reproducibility of results is a critical measure of sound science and essential for building the trust of other researchers. Replication is not the same as reproducibility. The former means that an independent group, trained in the discipline, can repeat the experiment; the latter means the results will come out the same. Many experiments are not replicated because there is no funding. However, when the results of an experiment are controversial, e.g., affects the bottom line of a corporation, a corporate stakeholder will replicate it to demonstrate that it is not reproducible. In one study, *Science Magazine* reported that rigorous replication efforts only succeeded in confirming the results of two of five cancer papers.[7]

Disagreements among scientists are hardly a new phenomenon.[8] In fact controversy is endemic to science. As far back as the Greeks, science was replete with competing explanations of the natural world. We refer to them as schools of thought. While the Greek scientists made observations of the world to reach their conclusions, they rarely did experiments. The period, beginning with the end of the Renaissance and the rise of the intellectual and social movement known as the Enlightenment, is cited by historians as the birth of the modern scientific revolution. From science as natural philosophy we witness the growth of new science as the merger of mathematics with experimental inquiry. Debates among scientists flourished.

Before scientists understood the role of oxygen in combustion, the Phlogiston theory prevailed. It was believed that a substance called phlogiston was contained within all combustible substances. When the substance was burned it released the phlogiston (the smoke and particles). First formulated by Johann Joachim Becker in 1667 and advanced by Georg Ernst Stahl in 1703, Phlogiston remained the dominant scientific theory until the 1780s, when experiments by Antoine-Laurent de Lavoisier discovered the role of oxygen in combustion.

The modern germ theory of disease was preceded by the Miasma theory, which held that diseases like cholera were caused by "bad air" (miasma) from contaminated substances such as rotting organic matter. This view of "bad air" can be traced back to antiquity continuing through most of the nineteenth century when it was replaced by the germ theory of disease upon the discovery of microorganisms.

As one sociologist of science notes, "Controversies are an integral part of the collective production of knowledge, disagreement of concept, methods, interpretations and applications are the very lifeblood of science and one of the most productive factors in scientific development." When two competing theories are proposed to explain an event, some clever investigators will devise an experiment to decide among the explanations. Competition within science and the ethos of skepticism helps to weed out false hypotheses.

To the general public, modern scientific controversy appears to be a matter of technical disagreements among competing schools of thought. What is often hidden from public view are the social, economic, and political factors that distinguish different outcomes from research findings. Historian of science Garland Allen wrote:

> [Scientific] controversies are often the focal point for social conflicts that are coming to a head for their own historical reasons. In such contests, scientific experts are used to buttress arguments within a controversy, giving legitimacy to one or the other side, helping to win public support for one special interest group or another, or in some cases camouflaging the real reason for the dispute, which are economic or sociopolitical in nature.[9]

In 1980 the U.S. Congress passed the Bayh–Dole Act (PL 96–517). Its goal was to create partnerships between business entities and universities. Global competitiveness was the Reagan Administration doctrine that prompted Congress to pass a series of laws and the Executive branch to issue regulations that were designed to foster technology transfer. It was argued that in the United States too few basic discoveries were turned into applied technology contributing to economic growth.

One of the key parts of the Bayh–Dole Act was that it gave universities, small business, and nonprofit institutions title to any invention that was discovered with federal research funds. Seven years later it was extended to industry as a whole, allowing large corporations to benefit from publicly-funded research.

The now defunct Office of Technology Assessment (OTA), a congressional body that published non-partisan evaluations of new technologies and legislation affecting science and technology, commented on Bayh–Dole and ancillary policies:

> The confluence of events and policies increased the interest of universities, industry and government in activities pertaining to partnerships between academia and business in all fields of science . . . .It is possible that the university-industry relationships could adversely affect the academic environment of universities by inhibiting free exchange of scientific information, undermining interdepartmental cooperation, creating conflict among peers or delaying or impeding publication of research results.[10]

While many of these outcomes have been documented during the 1980s, the one area omitted from OTA's list was the exponential rise in conflicts of interest among scientific and medical researchers. The reason why scientific and medical conflicts of interest are so important is that they are associated with bias in research methods, outcomes, or the interpretation of data. Studies have shown that corporate-funded research is more likely, compared to research funded by the government or independent nonprofit organizations, to yield results that are consistent with the financial interests of the sponsor. I have termed this the "funding effect" of science, also known as "sponsorship bias."[11]

At no time in U.S. history has the integrity of science become so important. Contested scientific beliefs have confused the public. Disagreements about the safety of chemicals, the effectiveness of medical therapies, the side effects of drugs, and the significance of anthropogenic causes of climate change permeate the media. The extent to which scientific studies and opinions are paid for by private interest groups is the basis of conflict of interest ethics. Transparency of the source of funding and any corporate involvement in a scientific publication has become one of the hallmarks of scientific integrity.

The tradition of preventing or managing conflicts of interest has a long history. Certain conflicts of interest of elected officials of government were prohibited by the U.S. Constitution. For example, the president is prohibited from receiving any addition to his salary from any other domestic source, or from accepting any gifts or salary (emoluments) from any other government. As stated in the Constitution: "No person holding any office of profit or trust under them, shall, without the consent of the Congress, accept of any present, emolument, office, or title, of any kind whatever, from any king, prince, or foreign state."

Government employees were covered by conflicts of interest in a series of laws, such as the federal criminal conflict of interest statutes found in the U.S. Code of Federal Regulations (U.S.C), Sec. 18, §§ 203, 205, 207, 208, 209. These laws going back to 1853 prohibit federal employees from representing private interests before the government and from seeking action from the government on behalf of private interests whether paid or unpaid. The 1962 amendments of 18 U.S.C. restricted post-employment activities of government employees, the Ethics in Government Act of 1978 restricts former senior government employees from contact with former agencies of past federal employment, and the Ethics Reform Act of 1989 amended the post-employment restrictions on the executive and legislative branches.

In late January 2018, after her recent appointment as director of the Centers for Disease Control (CDC), Dr. Brenda Fitzgerald was forced to resign because of her "complex financial interests that had imposed a broad recusal limiting her ability to complete all her duties as the CDC director."[12] Among them were investments in tobacco stock, which she traded even after taking the position at the public health agency.

While the federal government had long established conflict of interest requirements for its employees, the president, and for members of Congress, scientists and experts serving on federal advisory committees were not covered under those regulations prior to 1972. In that year the Federal Advisory Committee Act (FACA) was passed. Under FACA, scientists invited to serve on federal advisory committees are prohibited from holding substantial conflicts of interest on matters before the committee on which they serve. This applied to all federal agencies, including the Food and Drug Administration, the National Institutes of Health, and the Centers for Disease Control.

Up until the mid-1980s, there was little discussion about conflicts of interest within the scientific community. While scientists received funding, honoraria, equity and other forms of compensation from private corporations, the general consensus was that their professional ethics of pursuing the truth prevented them from being biased, regardless of their financial interests. Slowly, the premise behind this idea—that scientists were beyond letting financial interests affect their judgment—began to erode.

The medical sciences were the first discipline to identify financial conflict of interest disclosure as an integral part of publication. The *New England Journal of Medicine* introduced a policy in 1985 requiring transparency of financial interests for published articles. *NEJM* editor Arnold Relman wrote: "Lately, however, a new entrepreneurial fever has begun to affect the profession, and what was formerly on the fringe seems to be moving into the mainstream."[13]

The University of New Hampshire held an international symposium on Universities in the Twenty-First Century in 1985.[14] At that meeting I discussed the four personalities of the modern American university. I characterized them as distinct archetypes: classical, industrial, national defense, and public interest. In the classical form where "knowledge is virtue," university research is not designed to meet any economic or political purposes; rather it is focused on knowledge as an intrinsic value. In the industrial type, where "knowledge is productivity," the primary role of the university is to provide the personnel and intellectual resources for industrial development. For the third type, where "knowledge is security," universities provide the knowledge for developing military technologies for defense. Finally, the fourth type, where "knowledge is public interest," the function of universities is to produce the knowledge necessary to diagnose, treat, and cure disease, to protect the environment, and to ameliorate social pathology.

At the symposium, I concluded that "the classical image of the university as a sanctuary for the production of knowledge, set apart from the forces of the marketplace and the power brokers of political culture is a fading memory of a past era." But the dilemma with the new personality of the industrial university is that academic scientists who serve industry cannot at the same time serve society as disinterested experts. "One of the vital roles of universities—to provide independent expertise to government uncontaminated by the motive of financial association—is thwarted when industry and academia become indistinguishable."[15] Beginning in 1980, the university's industrial personality, where "knowledge is productivity," began to gain traction throughout American research institutions. With it came the rise of conflicts of interest in American science and medicine. When corporations began investing in academic science they expected a payback. The superposition of two value systems, namely, knowledge for its own sake versus knowledge to advance the bottom line, resulted in cases of corporate malfeasance leading to distortion of the scientific record with eventual harm to the public. In Chapter 2 "Tales of the Unholy Alliance" of my 2003 book, *Science in the Private Interest,* four cases are highlighted: Harvard ophthalmologist makes money off of his company's unproven drug; Scientist signs a research agreement with a drug company and loses control over the science; Government scientist oversees drug study while consulting for drug companies; Dangerous vaccine approved in process replete with conflicts of interest. Three other books provide additional examples of how public health has been damaged by conflicts of interest: *Doubt Is Their Product* by David Michaels; *Lead Wars* by Gerald Markowitz and David Rosner; and *Bending Science* by Thomas O. McGarity and Wendy E. Wagner.

This book outlines the development of conflicts of interest research as emerging from three distinct periods: early, middle, and late.

EARLY PERIOD: 1985–1995

U.S. science policy began largely after World War II. It is generally acknowledged that science was a critical factor in winning the war in Europe and Asia. This included electronic computers used to break the Nazi secret code, radar, and the discovery of atomic energy leading to the development of the atomic bomb. After the war there were debates about how to organize science and how to fund it. The principal challenge was to establish "an independent self-governing normative structure while at the same time becoming more fully integrated into the practical and economic life of the nation" (see Chapter 3).[16] The idea of multi-vested science grew first out of academic consultancies for private companies, then out of corporate funding of science and finally from academic-corporate partnerships for commercializing discoveries.

After passage of the Bayh–Dole Act molecular biologists were in high demand as they were the intellectual resource of these new venture capital companies seeking to exploit the commercial possibilities for gene splicing in the pharmaceutical and agricultural sectors.[17] I teamed up with a sociologist who specialized in network analysis and a recent Harvard graduate who worked for the *Multinational Monitor.* We were interested in understanding the pace of commercialization of university faculty as well as the growth of new biotechnology companies. The numbers spoke for themselves. Thirty-one percent of the biology and genetics faculty at MIT had commercial affiliations with the nascent biotechnology industry. Stanford and Harvard were at 19 percent. And it was learned from other studies that dual-affiliated faculty were four times as likely to report trade secrets resulting from their research (see Chapter 2). The implications of the data and the abuses in science were discussed in "The corporate capture of academic science and its social costs" (see Chapter 1).

MIDDLE PERIOD: 1996–2006

Medical and scientific refereed publications are not only the vital communication networks among scientists throughout the world and set the standards for what is publishable research, they also are primary sources for the science and medical writers who communicate results to the public. In 1996 I began to turn my attention to conflicts of interest in these publications. If financial conflicts of interest (FCOI) can bias outcomes in research publications, then how often do they occur, what is the extent of FCOI biases, and how much is disclosed? My colleagues and I examined about 1,100 university authors from Massachusetts institutions whose nearly 800 articles were published during 1992 in 14 scientific and medical journals. We learned that one-third of the articles had at least one lead author who had a financial conflict of interest. Yet, there were no disclosure statements describing these interests in the 267 articles. At the time we published our study very few of the journals we surveyed had FCOI policies. The journal *Science* introduced its FCOI disclosure policy in 1992 (see Chapter 4), whereas the federal government first established an FCOI policy for award recipients in 1995.

How many scientific and medical journals were beginning to adopt such policies? What were their editorial practices? Those were the questions I posed with my collaborator L. S. Rothenberg in Chapter 8. This was a bear of a study. We wanted to get a good representation of biomedical and science journals worldwide that were publishing in 1997. Estimates from various databases indicated the number of journals worldwide could be as high as 30,000. We used a measure called the journal impact factor to get a list of the highest impact journals. The journal impact factor is determined by how many times the articles published in the journal are cited in other articles. High impact factors are usually associated with prestige. We chose the top thousand journals in two categories related to citations: times cited and impact factor. After removing the overlaps, we were left with about 1,400 distinct high impact science and biomedical journals. Many of these journals were not accessible online so I had to spend considerable time at the Harvard Medical Library (Countway), which had the less popular journals in their stacks.

Of the nearly 1,400 hundred journals we examined, only 16 percent had a published policy on financial conflicts of interest. More remarkably, of the peer reviewed journals on our list that had a published policy during 1997, amounting to 327 journals, we examined over 61,000 articles and found 0.5 percent included a disclosure of financial conflicts of interest. We surveyed journal editors of these publications and we concluded: "Low rates of personal financial disclosures are either a result of low rates of financial interest (nothing to disclose) or poor compliance among authors to the journals' COI policies." Other studies suggested that the latter explanation was more probable. The *New York Times* article written about our study was headlined: "Scientists Often Mum about Ties to Industry" (April 25, 2001).

We used the data from these and other studies to support claims we made in an editorial in the *Journal of the American Medical Association (JAMA)*, which implored all journals to adopt FCOI policies. "While financial interest in itself does not imply any bias in the results of a paper and should not disqualify it from publication, readers and reviewers are the best judges of whether there is evidence of bias and whether that evidence favors those interests," (see Chapter 5). In a second *JAMA* editorial, I used the opportunity to reflect on the hidden industry-sponsored studies, which impact evaluations of the cost-effectiveness of drugs, especially high profile, expensive oncology drugs (see Chapter 7). One study found that industry-funded cost-effectiveness studies were eight times less likely to reach unfavorable qualitative conclusions than studies not funded by industry. Because such studies are important in establishing drug formularies, transparency of funding, openness of assumptions, and elimination of potential biases are critical in gaining the public trust.

The decade of the 1980s brought some major changes in intellectual property law and regulations. The Supreme Court, in a 5–4 decision, approved the patenting of a microorganism sui generis, independent of the process for which it was used. These changes brought new commercial opportunities for university discoveries, especially in patenting genes and GMOs. Multi-million-dollar industry contracts in academia became more common. These new intellectual property opportunities contributed to closing the "knowledge commons," the open and free exchange of ideas and the shared fruits of knowledge, in favor of secrecy. This came at the expense of communalism, and disinterestedness—two of the Mertonian norms of science. Chapter 6, written for the *Chicago Kent Law Review*, examines how intellectual property fostered the growth of researcher financial interests in biomedical research. "The commercialization of scientific research has compromised the traditional Mertonian norms on the dubious assumption that the appropriation of knowledge as intellectual property, and in its wake the erosion of communalism and disinterestedness, will yield a great public good in the long run."

Where did these norms of science come from? And will their erosion impact the objectivity of knowledge. That was the topic for the essay published in *Society* titled "Autonomy, Disinterest, and Entrepreneurial Science" (see Chapter 9). The distinguished British physicist John Ziman wrote that science can still be objective even if not disinterested, but will lose public trust. I introduce the "funding effect"—that corporate-funded research affects the outcome—to argue that financial conflicts of interest will compromise objectivity.

One day I received a call from a clinical psychologist Dr. Lisa Cosgrove, a professor at the University of Massachusetts, Boston, who studied the *Diagnostic and Statistical Manual of Mental Disorders,* more commonly known as the DSM. Published by the American Psychiatric Association

(APA), the DSM offers standardized criteria for the classification of mental disorders. It is widely used by clinicians, researchers, psychiatric drug regulation agencies, health insurance companies, pharmaceutical companies, and the legal system when therapies and reimbursements are contested. Expert panels are established by the APA to set criteria for the mental disorders listed in the DSM. Dr. Cosgrove wanted to study whether the members on these expert panels had financial ties to the pharmaceutical industry, which provided drug treatments for some of the disorders, like depression and anxiety. We formed a collaboration to answer the question, which resulted in our article featured in Chapter 10. There were 170 DSM-IV (fourth edition) panel members of which 56 percent had one or more financial ties to a company in the pharmaceutical industry. Not unexpectedly, the panels with the largest number of FCOIs had developed mental disorder criteria that were treated with pharmaceuticals. At the time there were no disclosures for the FCOIs of panel members in DSM volumes.

Cosgrove et al. and I wrote: "Transparency is especially important when there are multiple and continuous financial relationships between panel members and the pharma pharmaceutical industry, because of the greater likelihood that the drug industry may be exerting an undue influence on the DSM." The publication went viral. We stopped counting after 30 media outlets had covered the topic. Some examples: ***Chicago Tribune*** (April 20, 2006) Top mental health guide questioned; ***Washington Post*** (April 4, 2006) Writers of psychiatry "bible" have ties to drug companies; ***The New York Times*** (April 20, 2006) Study finds a link of drug makers to psychiatrists; ***Reuters*** (April 21, 2006) Mental illness writers had industry ties: Study; ***Newsday*** (April 21, 2006) Drugmakers, doctors linked; ***Bloomberg News*** (April 20, 2006) Psychiatric manual's authors have financial ties to drug makers; ***The Star Ledger*** (April 20, 2006) Study links money with drug experts. Among the foreign press were: ***Le Figaro*** (April 24, 2007) *Psychiatrie: des experts trop lié à l'industrie* and ***Panorama*** (November 5, 2006) *Manuale molto interessato.*

Research into scientific conflicts of interest had exploded in the 1990s. In 1974 there were no citations in Medline, the U.S. National Library of Medicine bibliographic database that contains more than 24 million references to journal articles in life sciences, on scientific conflicts of interest. In 2005, according to statistics from the former editor-in-chief of *JAMA*, it rose to 600 citations. And in 2017, for one year there were 750 Medline citations with "conflict of interest in the abstract" and 120 citations in the title. But what were the ethical foundations for requiring academic scientists to disclose their personal financial interests to journal editors? I felt I had to delve more deeply into the issue. That was the purpose of my paper featured in Chapter 11, "The Ethical and Legal Foundations of Scientific 'Conflict of Interest.'" After slogging through a number of ethical theories and literature on government ethics, legal ethics, and ethics in science I reached the conclusion that "integrity of academic science" was an important value and that "the ethical foundations needed for protecting the integrity of science demand measures that go beyond the mere disclosure of interests." One of the principles I arrived at was the separation of roles: "those who produce knowledge in academia and those stakeholders who have a financial interest in that knowledge should be kept separate and distinct." Under this principle, disclosure of interests is not the antidote to bias; rather prevention of FCOIs at the outset is the better approach.

LATER PERIOD: 2007–2017

While the DSM panel members had changed somewhat for the new issue scheduled to be released in 2013, there were already published reports that the FCOIs would not be diminished.[18] The fact that disclosure of conflicts of interest did not seem to be abating the conflicts suggested that transparency and potential bias were operating on different tracks. This led to a commentary co-authored with my principal collaborator on the DSM studies titled "A Comparison of DSM-IV and DSM-5 Panel Members' Financial Associations with Industry: A Pernicious Problem Persists" (see Chapter 18), published in the high visibility open access journal *PLOS Medicine*. We wrote quite assertively, "As an eventual gold standard and because of their actual and perceived influence, all DSM task force members [as the panel was known] should be free of FCOIs."

The next issue of the DSM (DSM-5) had financial interest disclosures of the panel members. In testing the hypothesis that disclosure of interests might reduce the FCOIs, we learned after comparing DSM-IV and DSM-5, and subsequently reported in the *New England Journal of Medicine*, that there were more FCOIs in the manual, which included such disclosures. Thus our hypothesis was proven false.[19] There were still more questions to ask about the new DSM.

Clinical trials are essential to obtaining the approval from the Food and Drug Administration for marketing a drug or medical device. The drug and device industry funds about six times more clinical trials than the federal government. And the financial stakes are very high. In 2008 industry funded 50 percent of the clinical trials listed in the public database clinicaltrials.gov and in 2014 about 36 percent.[20] It is reasonable to ask: What effect do conflicts of interest have on the outcome of clinical trials? Chapter 19 discusses a research study that examined 13 clinical trials for drugs designed for new DSM designated disorders such as "bereavement related depression" or "binge eating disorder." Since there were new DSM disorders, we questioned whether there were financial conflicts of interest between DSM-5 panel members who established these new disorder categories and the drug companies who manufactured the drugs for the clinical trials. "In all but 1 trial, FCOIs were found between DSM-5 panel members and the pharmaceutical companies that manufactured the drugs that were being tested for new DSM disorders." The paper addresses how these FCOIs could be minimized or eliminated emphasizing the importance of reducing multi-vested science.

Since faculty in research universities are encouraged, if not required, to obtain grants or contracts to fund their research, when they do receive a grant it is awarded to their university on their behalf. The conditions of the grant or contract must meet the university's standards, which are not uniform across academia. Some universities will not accept awards from certain organizations or will not accept awards that contain certain requirements or obligations. As an example, certain institutions prohibit faculty from soliciting awards from tobacco companies on ethical grounds. In other cases, universities will reject awards for a study that give the sponsor authority over the final publication of the results. The university may decline the award even if it is agreeable to the faculty member. Since faculty are generally given authority to select the research design and pursue funding to support it, there can be conflict between the individual investigator and the university over funding sponsorship. Chapter 12, "When Sponsored Research Fails the Admissions Test" explores the conflict between faculty autonomy and university values. Suppose a faculty member from a Jesuit college wishes to do research on safer abortion practices, which are inconsistent with the values of the institution. Or suppose a university has a prohibition against weapons research and a faculty member in physics seeks funding under the Strategic Defense Initiative called the "Star Wars" program. This paper proposes an ethical system to address these issues.

The medical community has had the lions' share of attention on conflicts of interest from Congress and the general public on financial conflicts of interest. It is understandable considering the importance of the medical sector in people's daily lives and the role played by the pharmaceutical companies. The public sensitivity to conflicts of interest in medicine is expressed through Congress' enactment of the Affordable Care Act in 2010 with a subsection 6002 titled the Physician Payments Sunshine Act, which requires medical product manufacturers to disclose to the Centers for Medicare and Medicaid Services (CMS) any payments (gifts, honoraria, travel) or other transfers of value made to physicians or teaching hospitals. Records of these payments for each physician are eventually placed on a publicly accessible database.

According to the Institute of Medicine, Clinical Practice Guidelines (CPG) are systematically developed statements to assist both practitioner and patient decisions about appropriate health care for specific clinical circumstances. A panel of specialists is convened by a government agency or a professional medical association to produce the CPG for a particular illness. The importance of CPGs as a guide to physicians who want to be sure they are using the standard of care cannot be overstated. A study of financial conflicts of interest in CPGs was a natural extension to the kind of work I and my collaborators—especially Lisa Cosgrove—had already accomplished. Chapters 13 and 21 cover this topic in depth.

Our team asked the question: If we looked at the Clinical Practice Guidelines (CPGs) issued by the American Psychiatric Association for DSM-5 disorders, what would the FCOIs of their authors look like? The APA issued 11 CPGs between 1998 and 2007 that correspond to specific disorders identified in the DSM. We chose three major CPGs (schizophrenia; bipolar; and major depressive disorder, MDD) and examined the relationships that their authors had with the pharmaceutical industry in a paper titled "Conflicts of Interest and Disclosure in the American Psychiatric Associations' Clinical Practice Guidelines" (see Chapter 13). There were twenty CPG authors for the three practice guidelines. We found that eighteen (90 percent) had at least one financial relationship with the pharmaceutical industry. None of the financial associations of the authors were disclosed in the CPGs. The APA had not yet reached the ethical standards of transparency for medical guidelines it issued that were subject to influence by the pharmaceutical industry. We proposed that: "In light of the increasing evidence that financial associations between biomedical researchers and the pharmaceutical industry may result in the publication of imbalanced—and sometimes inaccurate—results and recommendations . . . we recommend that the APA institute a more rigorous COI policy for the practice guidelines that the Association publishes and endorses."

The revision of the DSM in 2013 introduced new or expanded diagnostic categories, like "bereavement-related depression" or "binge eating disorder" about which there was some debate. The pharmaceutical industry had a financial stake in any new or expanded disease category for which there was an existing drug or a drug that was being investigated in clinical trials. Our research team asked the question: What, if any, were the FCOIs in the three-part relationship among DSM panel members, principal investigators (PIs) of clinical trials for new DSM-5 illness categories, and drug companies? There were no other studies that investigated this three-part relationship. Imagine a scenario where a DSM panel member was a consultant to a pharmaceutical company (one that manufactured a drug for the treatment of the panel's diagnosed disorder), while also serving as a principal investigator in the clinical trial testing the safety and efficacy of the drug. As will be seen in Chapter 19, we investigated thirteen clinical trials that were testing drugs for new DSM disorders. In twelve out of thirteen, FCOIs were found between DSM-5 panel members and the pharmaceutical companies that manufactured the drugs that were being tested for the new DSM disorders. Perhaps more surprisingly, in three out of thirteen (23 percent) of the trials the PIs were also DSM panel members. We concluded that this "raises questions about the potential of such multi-vested interests for implicit bias when making decisions about inclusion of new DSM disorders and their respective treatments."

The Institute of Medicine (IOM), now called the National Academy of Medicine, a part of the National Academies of Sciences, Engineering and Medicine (NASEM) issued a statement in 2009: "Financial ties between medicine and industry may create conflicts of interest. Such conflicts present the risk of undue influence on professional judgments and thereby may jeopardize the integrity of scientific investigations, the objectivity of medical education, the quality of patient care, and the public's trust in medicine."[21] Two years later the IOM recommended that FCOI policies be instituted for medical guideline development groups, that chairs and co-chairs should be free of commercial ties, and that ideally all participants in writing medical guidelines should be free of financial conflicts of interest.[22]

This prompted our research group to ask: Are the guideline groups following the recommendations of the Institute of Medicine? We decided to survey clinical guidelines (CPGs) issued by a variety of professional, governmental, nonprofit and commercial entities intended to treat depression. We found fourteen studies that fit such criteria. Chapter 21, published in *Accountability in Research*, reports the results of our study. Eight of the fourteen guidelines (57 percent) had at least one panel member who had a financial tie to at least one company that manufactures antidepressant medication. The most prevalent FCOI was being a member of a drug company's speaker's bureau. Of the 11 CPGs with identified chairs, 6 of those chairs (55 percent) had a clearly identifiable FCOI. The guidelines we studied were published from 2009 to 2016, some before the IOM recommendations were issued. There is some hope that the guideline developers will catch up with the IOM's recommendations and the future will see less conflicted and therefore more trustworthy CPGs.

I coined the term "funding effect" in science to describe the phenomena that has now been observed and replicated in a number of studies. It signifies that studies funded by for-profit companies compared to similar studies funded by government or nonprofit organizations tend to yield results that favor the financial interests of the for-profit sponsor. Writing for the *Stanford Law and Policy Review* (see Chapter 15) I connected the dots between the "funding effect," "conflicts of interest," "transparency," "ethics in research," "bias," and "objectivity" and "gifts to physicians." To highlight the limits of transparency I used a hypothetical and a statement by Judge I. D. Clair, who is quoted: "I will be sentencing the defendant, who has now been tried by his peers, to be incarcerated in a for-profit prison in which I have an equity interest. The extra money I earn from this partnership between my court and a reputable penal institution helps to compensate my low salary and allows me to serve the public interest and render more thoughtful and objective decisions." Why such a scenario would not be acceptable to the court but would to medical scientists who publish papers, establish disease criteria, and issue clinical practice guidelines is problematic.

While the scientific societies and journals were beginning to take financial conflicts of interest seriously, Congress held oversight hearings and made recommendations to federal funding agencies to adopt guidelines for the recipients of grants. The first federal guidelines were issued in 1995 (revised in 2011), by the Public Health Service under the Department of Health & Human Services, which includes the National Institutes of Health, the largest funder of medical research, and the National Science Foundation.[23] The federal FCOI guidelines were administered by the universities who were the recipients (along with their faculty) of federal grants. The goal of the federal guidelines was to promote objectivity in research by ensuring that recipient institutions managed FCOIs appropriately, collected information about faculty involvement in paid extra-university activities, and made such information available to the funding agencies on request. "This was a decentralized, locally managed system for addressing scientific COIs for investigators who received federal funds with federal walk-in rights to obtain information." The historical development of federal FCOI guidelines is discussed in Chapter 14. The role of Congress and the Inspector General in bringing the 1995 FCOI rules under the spotlight, highlighting gaps and limitations, and proposing revisions is analyzed in this paper published in the journal *Ethics in Biology, Engineering and Medicine.*

Every now and then a case arises that brings unusual clarity to a field of inquiry. When I learned that an external donor to a major university included in the gift's provisions requirements for the selection of faculty based on the funders' ideology, I felt compelled to write a commentary for *Nature*. It was titled "Beware of gifts That Come at Too Great a Cost" (see Chapter 16). This issue was a departure from the body of work on conflicts of interest among researchers. Here the conflict of interest was with the institution—namely a major research university. The federal guidelines on FCOIs mentioned but did not address the topic of institutional conflict of interest. A university might have an equity interest in a company, while its faculty is asked to do studies of the products the company manufactures.

In this case the donor was a charitable foundation with the namesake of the Koch brothers known for their conservative economic and political beliefs. The gift was for funding faculty positions in the economic department of Florida State University that had to meet specific requirements of the foundation, superseding the tradition of faculty governance and departmental autonomy in selecting colleagues. I wrote: "This agreement is a marked departure from the well-established separation between private academic philanthropy and faculty hiring decisions." In this case the continued oversight by the foundation for achieving its choice of economics to be taught subsumes the choice of curricula and economic theory to an external funder. "Compromising these [traditional] values, even under conditions of financial exigency, will turn a university against itself and corrupt its integral value to society."

A number of people have argued that FCOI policies are burdensome to faculty and do not reduce bias in research, or at least there is no evidence that it does. While there has been evidence for the "funding effect," that doesn't in itself support an outcome of bias. Industry drug studies might result in favorable outcomes more frequently than non-industry studies because of internal pre-testing that is done before clinical trials. Less effective drugs are eliminated before the published trials. In Chapter 17 I tackle the problem head on: "Do Financial Conflicts of Interest Bias Research?" I navigate the reader through the evidence for the "funding effect," explain what bias means, and reach a conclusion that even surprised me.

The National Academies has, over the years, issued some strong recommendations for protecting disinterested science and medical guidelines by trustworthy experts who at the very least disclose their financial interests to allow readers to make a judgment about the potential for bias. The Academies also had FCOI policies for their own publications. So much of the attention was directed at medical science, that far fewer studies of conflict of interest were addressed to other fields of science.

In 2016 I received an unpublished paper from analyst Tim Schwab of the Food and Water Watch—a national nonprofit environmental advocacy organization. The paper made claims about the FCOIs in a recent NASEM study of genetically engineered (GE) crops.[24] I collaborated with Mr. Schwab to investigate the NASEM report. We used criteria for FCOIs from the published literature and asked the question: Did any of the panel members who wrote the report have financial interests with the biotechnology industry whose products they were examining on issues of safety and quality? The NASEM report on GE crops, which listed twenty panel members, wrote in the report that there were no financial conflicts of interest to disclose.

Our inquiry, which was accepted for publication in the journal *PLOS ONE* (see Chapter 20) revealed another story. "Our results showed that six panel members had one or more reportable financial COIs, none of which were disclosed in the report. We also report on institutional COIs held by the NASEM related to the report."

NASEM responded to our study in a news release: "The National Academies of Sciences, Engineering, and Medicine have a stringent, well-defined, and transparent conflict-of-interest policy, with which all members of this study committee complied. It is unfair and disingenuous for the authors of the *PLOS* article to apply their own perception of conflict of interest to our committee in place of our tested and trusted conflict-of-interest policies." Our study was carried in *Le Monde* and the *Chronicle of Higher Education*. In the *Chronicle* story NASEM's response to the *PLOS ONE* study was reported:

> The National Academies initially responded to the criticism by defending the rigor of the process by which they produce some 200 reports each year, including an extensive system of iterative, blind peer review. The academies faulted Mr. Krimsky and Mr. Schwab for applying some measures of financial conflict that were stricter than current National Academies policies. But Mr. Hinchmanm [of NASEM], in an interview, conceded that it may be time for the National Academies to update those measures and make other changes, including giving the reports' readers clearer information on when conflicts are identified.

The Committee on Genetically Engineered Crops of NASEM responded to the *Chronicle* report with sharply worded criticism of the *PLOS ONE* study.[25] "You recently published an article titled "Under Fire, National Academies Toughen Conflict-of-Interest Policies" (*Chronicle*, April 25, 2017). A premise of the article is that this toughening was triggered by claims of personal conflicts of interest "tainting" the Academies' report on genetically engineered crops (GE crops) that we authored.

> The Academies have already gone on record to dispute this false claim, but our concern is that you have specifically sown undeserved doubt about the credibility of the GE crops report by uncritically conveying claims made in one article (*PLOS ONE* 12(2):e0172317) without considering a larger body of evidence contradicting that article's conclusions. For example, no mention was made of leaders of 15 academic societies (ecology, sociology, economics, toxicology, science ethics, and agriculture) who independently examined the report and published an article titled "National Academies Report Has Broad Support" (*Nature Biotechnology* 35, 304). That strong endorsement is inconsistent with your conclusion that the National Academies "reputation has been challenged" because of the GE crops report.

The Committee went on to state that some of its members were actually critical of GE crops and therefore implying that they need not declare any FCOIs.

> You echo the *PLOS* article in stating that six members of our committee had "grant support or patentable discoveries [that] suggested alliances with producers of genetically modified organisms." We wish that before making such a blanket statement, you had done due diligence to assess this conclusion. As an example, one of the six members with alleged "alliances," Dr. Carol Mallory-Smith, has long been critical of measures taken by companies to guard against gene flow from GE crops to wild plants. The only reason she had a grant from a corporation was because that

> corporation was required by USDA regulators to have surveys performed to check for gene flow. Does this grant suggest an alliance with the corporation?
>
> Similarly, before accepting the blanket statement that "None of the 20 panel members could be found to have any significant alliances with groups skeptical of GMOs," we wish you would have investigated the chair of the committee who helped the Union of Concerned Scientists by writing a substantial section of their report that was critical of company and government approaches for deploying GE crops. Other committee members have published criticisms of risk assessments and past socio/economic analyses of GE crops.

We sent a letter to the *Chronicle* in response to the Committee's claim, but it did not get published. Here is our unpublished response.

> To the Editor,
>
> The letter submitted by the Committee on Genetically Engineered Crops, National Academies of Sciences, Engineering and Medicine (NASEM) (June 12) takes umbrage with *The Chronicle* article "Under Fire, National Academies Toughen Conflict of Interest Policies" (April 25), which refers to a recent study we published in *PLOS ONE* (Feb 28). We applaud the National Academies' announcement that it is overhauling its conflict-of-interest policy, but we remain baffled that the Committee on Genetically Engineered Crops pushes back against our recommendations for transparency and full disclosure.
>
> The Committee's letter suggests that determining whether an individual has (or should disclose) financial conflicts of interest is a complex endeavor, open to widely varying interpretations. This represents a misunderstanding of current scientific norms, which warrants a response. Research institutes, government agencies, academic journals and even

member Academies have long established guidelines and definitions of financial conflicts, which we drew from in creating the criteria used in our analysis. We found six Committee members who had undisclosed financial conflicts, including substantial research funding from corporate sponsors with financial interests in agro-biotechnology, and patents in GMO crop related areas. We also found that the National Academies failed to disclose in its report the millions of dollars it receives from companies with a financial interest in the GMO topic.

Whether or not panel members have historically been critical or supportive of GMOs is a separate issue from a member's obligation to disclose—and NASEM's obligation to report—financial conflicts of interests. And, to set the record straight, our study never discussed nor demeaned the findings of the Academies' scientific report. We acknowledge the importance of NASEM and its efforts to remain an objective and independent, preeminent scientific body. If anything, our findings and recommendations were designed to elevate the standing of NASEM by having it adhere to the highest standards of transparency with respect to financial conflicts of interest.

—Sheldon Krimsky and Tim Schwab

The above interchange with the nation's preeminent scientific body brings into sharp focus the dilemma of our times. Research universities depend on external funding and in many fields, much of that funding comes from private corporations. This trend is not likely to change as government funding has not kept pace with the expansion of academic science. One cannot make a blanket judgment that corporate external funding always biases research. However, in certain fields where the commercial stakes are high, such as agricultural chemicals, tobacco, pharmaceuticals, and climate change studies, we have seen examples of the biasing effect of the sponsoring entity, ghostwriting, withholding of public health data, and fabricated attacks on responsible scientists. Disclosure of conflicts of interest is not a panacea for protecting the scientific record or the public health. But it is an essential part of the process. In some cases, such as in advisory committees of government, professional societies and the National Academies, the gold standard should be zero tolerance for panel members with the actual or even the appearance of financial conflicts of interest.

In the final chapter (22) my coauthor, journalist Carey Gillam, and I show just how far a corporation will go to protect its product from scientific findings that affect the company's bottom line. The article titled, "Roundup litigation discovery documents: implications for public health and journal ethics" was based on the analysis of litigation discovery documents that came from Monsanto's own files.

The volume of twenty-two essays in this book covers thirty-three years in my career involving investigations on conflicts of interest in science and medical research. The articles raise the question of whether scientific inquiry is independent of financial interests, and if not, whether those interests affect the objectivity of the research findings. Society's trust in the integrity and independence of the academic research community has never been more important, as false idols of truth have emerged denying credible scientific results. Avoidance of conflicts of interest in science will ensure its higher quality and secure greater public trust. Awareness of persistent conflicts of interest will at least help readers identify these potential biases.

Chapter 1

THE CORPORATE CAPTURE OF ACADEMIC SCIENCE AND ITS SOCIAL COSTS[1]

SHELDON KRIMSKY

Tufts University, Department of Urban and Environmental Policy

Commercial applications of molecular genetics and cell biology have resulted in a flurry of entrepreneurial activities among academic biologists and universities eager to cash in on the financial side of this technological revolution. The situation is not unique to biology. It is following the path of other disciplines that have formed close partnerships with industry, including nuclear and petroleum engineering, computer sciences, nutrition, electronics, and chemistry. Nevertheless, the current debate that has centered on the commercial ties of academic biologists has been more widely publicized than at any time in the past. Several hypotheses may be offered to explain this phenomenon.

The commercialization of biology occurred rapidly, and considerable media attention was given to the discoveries and the personalities involved. By the time many biologists developed commercial interests, a widely publicized controversy over the safety of recombinant DNA techniques had already taken place. The social and ethical issues associated with gene

1Originally published in: Aubrey Milunsky and George Annas, *Genetics and the Law III: Proceedings of the Third National Symposium on Genetics and the Law*, April 2–4, 1984. (New York: Plenum Press, 1985), 45–55.

splicing provided grist for the public's concern over the commercial activities of its pioneers. The confluence of social and ethical debates with commercialization of science generated a larger public reaction to the latter.

A second explanation centers on the perceived role of biomedical science in society. Unlike other scientific and engineering fields that have developed linkages with the private sector, biological research has been closely associated with public health. The public expectations of this area of research are greater than they are of such areas as chemistry or computer sciences. Moreover, the preponderance of funding for biomedical research comes from social resources. Consequently, in the public consciousness, the conjunction of these factors—namely, the sources and goals of funding—makes the academic entrepreneurs in biomedical science accountable for their commercial activities in ways that other scientists are not.[1]

However, I would conjecture that the distinction between the goals and funding of biomedical research and those of other commercialized disciplines provides only part of the answer. The types of university—industry relationships in biology are more varied, more aggressive, more experimental, and more indiscreet than they have been in similar historical circumstances. A significant number of new firms in biotechnology have sprung directly out of academia. By contrast, in the microelectronics field, most firms were spawned directly from industries that were recipients of U.S. Department of Defense contracts.[2]

Another explanation that sets biotechnology apart from other academia—industry partnerships was advanced by Congressman Albert Gore (D. Tenn.). According to the Congressman, in the past there has always been a distinction between pure and applied research in the means by which technology is transferred from academe to industry. Gore observed that, in genetic engineering, "there seems to be no phase of applied research: the discovery of the basic scientists may go directly and swiftly from the laboratory bench in the university into a profit-making venture."[3] As a result of the omission of the intermediate stage, Gore believes that an unusual set of ethical issues results.

An additional factor that helps to account for the vehemence of this issue is that our society has changed. In the post-Watergate period, we have become more sensitive to conflicts of interest. Laws have been passed to protect society from unsavory kinships between the public and the private sectors. Public interest groups monitor corporate influence on government agencies. As a consequence, the public and the media are more sensitive to allegations that public funds are being misused or that private interests are exploiting social resources. Public universities have been sued for violating their mandate to serve the general interest.[4] Recent attention has also been directed at faculty misuse of federally supported projects.

The public perception of science, by and large, still portrays the contemporary scientist as a selfless discoverer of truth, despite efforts on the part of sociologists and the media to show otherwise. The marriage of science and Wall Street portends an illicit affair to most people. Perhaps this attitude is an outgrowth of the American Puritan tradition that financial gain distorts truth and values.

Why should we be concerned about what our universities or their faculties do to raise money? For one thing, universities and their faculties are a national resource. Our government depends on the expertise in academe for public policy formation. Second, universities are recipients of substantial government support, which implies some responsibility and accountability.

Much of the debate on the commercial ties of university faculty has centered on a number of issues involving the conflicting missions of business and academe. These include the control of intellectual property, the openness and accessibility of scientific and technical knowledge, the commingling of funds, the ownership of tangible research property, the use of public research funds for private business interests, and the influence of entrepreneurial faculty on the education of students. These are serious issues, and they have been aired to some extent in media coverage, university debates, and congressional hearings.[5] Several leading universities have issued guidelines for faculty pursuing commercial interests and have established policies on contractual agreements between the university and the private sector.[6]

Notwithstanding these initiatives, there is an important side to the problem that the current debate has totally neglected. Even if all the aforementioned problem areas are satisfactorily resolved and the conflict of interests is removed, intense commercialization of biology could result in an enormous social liability. Let me summarize the principal argument in the form of a conjecture.

If a sufficiently large and influential number of scientists or engineers become financially involved with industry, problems related to the commercial applications of the particular area of science or engineering are neglected. The scientific community becomes desensitized to the social impacts of science. This desensitization leads to a conservative shift in attitudes and behavior. The new values emphasizing science for commerce become internalized and rationalized as a public good. The disciplinary conscience becomes transformed. This transformation happens incrementally, without conspiracy or malice. Scientists or engineers with a stake in the commercial outcome of a field cannot, at the same time, retain a public interest perspective that gives critical attention to the perversion of science in the interests of markets.

We are not dealing with a threshold phenomenon. There is no clear stage in the growth of academic-corporate partnerships when the effects I have outlined suddenly become observable. That this phenomenon exists must be inferred from the psychology of individual behavior and from our knowledge of how people's values are shaped by their institutional affiliation and financial associations.

When the number of faculty involvements are small, the effects on public interest science are not likely to be important. As long as a sufficient number of scientists remain free from corporate influence, there will be a disinterested intelligentsia to whom the public can turn for a critical evaluation of technological risks, goals, and directions.

If my argument is correct, then the individual instances of faculty—industry ties are far less important than the aggregate corporate penetration into an academic discipline and the degree to which the major institutions and leading faculty are involved. It is my contention that, unless we have some quantitative information about the degree of corporate-academic interaction, we cannot appreciate the gravity of the problem.

What can be learned from a study of dual relationships in the area of biotechnology? Suppose we had before us perfect information about the commercial affiliations of academic scientists. What questions would we ask? We might want to know what percentage of faculty at leading universities has a substantial involvement in commercial enterprises. We might look at the extent to which the dual-affiliated academic population participates on public advisory committees or study panels. Unexpected results may be interpreted in several ways. Imagine that the participation is heavy. A skeptic might question whether the peer review process is being compromised by having a substantial number of commercially affiliated scientists reviewing grant proposals. It is inescapable that a diffusion of ideas will take place between reviewers and the institutions with which they are associated. On the other hand, suppose the participation is low. The same skeptic might explain this low participation by arguing that business interests have taken priority over the responsibility of scientists to participate in the peer review system.

Alternatively, there will be those who interpret the results, whatever they might be, as irrelevant to the effects mentioned. Notwithstanding problems of interpretation, I believe that there is some value in understanding the degree to which academia has financial interests in biotechnology.

To investigate the corporate relationships of academic scientists, I developed a data base of university faculty in biology, biochemistry, molecular genetics, and medicine who meet one or more of the following criteria with respect to biotechnology firms: (1) they serve on a scientific advisory board; (2) they hold substantial equity; (3) they serve as a principal in a company. Armed with this data base, which represents a lower bound of involvement because faculty connections to private firms are not ordinarily available, one can correlate this population group with other academic populations comprising leading biology departments; service on committees of the

National Institutes of Health, the National Science Foundation, and the U.S. Department of Agriculture; and membership in the National Academy of Science.

This inquiry is not designed to test a hypothesis. Rather, it is designed to suggest a research program. If the degree of the corporate penetration of academic biology is sufficiently high, the next obvious question is: What are the effects of this penetration? I conjecture that these effects will reveal themselves as a shift from a public orientation of science to science for private profit. In the long run, this shift will result in the social neglect of technological abuse. Before I turn to the data, I shall offer some qualitative and historical examples that support the conjecture.

SCIENTIFIC OBJECTIVITY AND INDUSTRIAL INTERESTS

Public policy formation in a highly industrialized society such as ours is a complex affair. It frequently involves input from many areas of expertise. Scientists serve on a labyrinth of public advisory committees, review boards, and risk assessment panels throughout all levels of government. How do we ensure objectivity in the contributions of scientific experts to public issues, particularly where consensus is difficult to find? Recently, the Office of Technology Assessment (OTA) issued a report on biotechnology that made the argument that the dual affiliation of scientists in the academic and commercial worlds is actually more desirable from a public policy standpoint when expertise is needed:

> An argument could be made that because the public has supported research in universities, it has a right to know whether a particular university faculty member who is giving testimony, for example, has a consulting relationship with a company that manufactures a particular harmful chemical. The negative side of the disclosure policies is that "objective" information may be judged "subjective" because of guilt by association. If a faculty member's consulting arrangement with industry is declared

> openly, it is not necessarily the case that his or her testimony is biased. In fact, the expert may have a more objective view because he or she understands both the research and development aspects of the technology.[7]

There are two arguments here. The first is that the veil of confidentiality on the commercial affiliation of a scientist testifying before a governmental body would prevent bias against the individual's presentation. According to the OTA, if the disclosure is required, testimony would not be taken on face value but would be dismissed for reasons of association. The second argument interprets objectivity to mean "multidimensionality." The implication is that the more affiliations a person has, the more objective that person can be.

The OTA analysis confuses objectivity with eclecticism. There are many advantages in having faculty link up with the private sector. Those advantages include a greater awareness of the full life cycle of science, from discovery to manufacture. But the OTA makes a serious error when it describes the financial involvement of academic scientists in commercial ventures as a contributor to objectivity. The argument fails because of the financial interest. A form of eclecticism that is independent of pecuniary interests could indeed enhance objectivity. Proposals for a disinterested and eclectic intelligentsia have been advanced by a number of social theorists, including the Greek philosopher Plato and the German sociologist Karl Mannheim, both of whom were aware that knowledge is subject to the control of economic interests.

The history of technology provides an abundance of examples illustrating the distortion of objectivity when scientific expertise is beholden to the industrial sector. The causal relationship need not be absolute. We are dealing with a statistical phenomenon that is guided by factors of social psychology. Our conflict-of-interest laws are based on assumptions of human frailty as exemplified by the aphorism "Don't bite the hand that feeds you." It is a mistake, however, to view conflict of interest in terms of conspiracy of conscious design. It is my hypothesis that a sizable academic-industrial association will slowly change the ethos of science away from social protectionism and toward commercial protectionism. The aggregate of isolated individual decisions to go commercial creates a qualitatively new effect. It should be emphasized that the discovery of a problem tells us nothing about the solution. In some cases, it might be wise to live with the problem, to understand its social consequences, and to avoid draconian measures. For the problems outlined in this chapter, I will offer a few modest social antidotes.

Let me begin with a phenomenological exercise to illustrate my thesis. Imagine that you are heavily funded by a company to engage in research. Is it likely that you would publicly embarrass the company by revealing information or posing questions about its technological direction? Most scientists with a conscience would make their viewpoints known to the firm's directors. But who wants to jeopardize his or her funding by making an issue public? The closer the relationship one has to a firm, the more propriety and self-interest dictates keeping criticisms within the corporate family.

A few years ago, I supervised a policy study involving the chemical contamination of a town's water supply. The actors included a multinational corporation; town, state, and federal officials; a public advocacy group; and technical people. I chose to do the study for several reasons. First, it served the public interest. Second, it was a useful case for instructional purposes. Third, from a public policy standpoint, it represented a milestone for the implementation of a major federal law. If I had been funded by the corporation in question, that research study would never have entered my mind because of the likelihood the company would not be shown in the best light. If my department had been heavily funded by the company, including possibly graduate student stipends and multiyear grants, it is extremely doubtful that any faculty member would have chosen to study how the department's corporate benefactor was implicated in the contamination of a water supply, unless there was reasonable assurance that the outcome would not be an embarrassment.

As the financial connections become more remote, the psychological and social factors that limit or restrict freedom of inquiry become less important. A corporate representative on the university board of trustees might have an effect on the choice of a research program or even on its outcome. However, the strength of that influence would be severely weakened as it was mediated through university channels.

When our policy study on the chemical contamination of the town's water supply was complete, a vice-president of the corporation made a personal visit to the president of my university and asked to have the study suppressed or totally disassociated from the university. It is gratifying to report that my university made no efforts to restrict my academic freedom. But the direct political influence on research has become less of a problem since the introduction of the tenure system in the aftermath of the McCarthy period.

However, the economic determinants of research and their influence on the latitude of inquiry are far more pervasive and subtle. Sometimes, this influence manifests itself in the distortion of science. At other times, it is expressed in the control of information. Most frequently, it is felt by the kinds of questions that are pursued in the areas where science and social policy intersect.

SCIENTISTS AND THE PUBLIC TRUST: SOME HISTORICAL ABUSES OF DUAL AFFILIATIONS

Periodically, a story appears in the media about an academic scientist who expresses views sympathetic to an industry position on a controversial health or environmental policy. The article may then mention the financial association between the scientist and the company that has a stake in the outcome. Considering the amount of industry consulting that takes place, the public learns only about the proverbial "tip of the iceberg" of the associations. Because the documented cases may be small in number, there is no clear way of knowing the aggregate effect that these individual associations will have on social policy formation. Given the choice, the public sector would rather place its trust in scientific experts who are not linked financially to industry. Problems arise when the pool of experts in molecular genetics who are unaffiliated with industry becomes vanishingly small.

A situation like this occurred in 1969 when close ties between the oil industry and university experts in academic disciplines such as geology, geophysics, and petroleum engineering incensed California officials and federal authorities. State and federal agencies were responsible for the environmental problems arising from massive oil leaks of the Union Oil Company's offshore well in the Santa Barbara Channel. According to the report in *Science*:

> California's chief deputy attorney general . . . publicly complained that experts at both state and private universities turned down his requests to testify for the state in its half-billion dollar damage suit against Union and three other oil companies.[8]

The explanation offered by state officials for the difficulty they had had in getting testimony from experts was that petroleum engineers at the University of California campuses of Santa Barbara and Berkeley and at the privately supported University of Southern California indicated that they did not wish to risk losing industry grants and consulting arrangements.

It was reported in *Science* that most petroleum engineers in academia did extensive consulting for oil companies and formed part of the university–industry "oil fraternity."

> Consulting is regarded not simply as a lucrative perquisite of the profession but as a necessary way to establish and maintain a departmental reputation and create job opportunities.[9]

Another obstacle facing public officials attempting to getting objective advice from the experts who serve on public service panels is that many own stock in the companies that are affected by their decisions. The lesson illustrated by this case was not that petroleum engineers refused to testify. They were probably acting ethically in not testifying, as their corporate ties might have compromised their objectivity. The real problem was the scarcity of academic experts who were not affiliated with the oil industry and who could thus provide a disinterested perspective.

In some situations, research is so highly specialized that only a few scientists in the entire country may have the information necessary to render a decision on the health and safety of a new substance. Several decades ago, it was common practice for scientists to sign restrictive publication agreements with companies. Such agreements are still used today in the biotechnology industry. In one important case, information withheld from publication could have prevented a toxic pesticide from being marketed. A clinical professor of occupational and environmental medicine at the University of California at San Francisco was engaged in toxicological research on the pesticide DBCP for the Shell Development Corporation in the 1950s. He discovered that the chemical caused severe cases of testicular atrophy in test animals. As was common practice at that time, the research results were kept out of print for a period of time to protect trade secrets. Although a brief abstract of the toxicological study was published in 1956, the full results were held back from publication until 1961, six years after the pesticide had been approved for marketing.

In the late 1970s, workers in a DBCP plant were monitored. An unusually high incidence of male infertility was reported. At state hearings on DBCP, it was noted that the scientist who had studied the pesticide had testified at public hearings on other environmental health matters without disclosing his consultant work with firms that had a financial interest in the subject matter under investigation. The chairman of the panel stated:

> . . . it is difficult to know in the cases of [such scientists] with 30 years of dual relationships with the university and with Shell where advocacy on behalf of private interests ends and where responsibility as an "objective" professor begins.[10]

A special feature of the journal Business and Society Review reported cases where the public received expert testimony from scientists with undisclosed relationships to companies that stood to gain from the recommendations. Michael Jacobson, Executive Director of Science and the Public Interest, described conditions in the field of nutrition:

> In the area of food safety and nutrition . . . a large percentage of experts has received industry money. Rare is the expert who accepts such funds and is an ardent defender of the public's interest.[11]

Similar examples can be found in nuclear engineering, occupational health and medicine, and ecology. Ultimately, it is socially desirable that there be a balance in the academic community. For any discipline that has a commercial offspring, it is vital that a critical mass of experts remain disassociated from industrial ties in areas related to their field of expertise. And when scientists maintain such ties, it is essential that the public understand the nature of the relationships when their expertise is sought in setting policy. But just how extensive is the problem in biotechnology?

ACADEMIC-CORPORATE LINKS IN BIOTECHNOLOGY: SOME QUANTITATIVE RESULTS

To evaluate the degree of the link between academe and the biotechnology industry, I developed a data base of scientists who are formally affiliated with biotechnology firms either as members of scientific advisory boards, as consultants on retainer, as principals in the firm, or as large stockholders. I shall label this subpopulation of dual-affiliated scientists with the term ACIND for "academics in industry."

The primary source for the data base was fifty public biotechnology companies. These were drawn from a pool of 250 public and private firms that have been inventoried by trade organizations in the United States. Each of the public corporations reviewed for the study publishes a list of its scientific advisory board, its management, and major stockholders in its company prospectus (the 10K report required by the Securities and Exchange Commission). Additional entries into the data base were gleaned from the trade literature and media reports. However, as private firms are not legally obligated to file reports in the public domain, it is more difficult to obtain this information.

For this study, only 20 percent of the total number of biotechnology firms has been systematically surveyed, although these represent the largest and most active firms. Therefore, the actual number of academic scientists involved in the biotechnology industry could run three to four times the figure of 345.

Another consideration in interpreting the data is that the number of biotechnology firms has increased at an exponential rate within the past few years. The trade magazine Genetic Engineering News reported that there was a handful of biotechnology companies before 1981.[12] By the next year, 184 companies were listed in its registry. That figure climbed rapidly to 220 by November 1983, and current estimates place tae number of firms at 250.

One of the goals of the study is to examine a number of assumptions about the extent of corporate affiliations in biomedical science. For example, in 1982 Barbara Culliton, writing for *Science*, stated that most of the country's leading biologists are associated with biotechnology companies.[13] Under what criteria can we evaluate such a statement? I looked at corporate affiliation as a function of membership in the National Academy of Sciences (NAS), choosing membership as a proxy for "leading biologists."

Another area of inquiry is the relationship of ACIND scientists to participation on study panels and public advisory committees in major government funding agencies. The National Science Foundation (NSF) has a rigorous criterion for weeding out potential conflicts of interest on proposal review. An individual's commercial relationship is a relevant input in the review process according to NSF staff. It is expected that such information will be disclosed by mail reviewers as well as by members of study panels. The obligation of disclosure rests with the prospective reviewers and is used by the NSF in determining whether a conflict of interest or its appearance exists. In this study, we are at the stage of developing aggregate statistics. The inquiry into NSF affiliation is still in progress.

The quantitative information that we have compiled thus far can be summarized as follows. This information is based on a total data base of 345 dual-affiliated scientists in 50 biotechnology corporations and various private companies.

1. Sixty-two scientists of the 345 are members of the NAS (18 percent).

2. The four most relevant categories of NAS membership for biotechnology are biochemistry, cellular and developmental biology, genetics, and medical genetics. The ACIND entries in our data base constitute 25 percent of the NAS members in the four categories. The Percentage of NAS scientists with industry affiliations revealed in the study represents a lower bound for the profession as a whole. The scientific affiliates of the additional 200 firms could bring the number of corporate–affiliated NAS members well beyond 50 percent. This percentage is particularly significant in that the NAS is frequently called on to render decisions on the social uses of science and technology.

3. In another correlation, we looked at the number of ACIND scientists who participated on NIH study panels or in public advisory groups. In this category, forty scientists were identified in a listing of NIH groups that covered a twelve-month period. That figure represents 12 percent of our data base.

These statistics must be understood in the context of other questions. For example, is there an appearance of conflict of interest if an ACIND scientist serves on a study panel? If the answer is affirmative, then forty is a significant number. Alternatively, are ACIND scientists less likely than their nonaffiliated counterparts to serve on study panels? Are they being self-selected out of this responsibility because of commercial interests and are the numbers large enough to have an effect on the peer review system? To date, from our data base, ACIND scientists make up 10 percent, at most, of NIH study panels. Further studies are needed to interpret the significance of the data. Trends should be followed over a longer time period to disclose the impact of the biotechnological revolution on the social and scientific character of academic biology and the health sciences.

In addition to quantitative data, the preliminary results reveal interesting qualitative information. A number of the biotechnology firms offer stock to members of their scientific advisory boards as one means of remuneration, giving scientists direct equity in the business. Among the ACIND scientists, we found many heads of departments, chairpersons, a college president, and a former director of the National Institutes of Health (which has established the principal regulatory apparatus for the industry use of recombinant DNA techniques).

Our inquiry also revealed several types of restrictions placed on scientific advisors by the companies with which they are associated to protect proprietary information. One prominent firm requires that its academic affiliates "not perform research in competition with the Company for a period of three years after termination of the consulting arrangements."[14]

Another company states that "there is no assurance that the company's business will not conflict with the business of the institutions with which the various consultants are affiliated."[15] These are particularly troublesome signs. Several universities that have acted on potential conflicts of interest among ACIND individuals have introduced disclosure procedures that are supervised by departmental chairpersons. But if the degree of faculty linkage becomes significantly high, many chairpersons will be part of the same reward system.

In conclusion, there is much that needs to be done to improve the public's attitude toward the role of science in social policy, and particularly to enhance the image of scientific objectivity. One contribution to this end is to promote disclosure. The commercial connections of scientists with dual affiliations should be part of their résumé and open to the public record when they enter the policy realm or when they serve on public advisory committees. This is not a difficult or burdensome requirement.

A second recommendation, which is more difficult to implement, would reward scientists who maintain an independence from commercial activities. Such independence might be factored into appointments on prestigious commissions and other policymaking activities, including service on study panels, as well as preference in the competitive grants program.

Without some incentives to reverse the momentum of the ACIND phenomenon, the pure biomedical scientist may become the vestigial remains of a past generation. At risk is the foreclosure of an important agenda: the social guidance of a technological revolution and the increasing erosion of public confidence in scientific objectivity.

Chapter 2

ACADEMIC-CORPORATE TIES IN BIOTECHNOLOGY: A QUANTITATIVE STUDY[2]

SHELDON KRIMSKY
Tufts University, Department of Urban and Environmental Policy

JAMES ENNIS
Tufts University, Department of Sociology

ROBERT WEISSMAN
Multinational Monitor

Rapid commercialization of the biological sciences began several years after the 1973 discovery of recombinant DNA (rDNA) molecule techniques. Potential applications of "gene splicing" to a wide range of industrial, agricultural, and pharmaceutical products stimulated the founding of hundreds of new firms (see Figure 2.1). Billions of dollars of venture capital were invested in just a few years.[1,2] Academic scientists were centrally involved in the birth and development of many of these firms. While dual affiliation of scientists with universities and with for-profit companies is common in some academic fields, there is reason to believe that the practice is especially pervasive and significant in biotechnology. These linkages merit consideration for several reasons. Some possible negative consequences include

2Originally published in *Science, Technology and Human Values* 16, no. 3 (Summer 1991): 275–287.

potential conflicts of interest, redirection of research from basic to applied areas, erosion of openness of scientific communication, and detrimental effect on graduate training. On the other hand, positive consequences of such ties might include new and necessary sources of research funding, increased incentive for scientific innovation, increased yield of beneficial new products, more rapid technology transfer from universities and government labs to industry, and competitive advantage in international markets.

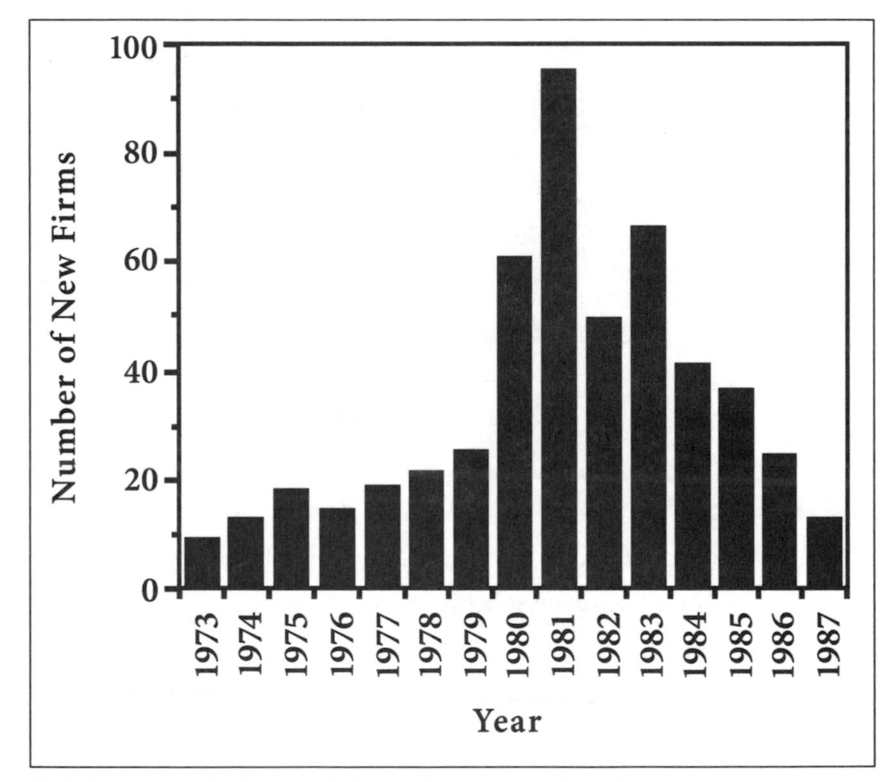

FIGURE 2.1: New biotechnology firms, by founding dates: 1973–1987 (n=493).

University-industry relations were the subject of intense media scrutiny and congressional oversight between 1980 and 1983. Subsequently, the commercialization and possible politicization of the biological sciences became the focus of new scholarly studies. Etzkowitz[3] finds the historical roots of academic entrepreneurship in early nineteenth-century German science and traces its evolution and development in pre–World War II America. Weiner[4] presents historical evidence that spirited debates over patenting medical discoveries raged early in this century. Blumenthal, Epstein, and Maxwell[5] describe the success of the Wisconsin Alumni Research Foundation (WARF), which provided a commercial outlet to university scientists for more than a century. The works of Kenney[6] and Dickson[7] describe how academic institutions have sought new sources of funding and a more favorable political climate for university-industry partnerships in biology. And through a series of in-depth interviews, Etzkowitz[8,9] explores the values, motivations, and changing norms of entrepreneurial scientists and university administrators across disciplines.

Until 1984, there were many conjectures about how commercialization of applied genetics was affecting universities and their faculty, but as yet no systematic effort had been made to study these effects. In that year Harvard University's Center for Health Policy and Policy Management initiated a study that surveyed over 100 firms and more than 1,200 biomedical faculties in over 50 U.S. universities. Two published papers from that study provide the best current data on the extent and impacts of faculty-industry research relationships.[10,11] Also in 1984, the Office of Technology Assessment (OTA) published a study on commercial biotechnology with a chapter dedicated to university-industry relationships.[12] In connection with its study, OTA sponsored a survey of six universities with strong biomedical research facilities and of fifteen biotechnology companies to gain information on the factors responsible for and the nature of the new relationships in biology.

Academic-industry relations are only partly expressed through research relationships. Other types of relationships include shared patent rights, equity interests, consultantships, and managerial roles. The question we posed for our study is: Can a quantitative and objective measure be developed that exhibits the structure of faculty-industry relations across American universities but does not depend on faculty self-reports? The advantage of such a measure is that it can reveal demographic patterns while also showing the fine structure of industry-faculty associations within individual institutions. These results are important for gaining a better understanding of the potential for conflicts of interest, shifts in the research agenda, and the potential obstacles to intellectual exchange.

METHODS

The goal of our study is to exhibit the linkages between university faculty and the biotechnology industry during the industry's early stage of development. The years 1985–1988 were chosen as the test period. The method consists of three elements. First, the term faculty-industry linkage was defined. Second, a system for quantifying the linkages was developed. Finally, a linkage map was constructed of commercially active faculty and the new biotechnology industry for North American colleges and universities. The university-industry relationships we chose to measure are those in which faculty has a formal association with a firm. Faculty are considered to have formal associations if they satisfy one or more of the following conditions: (a) serve as a member of the firm's scientific advisory board (SAB) or as a standing consultant; (b) hold managerial position in the firm; (c) possess substantial equity in a firm (i.e., sufficient equity to be listed in a public firm's prospectus); or (d) serve on the board of directors of a company. University biological/biomedical faculty who satisfy the criteria listed above will be called dual-affiliated biotechnology scientists (DABS).

Industry-university linkages as defined above represent the highest degree of scientist involvement in the commercial sector, namely, formal associations with a firm. As such, they are most likely to have consequences of concern. While the Harvard study encompasses industry-funded research relationships of any duration, the focus of this study is on those ties likely to be more enduring and involving. Our faculty sample excludes individuals with small equity interests in a company who otherwise do not meet any of the four conditions listed above. While equity interests of any magnitude may influence behavior of faculty, small equity interests were excluded from our study because they could not be systematically identified by our method and because we sought an indicator of active and substantial firm involvement.

Between 1985 and 1988 we constructed a list of 889 U.S. and Canadian biotechnology firms. Two criteria guided the selection of firms. First, a firm must be involved in the microbiology, genetics, or biochemistry of cells. Companies that specialize exclusively in bioprocessing, fermentation, large-scale purification, and instrumentation were excluded. Second, we included only newly established (post-1973) companies, their subsidiaries, or established companies that formed new research and development divisions in biotechnology in the aftermath of the genetics revolution. Firms were located by using standard industry directories, trade association lists (e.g., Industrial Biotechnology Association), publication inventories *(Genetic Engineering News)*, government studies, personal contacts, media accounts, and other sources. Under the U.S. Securities and Exchange Act, firms that issue public securities are required to file prospectuses and annual financial disclosure statements (10K reports) that include information about management personnel. By reviewing these documents, we obtained the names of university faculty who were founders, board members, major shareholders, standing consultants, or members of the firm's SAB. For private companies that are not required to issue public prospectuses, we sent a one-page questionnaire requesting the founding data, the public/private holding of the firm, whether it employed a SAB, and the names of its board members.

Pretest samples were sent first to 51 and then to an additional 53 firms during the spring of 1985 another 289 private firms were queried in March 1987. Firms that did not respond to the mail survey were contacted by telephone and asked the same questions. In August 1988, 507 remaining firms on which we had no information were sent the survey. Overall, we obtained usable information on 539 firms or 60.6 percent of the total list of 889.

Of the firms for which we obtained information, 54.0 percent (291) reported having a SAB, and 46.0 percent (48) responded negatively. Response bias may inflate the yes percentage somewhat, since firms with SABs might view the question as more pertinent and, therefore, be more likely to respond. Nevertheless, the overall yes percentage can be no lower than 32.7 percent (i.e., 291/889) Thus the proportion of all biotechnology firms with SABs is roughly between one-third and one-half. In order to determine whether non-respondents in the final sample were systematically different from respondents, we randomly selected 95 non-respondents from the final sample and contacted them by telephone. Of this group, 22 (23 percent) had SABs and 18 (19 percent) did not; 24 (25 percent) had gone out of business or were no longer at their listed address or phone number. Eight firms refused to respond to the survey. The remaining 23 (24 percent) did not answer repeated calls. Those firms that provided information had a 55 percent SAB rate (22/40), which is comparable to what we found in the mail survey. In a rapidly moving field like biotechnology, there is a substantial turnover of firms (new arrivals and bailouts) within a short time period. If we assume, conservatively, that one-half of the 350 firms that failed to respond are still in business and that 55 percent of these have SABs, then there are roughly another 96 SABs that we have not located.

The final data base has two parts. The first consists of 889 biotechnology companies, and the second consists of 832 scientists (including plant pathologists, microbiologists, geneticists, and biochemists) with formal relationships to biotechnology firms. As a small number of scientists hold affiliations with more than one company, they therefore appear more than once in the data base. The search yielded 927 linkages with a median SAB board size of 3 (mean = 4.3, maximum = 29; minimum = I). If the estimated number of missing SABs (96) is multiplied by the median SAB size, then we estimate roughly that an additional 288 "missing" scientists serving on biotechnology advisory boards are not recorded in our data base.

Of the 889 firms, 286 (32 percent) are public, 406 (46 percent) are private, and 197 (22 percent) have an undetermined status. More than one-half the total number of firms are composed of newly established companies (post-1973), subsidiaries, or new biotechnology divisions of established companies. The data base represents aggregated entry points in the biotechnology field for firms and DABS. Although there has been some winnowing of firms during the four-year period and some scientists have retired from academe or disaffiliated from a firm, these data points remain included so long as a scientist or a firm meets the criteria at any time during 1985–1988.

PATTERNS OF LINKAGE

As noted, our survey of biotechnology firms revealed that 291 (32.7 percent) have SABs. New firms are more likely than established firms to have SABs, since displaying intellectual capital can help attract financial backing. Many private firms treat their SAB membership as proprietary information. Moreover, new firms are constantly forming, and some may have been missed in our surveys. Despite these limitations, a data map with over 800 DABS reveals the structural relationships between a nascent industry and the academic sector.

Our demographic analysis of the biotechnology industry shows that the highest concentrations of new biotechnology firms are in California, Massachusetts, Maryland, New Jersey, and New York, in descending order. Nearly 40 percent of the firms are located in California[13] (see Figure 2.2).

Start-up companies frequently sought financing based on the promise of the techniques, in some cases without explicit product ideas, prototypes, or patents in hand. Having prominent scientists on the firm's SAB can bolster the confidence of prospective investors. The data indicate that faculty from large, prestigious universities were most heavily involved in the promotion and development of new firms. Table 2.1 shows the distribution of DABS and their linkages to firms for 24 universities with the largest number of faculty involved in the commercialization of biology. Thus, from our database, Harvard has 69 DABS from its Faculty of Arts and Sciences and its Medical School, representing 83 linkages (formal ties) to 43 companies. The higher figure for linkages indicates that some faculty has formal ties to more than one firm. Also, some companies sign up multiple faculties from a given institution.

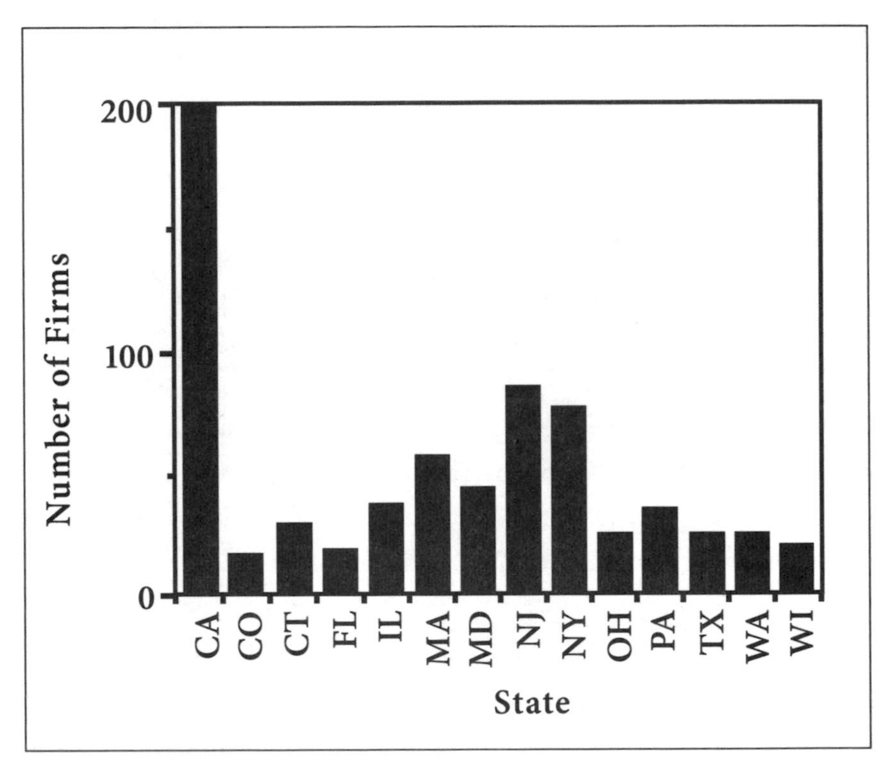

FIGURE 2.2: States leading in the formation of new biotechnology firms: 1973–1987.

The closest competitor to Harvard in faculty-industry links is Stanford, with 40 DABS, 51 linkages, and 25 firms. MIT, which does not have a medical school, shows a very commercially active biology department with 35 DABS, 50 linkages, and 27 firms. The ratio of DABS to the total biotechnology faculty of an institution or department is an indicator of university penetration by the new industry. However, the total number of DABS for a given university may be distributed over many departments, some of which may be only marginal to the commercial development of biotechnology. Thus, the ratio of DABS to total biomedical faculty is not a useful indicator of the degree to which departments have been affected by commercial affiliations. Instead, we have calculated the percentage of DABS in select departments for ten universities. The total number of faculty in these key departments was obtained from university catalogs and Peterson's Guide (see Table 2.2). Twenty-three of MIT's 35 DABS are in the Department of Biology, which lists 74 members. The rate of commercialization penetration (de minimis) for this one department at MIT is 31.1 percent.

In contrast, Stanford and Harvard, with medical schools, have, respectively, four and six commercially active departments with de minimis penetration rates of 195 percent and 19.2 percent, respectively. At the University of California, Davis; the University of California, San Francisco; and the University of California, San Diego, most of the dual-affiliated faculties are located in one or two departments, with penetration rates between 11 percent and 15 percent.

Blumenthal, Gluck, Louis, et al.[14] found that biotechnology faculty with industry support were four times as likely to report that trade secrets resulted from their research than other biotechnology faculty were. Another factor that may impede open communication is the number of firms represented within one university or within a single department. In settings where many professors are linked to competing firms, there may be more restrictions on scientific interchange. Table 2.1 shows the number of firms represented by the DABS at selected high-profile universities. At Harvard, at least 43 independent firms are represented by the formal affiliations of its biomedical faculty. A small percentage of these firms were started by Harvard scientists. At MIT and Stanford the figures are 27 and 25, respectively.

TABLE 2.1: Linkages of Biomedical Faculty to Firms

	Number of DABS	Number of Links	Number of Firms
Harvard	69	83	43
MIT	35	50	27
Stanford	40	51	25
University of California, San Francisco	24	28	14
Yale	22	26	21
University of California, Los Angeles	26	30	19
University of California. Berkeley	22	24	16
University of California, San Diego	22	22	11
Johns Hopkins	20	24	16
Columbia	15	18	15
University of California, Davis	17	17	12
New York University	14	15	12
California Institute of Technology	12	15	11
Baylor	17	18	11
Cornell	20	20	15
University of Texas	21	27	22
University of Wisconsin	24	24	19
University of Washington	21	22	18
University of Colorado	12	15	10
University of Michigan	11	12	12
University of Minnesota	16	16	12
University of Pennsylvania	15	17	11
Rockefeller University	10	12	12
Tufts	11	12	11

TABLE 2.2: Rate of Commercial Penetration into Select University Departments

	Number of Key Departments	Number of Faculty	Number of DABS[a]	% Penetration
MIT[b]	1	74	23	31.1
Stanford[c]	4	82	16	19.5
Harvard[d]	6	156	30	19.2
University of California, Davis[e]	2	38	6	15.8
University of California, San Francisco[f]	1	61	9	14.8
University of California, Berkeley[g]	5	103	14	13.5
University of Washington[h]	2	79	10	12.7
University of California, Los Angeles[i]	4	115	14	12.2
University of California, San Diego[j]	1	77	9	11.7
Yale[k]	4	126	14	11.1

a. DABS=Dual-affiliated biotechnology scientists. Department of Biology.
b. Medical School—Departments of Biological Chemistry, Genetics, Microbiology and Immunology, and Biological Sciences.
c. Arts and Sciences—Departments of Biochemistry and Molecular Biology; Division of Medical Sciences—Departments of Biological Chemistry and Molecular Pharmacology, Cellular and Development Biology, Genetics, Microbiology and Molecular Genetics, and Medicine.
d. Departments of Plant Pathology and Department of Biochemistry and Biophysics.
e. Department of Biochemistry and Biophysics.
f. Departments of Biochemistry, Microbiology and Immunology, Plant Biology, Plant Pathology, and Molecular Biology and Genetics.
g. Medical School—Departments of Biochemistry and Microbiology.
g. College of Letters and Science—Department of Cell and Molecular Biology and Department of Microbiology; School of Medicine—Departments of Biochemistry and Microbiology.
h. Department of Biology.
i. Department of Biology and Department of Molecular Biophysics and Biochemistry. Medical School—Departments of Human Genetics and Cell Biology.

Prospectuses of some firms stipulate proprietary covenants with SAB members. Some may view the diversity of corporate affiliations at a single institution as a positive sign that universities are not subject to the dominance of a single firm. Nevertheless, the magnitude of firm representation within the university helps us to explain the emergence of a new climate in biology in which limited secrecy[15] replaced free and open communication.

NATIONAL ACADEMY OF SCIENCES

Our data show that the new biotechnology industry was actively supported by academic scientists in the nation's leading universities. Participation in the commercial applications of molecular genetics by the nation's elite scientists is reflected in the membership of the National Academy of Sciences (NAS). The academy plays a major consultative role for Congress and other government bodies on a wide range of issues of major concern to society. Academy members frequently serve on panels that issue health and safety recommendations for new products or technologies. The data base was used to determine a lower bound of DABS who were members of the NAS. Of a total of 359 academy members who may be classified as biologists and biomedical scientists (as of 1988), 132 (37 percent) were identified in the data base with formal ties to companies. Since membership in the academy is lifelong, the effective percentage of currently active NAS members with industry affiliation may be significantly higher. One NAS member estimated that for active members, the number of DABS is well over 50 percent.[16]

PEER REVIEW

Peer review is an essential part of the international system of science. It is difficult to imagine the organization of science as we know it without a peer-review process. Not only does it help improve the quality of published papers, but it also plays an invaluable role in allocating federal research funds. Reviewers of grant proposals, where innovative ideas germinate, are bound by a code of ethics. A reviewer is expected to protect unique unpublished ideas in a funding proposal from precipitous disclosure and exploitation.

Many of the peer reviewers today have commercial ties. Cutting-edge research may be essential to a firm's competitiveness. Consequently, there is more incentive to circumvent the norms of peer review and channel innovative ideas of grant applicants directly to the commercial sector.

We used the DABS data base to test the relationship of peer reviewership with commercial affiliation of scientists. The National Science Foundation (NSF) provided a complete list of potential and actual peer reviewers for fiscal years 1982 and 1983. The list includes the number of proposals reviewed by each individual. We compared that list with the DABS data base. Forty-nine percent of the dual-affiliated scientists appeared on the NSF peer-review list as potential reviewers in the biomedical sciences. Of the 832 scientists in the data base, 343 (41.2 percent) reviewed one or more proposals during the two-year period.

It is very difficult to prevent people who are so inclined from pilfering ideas while they serve as peer reviewers. The integrity of the system depends upon the adoption of informal ethical norms by members of the scientific community. Stigmatization and moral opprobrium are important disincentives for violating the norms. But the opportunity to channel innovative ideas in funding proposals to selected commercial enterprises may exacerbate the pilfering of innovative ideas. This may lead some scientists to seek commercial funds for their ideas rather than risk having them stolen through the peer-review process.

CONCLUSION

In less than a decade, the fields of molecular biology, genetics, and biochemistry in the United States have experienced a dual transformation. First, they have been transformed as basic sciences in the aftermath of the discoveries of gene splicing and gene synthesis. Second, they have been transformed as social institutions as the marriage between academic and industrial science was consummated.

This study has generated the first quantitative map of university-industry linkages across the biological disciplines arising from these transformations. The results indicate that such ties are widespread. U.S. biologists, en masse, have responded to the opportunities of entrepreneurial science. These linkages have appeared rapidly, as university policies have changed and as norms of behavior among biological scientists have shifted. In addition to the overall rate of affiliation, however, the differential location deserves attention. To the extent that these ties are focused on central, elite universities, their consequences are likely to be profound. These institutions not only are the wellspring of future path-breaking discoveries, they also are vital as the training grounds of the next generation of leading scientists and academic faculty. For these reasons, the heavy concentration of faculty-industry ties in first-rank institutions is likely to magnify their possible consequences. Faculties with university and industry affiliation in the biological sciences are becoming the rule rather than the exception in the United States and Canada. Many leaders in the field of molecular biology paved the way to entrepreneurship and serve as role models for younger faculty.

These data focus on relatively enduring forms of corporate involvement. While we have not examined less involving, more transitory associations such as contract research relations, we believe that they will magnify the pattern seen here. Most established large biotechnology firms do not have SABs, but they do have networks of consultants. These relations may also affect the behavior of academic scientists. Therefore, our data understate the full extent of faculty-industry linkages by emphasizing SAB membership or managerial participation, exclusively. The data base of DABS offers the possibility of testing additional hypotheses about the effects of industry associations among academic faculty, particularly the potential influence of commercial affiliation on research agendas, conflicts of interest, and norms of scientific communication.

Chapter 3

SCIENCE, SOCIETY, AND THE EXPANDING BOUNDARIES OF MORAL DISCOURSE[3]

SHELDON KRIMSKY

Tufts University, Department of Urban and Environmental Policy

MULTI-VESTED SCIENCE

(AN EXTRACT FROM *SCIENCE, POLITICS AND SOCIAL PRACTICE*)

Among the four institutional norms of science cited by Robert Merton, the shared value of disinterestedness embodies the moral commitment to leave aside one's personal interest when investigating the laws of the natural world. In the idealized situation, scientists have only one interest in the outcome of inquiry and that is to ascertain the truth or falsity of a hypothesis. Fraud, Merton noted, was virtually absent in the annals of science as a consequence of this widely shared norm. If the sole interest of the scientist is the pursuit of truth, conflict of interest among scientists must be an oxymoron. As in cases of fraud, such conflicts were traditionally viewed as deviant cases that do not negate the near universality of the norm.

Fifty years after Merton published his thesis, the social relations and structure of science research has undergone significant change. In particular, it is no longer uncommon for academic scientists in certain fields to

3An extract from an originally published chapter in *Science, Politics and Social Practice, Boston Studies in the Philosophy of Science* (Dordrecht: Kluwer, 1995).

have consulting relationships with industry, to hold stock in companies related to their work, to hold multiple patents from their research, or to be involved in the development of a new company. These changes in the biological sciences started almost immediately after the discovery of gene splicing techniques just as industrial genetics had gotten underway. And with these changes came a rapid transformation of the norms of the biological sciences.[1]

For several reasons, entrepreneurship in the biological sciences created a more intense public response and media reaction than similar activities in other disciplines. This was surprising to many biologists who viewed the commercial possibilities in molecular genetics as analogous to what had happened in chemistry, physics, computer sciences, and many other disciplines that had something to offer industry. The public seems to expect a higher standard of ethical conduct in the health sciences than it does in other fields. Since the major portion of health sciences research in universities is publicly funded, scientific investigators are considered more directly accountable to the public interest than other fields.

Scientific fraud and conflict of interest were the subjects of numerous media accounts in the 1980s. After a series of hearings, Congress highlighted these issues in a report titled "Are Scientific Misconduct and Conflicts of Interest Hazardous to our Health?"[2] The report cited scientific studies compromised by financial conflicts of interest and questioned the ability of the scientific community to set its own standards of professional conduct.

Once remote from commercial linkages, the biological sciences witnessed a rapid and aggressive rush among leading scientists and their universities to capitalize on the financial expectations of biotechnology.[3] Many new firms were started by scientists who retained their full academic appointments. The notion of "disinterestedness" in science was subject to a new critical inquiry. Can a scientist whose research is funded by a drug company be disinterested in the outcome? What about a scientist who is evaluating the safety or efficacy of a product that his company is poised to manufacture? Should a scientist who has equity in a company be required to disclose the relationship in a scientific publication related to the firm's commercial interests?

These questions, once the purview of professional ethics, now are at the centerpiece of public policy. The suspicion of the "disinterested" scientist was expressed best by the action of two major scientific journals: The *Journal of the American Medical Association* and the *New England Journal of Medicine* that require authors to disclose consulting and equity relationships related to their research.

The changing social structure of science and the growth of reciprocal and symbiotic relationships between academic research and commercial development has resulted in a new public examination of the moral status of science. Despite efforts within the scientific and medical community to retain internal control of the ethical behavior of scientists, the scope of public accountability has been widening. Issues once deemed to be in the domain of professional ethics, have become issues of social ethics. Swazey notes: "there has been a progressive shift in the locus of control from within clinical investigation to extra professionally or bureaucratically mandated laws and regulations."[4] Scientific fraud, conflict of interest, intellectual property, experiments with animals and fetal tissue, and genetic engineering are among new areas where governments have begun to take a more proactive role in monitoring the behavior of scientists. These trends do not imply that scientists are less moral than their predecessors. Rather, scientists are embedded in a new system of social relations that blur the boundaries between pure and applied research, between private versus public science, and between inquiry for knowledge and inquiry for profit. Scientists are faced with more choices and are situated in more varied contexts than they once were. Changes in the normative structure of science are reflections of changes in the nature and organization of research.[5]

It is a popular misconception that the autonomy of science is narrowed as external events encroach on its self-governance.[6] The concept of "scientific autonomy" in this context is treated as an entitlement that comes along with being a member of the professional guild. Not only is science treated ahistorically in this way but "autonomy in science" takes on an essentialist status somewhat like a theory of natural rights. I would argue that the moral autonomy of science expressed in such terms as "freedom of inquiry" does not precede the social context within which science is carried out but is derivative of it. This means that the normative conditions of science do not make sense apart from the political and economic context within which science is embedded.

Once remote from the affairs of public life, there is much more overlap between the moral domains of science and society. As scientists and their institutional cultures have become more deeply woven into the fabric of society through military research, federal grants, and entrepreneurial affairs, the spheres of normative behavior have been pulled closer together resulting in a recalibration of the boundaries of self-governance. The independence of moral spheres is no longer functional in the new system of relations. Thus, it is not as if the moral autonomy of science is threatened, rather a new concept of scientific autonomy and public responsibility is emerging.

Chapter 4

FINANCIAL INTERESTS OF AUTHORS IN SCIENTIFIC JOURNALS: A PILOT STUDY OF 14 PUBLICATIONS[4]

SHELDON KRIMSKY
Tufts University, Department of Urban and Environmental Policy

L. S. ROTHENBERG
University of California, Los Angeles, Division of Medical Genetics, Department of Medicine

P. SCOTT
Tufts University, Fletcher School of Law and Diplomacy

G. KYLE
Rutgers University, Office of Corporate Liaison and Technology Transfer

INTRODUCTION

There was a time decades ago when it would have been problematic for an academic scientist in most fields to acknowledge publicly a relationship with private industry. So it would have surprised—if not shocked—most attendees at the 1951 annual meeting of the American Association for the Advancement of Science if they had wandered into its Section on Science and heard a vice-president of Georgia Tech predict:

4Originally published in *Science and Engineering Ethics* 2, no. 4 (1996): 395–410.

> We may envision the time when industry and education are so closely related and so interwoven that it may be difficult to tell whether an individual is a worker in industry, or a faculty member in college, or whether he is a teacher or a student, and as a matter of sober fact, he may be all four over a year's time.[1]

Considerable evidence exists that attitudes among academic scientists toward private industry have changed, particularly for those in the biological and biomedical sciences. Two 1986 reports[2,3] found that in the United States, university-industry research relations had become increasingly important to both parties. Almost half of the biotechnology firms surveyed had such relationships,[4] and their financial support "may account for 16 to 24 percent of all external support for university research in biotechnology."[5] Other studies also suggested that a sizable number of faculty members working in the life sciences at leading universities are involved in the commercialization of biotechnology.[6]

The Human Genome Project has generated an unusually large number of collaborations between academic scientists at universities or publicly funded research institutes and newly created biotechnology companies. Many leading scientists searching for genes linked to human diseases have been identified publicly as being associated with biotechnology and biopharmaceutical companies seeking to commercialize such technology.[7]

Since the U.S. Congress in 1980 passed the Bayh–Dole Act (Public Law 96–517) that gives educational institutions ownership of inventions created with federal grant funds, an increasing number of universities and colleges have been actively engaged in patenting and licensing such inventions. In fiscal year 1993, royalty revenues for the top ten universities or state college systems amounted to $170 million, and faculty scientist-inventors shared a varying percentage of those royalties.[8] As U.S. researchers compete for decreasing levels of Federal research support and as major universities themselves create start-up biotech companies or engage in joint ventures with existing companies, the incentives for further academic-industry relationships increase.

While there are many positive benefits (e.g., financial, scientific, educational, and societal) in academic-industry relationships, particularly in biotechnology and bio-pharmacology, some observers have begun to see potential hazards that must also be considered. Blumenthal[9] suggests that to the extent that academic-industry relationships or individual university investigators' financial relationships are perceived as influencing the conduct and publication of research, "observers will cease to see universities as sources of impartial, disinterested knowledge that deserve public support and the freedom to use that support as they see fit." Similar concerns have been expressed regarding European, especially British, universities.[10] Djerassi says that this involvement is of "a magnitude that invites instant suspicion and criticism."[11]

Concerns over the potential for conflict of interest (defined as "situations in which financial or personal considerations may compromise, or have the appearance of compromising, an investigator's professional judgment in conducting or reporting research")[12] in science were heightened in the 1980s in response to the rapid commercialization of molecular biology and the overall growth of academic entrepreneurship.[13–16] University scientists and their institutions took on expanded roles in commercializing discoveries.[17–21] Many professional organizations and journals have given voice to these concerns.[22–29] In the aftermath of a series of Congressional hearings, scientific journals, universities, and most recently federal agencies responsible for funding have issued guidelines for disclosure of possible conflicts of interest.[30–33]

Three important areas where disclosure has been discussed are federal grant proposals, articles submitted to journals for publication, and peer review.[34–40] The most far-reaching regulations for preventing and managing financial conflicts of interest in federally sponsored research were issued by the Public Health Service (PHS) in July 1995.[41] These regulations apply to all research funded by the PHS, including the National Institutes of Health. Under these new rules, universities are required to establish review procedures to evaluate whether the mandatory disclosure of financial interests by a candidate for a grant constitutes a conflict of interest, and if so what steps should be taken.

These regulations define a "significant financial interest" as "anything of monetary value, including but not limited to, salary or other payments for services (e.g., consulting fees or honoraria); equity interests (e.g., stocks, stock options or other ownership interests); and intellectual property rights (e.g., patents, copyrights and royalties from such rights)." Investigators need to disclose only those significant financial interests that "reasonably appear to be affected by the research proposed for funding by the [Public Health Service] including the investigator's financial interest in entities whose interest would be affected."[42] A series of exemptions are also provided, including equity or salary interests that do not exceed $10,000 in value or a 5 percent ownership interest in any single entity.[43] The National Science Foundation has a similar policy[44,45] whereas the Food and Drug Administration has proposed a different approach.[46]

Several journals have developed disclosure policies for contributors in reaction to public concerns over the appearance of bias in their publications.[47] The International Committee of Medical Journal Editors (ICMJE) passed a resolution in 1988 that authors should acknowledge any financial relationships that "may pose a conflict of interest,"[48] but the resolution is not obligatory for the twelve journals participating in the organization. A few medical journals, including the *Journal of the American Medical Association*, *The New England Journal of Medicine (NEJM)*, and *The Lancet* have established policies that request authors to disclose to the readers any commercial associations they or close family members have that might pose a conflict of interest in connection with a submitted article.[49,50] Contributors to these journals are expected to disclose consultantships, stock ownership, honoraria, and substantial gifts. In 1993, the ICMJE reinforced its previous resolution urging that authors who submit a manuscript, whether an article or a letter, acknowledge all financial support and other financial or personal connections to the work.[51]

This reports on the results of a pilot study that measures the frequency of one set of verifiable "financial interests" (as defined for this study) in the subject matter of the articles published in fourteen scientific journals that are linked to the principal authors of those articles. The objectives of this study were: (i) to select a set of published and observe the degree to which a sample of authors hold a financial interest in areas related to their research that are reportable under current standards; and (ii) to examine the hypothesis that significant numbers of authors of articles in life science and biomedical journals have verifiable financial interests that might be important for journal editors and readers to know.

These objectives were applied to a pilot study of Massachusetts academic scientists who were cited as first or last author in at least one article published during 1992 in fourteen leading journals of cell and molecular biology and medicine. Massachusetts was chosen for the pilot study because it has significant biomedical activity, it is the home of many new biotechnology start-up companies, and it has strong academic institutions with faculty that are likely to have involvement in the biotechnology industry.

METHODS

To achieve these objectives we first adopted an indicator of "possessing a financial interest," following recent federal policy guidelines, that applies to scientific authors. Second, we constructed a database of every published in 1992 by fourteen leading life science and biomedical journals that had a first or last author whose affiliated institution was located in Massachusetts. Third, we applied the indicator to determine the frequency with which authors had satisfied the condition of "possessing a financial interest." Fourth, we examined the articles for any disclosure of financial interest.

Indicator of Possessing a Financial Interest

For this study a scientific author is said to possess a financial interest in his/her published work if he/she meets one of the following conditions: (i) is a member of a scientific advisory board of a company that develops products related to the scientist's expertise; (ii) is listed as an inventor on a patent or patent application for a product or process closely related to the scientist's publication under review; (iii) serves as an officer, director, or major shareholder in a for-profit corporation involved in commercial activities related to the scientist's field of expertise. This indicator is not meant to exhaust the meanings of "possessing a financial interest." Other possible criteria, which we were unable to check, include personal and familial investment holdings, consultantships, and honoraria[52,53] Note that many scientists have consulting relationships with biotechnology firms which are not in the form of membership on scientific advisory boards, but because we did not have data to independently verify these relationships, these financial interests were not included as indicators in this pilot study.

Scientists as Advisors to Companies

We used two data sets: 1) Massachusetts biotechnology firms, including their officers and scientific advisory boards (SABs), and 2) scientists listed as inventors on patents or patent applications registered with the World Intellectual Property Organization (WIPO). The methodology for developing an inventory of SABs for a population of companies was adapted from Krimsky et al.[54]

An inventory of Massachusetts biotechnology firms was developed in two stages. First, a comprehensive list of firms was derived from four sources: (i) unpublished data compiled for Krimsky et al.;[55] (ii) the 1994 Massachusetts Biotechnology Council membership list; (iii) The *Genetic Engineering News Guide to Biotechnology Companies*, 1994;[56] and (iv) an inventory of Massachusetts biotechnology companies prepared in June 1993 by Lyndon Lien for the Biotechnology Center of Excellence Corporation, Boston, Mass. We found a total of 149 biotechnology companies (i.e., the comprehensive list) with a Massachusetts address derived from the four sources.

Second, a subset of the comprehensive list (i.e., the dedicated list) was compiled by selecting from the comprehensive list those companies that use genetic or cellular techniques to manipulate genes or organisms, that work with genes or proteins, or that use cells to clone genes or reagents. Excluded from the dedicated list were those firms that are primarily instrument manufacturers or that engage in large-scale fermentation from source materials provided by another company. The list of dedicated Massachusetts biotechnology companies (DMBC) consists of eighty-four entries. Scientific advisory boards and company officers were obtained from DMBCs through sources that included corporate annual reports and federally mandated financial disclosure statements of public companies. Corporations that are not required to file periodic reports with the U.S. Securities and Exchange Commission were surveyed by letter and phone. Using this method, we compiled a list of 370 different scientists on SABs of Massachusetts firms.

Author Database

To select our study population of journal authors against which to apply our indicator "possessing a financial interest," we chose a base year of 1992 and fourteen leading journals in cellular and molecular biology, and medicine. We chose 1992 because it was the most recent year for which complete patent information could be obtained from the WIPO (the information becomes public eighteen months after filing), it coincided with available information on SABs and it was a year that saw considerable commercial activity in biotechnology and a heightened discussion about conflict of interest.

The fourteen English-language scientific journals represented a sample of the leading biologically-oriented science and biomedical journals based on the 1992 journal impact factors calculated by the Institute for Scientific (ISI).* [5]We sought journals that were publishing articles of potential commercial interest to biotechnology and biopharmaceutical firms, both in the clinical and basic sciences (see Table 4.1).

5The journal impact factor has been defined as "a measure of the frequency with which the 'average article' has been cited in a particular year."[57]

The journals [and their ISI impact factors] represented the **general sciences** (*Nature* [22.139], *Science* [20.967], and *Proceedings of the National Academy of Sciences of the U.S.A.* (PNAS) [10.480]); **biochemistry and molecular biology** (*Cell* [33.617], *EMBO Journal* [12.634], *Journal of Cell Biology* [11.118], *Molecular and Cellular Biology* [8.ins291], *Journal of Biological Chemistry* [6.733], and *Plant Cell* [6.342]); **genetics and heredity** (*Genes and Development* [14.270] and *American Journal of Human Genetics* [9.076]; and **general and internal medicine** (*New England Journal of Medicine* [24.455] and *The Lancet* [15.940]). A new genetics journal, which only began publishing in 1992 and produced only nine issues that year, *Nature Genetics*, was not rated by ISI but was included on the basis of its subsequent reputation.

From these journals, all original articles (10,148) about cellular and molecular biology and genetics were selected. From these we selected a subset on the criteria that either the first or last author was affiliated with a Massachusetts nonprofit academic or research institution (812 or 8.0 percent of the original set).

We chose first and last authors to set boundaries on the size of the author database while insuring that it included the significant contributors to the research publications. We assumed that one or both of these two authors would likely have had primary authorship responsibility, as is common practice in the biological sciences for multiple-authored papers. The number of unique Massachusetts-based authors derived from the screening criteria was 1,150. The total number of authors on all articles screened is likely five to six times larger than that figure of 1,150. As expected, many of these authors were listed on multiple papers. In our analysis, we deleted from the reference group forty-five authors who listed a Massachusetts biotechnology company as their address since that constitutes a disclosure of financial affiliation. This left 1,105 authors who gave as their affiliation a nonprofit academic or research institution.

Patent Applications and Patents Issued

One of the objectives of the study was to determine the percentage of authors who were listed as inventors on patent applications, or were issued patents on products or processes that closely resemble the content of their scientific papers. In this respect, the inventorship status on a patent/patent application meets one of the criteria for having a financial interest. Patent applications filed in the U.S. Patent and Trademark Office in Washington remain confidential until the date they are issued as patents, a process that can take two to three years or longer. Thus, at the time of the study, the patent application system in the United States was essentially a secret one, and there were no industry or academic databases containing information on filed U.S. patent applications. Therefore, we chose to review the patent applications of U.S. origin filed under the Patent Cooperation Treaty (PCT), which was signed in 1970 and came into effect in June 1978 under the supervision of the WIPO in Geneva. Under the terms of the treaty, WIPO is required to publish the patent applications it receives exactly eighteen months after the date of their submission to the Patent Office. Using the PCT filings, we were able to identify the patent applications of U.S. origin on which authors were listed as inventors, which are otherwise required by U.S. law to treated be as confidential information by the Patent Office.[58]

We matched the list of authors selected from the fourteen journals with WIPO-listed patents and patent applications for 1992. The WIPO information was available on PCT Patent Search, a CD-ROM available from MicroPatent in East Haven, Connecticut, and Cambridge, England. We checked every author in the author database against the names of inventors on patent applications and patents listed on PCT Patent Search. Then we screened for those patent applications that listed the relevant author's name as an inventor and whose subject matter was closely related to the author's journal article. All four members of the study team reviewed the match between the subject matter of the patent and the subject matter of the journal article.

One of these reviewers (G.K.), a scientist with extensive experience with university intellectual property and technology transfer, served as the final arbiter of whether the patent application was indeed based on the research article in question. This process relied on the frequent similarity in tables and graphs used in both articles and patent applications, as well as a non-mechanistic visual search for similarity of language in the examples used to describe processes and findings. We reviewed the abstracts of patent applications for a match, and, in certain circumstances, obtained the full text of the applications in order to resolve uncertainties.

Corporate Officers, Directors, or Major Shareholders

To identify authors who are officers, directors, or major shareholders of biotechnology companies, we used a database of information on public corporations created by analyzing filings with the U.S. Securities and Exchange Commission (SEC). The database titled Compact D/SEC (1995) is owned by Disclosure, Inc., Bethesda, Maryland, and is updated quarterly from July 1990. About 50 percent of the biotechnology firms in Massachusetts, and the overwhelming majority of biotechnology firms in the United States, are privately held, and therefore are not required to report information to the SEC. For other sources of information on private companies we used surveys, news reports, and published materials from companies, but this information was very spotty and not very helpful.

RESULTS

From the 1,105 journal authors we found that 112 or 10.1 percent were listed as inventors on patents or patent applications on file with the WIPO that correlated with published articles in our study sample. There were 69 authors who were SAB members in Massachusetts biotechnology companies (6.2 percent). There were 15 authors who serve as company officers, directors, or major shareholders (1.4 percent). The frequency with which an author who does not give a firm affiliation is associated with one or more of the three categories of financial interest under study is calculated by the union of the three sets, which have overlapping members. This condition is satisfied by 169 authors, which indicates that 15.3 percent of the author population had at least one financial interest in their published articles (see Table 4.2).

We also calculated the frequency that an article selected from the reference population (n=789) has one of its lead authors identified with one or more of the three categories of financial interest. Twenty percent of articles have a lead author on a Massachusetts SAB of a biotechnology company (n=160); 7 percent have a lead author who served as an officer or major shareholder in a biotechnology company (n=57); 22 percent have a lead author who is listed as an inventor in a patent or patent application closely correlated with a publication in our study sample (n=175). Thirty-four percent of the articles in the study sample (n=789) meet one of the three criteria satisfying the condition of having at least one lead author with a financial interest (n=267).

After reviewing the 267 identified as having a lead author (first or last) with at least one financial interest closely related to their publication, we could find no statements of disclosure for any of the three indicators of financial interest linked to a lead author who gave an academic affiliation. There was a disclosure of stock ownership in one article in the *NEJM* but the authors cited were identified as employees of the corporation in which they held stock and were deleted from our analysis.

Patent applications are often filed as an afterthought by universities and nonprofit research centers that own the patent rights and require researchers to disclose their inventions in order that these institutions might benefit from successful technology transfer. The federal and state governments, to the extent that they have research laboratories, are doing exactly the same thing, and the U.S. Congress and state legislatures encourage such activities, public and private, to spur economic development.

DISCUSSION AND LIMITATIONS OF DATA

Scientific Advisory Board Membership and Corporate Officers/Significant Shareholders

Our database included active SABs only, but occasionally members do cycle off. Membership information on SABs obtained from company reports and federally mandated documents does not stipulate the dates a person begins or terminates service on the board. Because our SAB data were collected in 1994 for companies founded up until 1992 and because SABs do not change very frequently, we felt it was reasonable to assume that the scientific advisers we identified in 1994 were active in companies that existed in 1992. However, our assumption may not be correct.

Our inventory of SABs was based on Massachusetts companies because we had access to the most complete information on SAB membership in that state. It seems likely that the profile of academic industry ties found in Massachusetts is similar to that of other biotechnology rich states that contain high concentrations of academic researchers such as California, New York, Maryland, New Jersey, and perhaps Texas. However, without a national study of financial interests in publications or other state profiles, it is premature to extrapolate these results to other parts of the country. Because many Massachusetts scientists are on the SABs of out-of-state U.S. companies or international biotechnology companies, our data significantly undercount the corporate SAB affiliations of journal authors. Also, three private companies in Massachusetts declined to report their SAB composition.

A national data set of company officers, directors and major shareholders (beneficial owner of 10 percent or more of stock issued) was used to identify journal authors who have a university affiliation and fall into one of those categories. However, this database only applies to public corporations, which comprise about 50 percent of our Massachusetts firms but a much smaller percentage of all biotechnology companies in the United States. Thus, our sources underestimate the number of academic faculty who are corporate officers, directors or major shareholders. Because the number of such corporate affiliations among academic faculty in our sample is small (n=15) compared to patent/patent application inventors and SAB members, undercounting in this category is unlikely to affect the outcome significantly.

In general, our data underestimate financial interest because we only considered three factors in measuring it. Other circumstances such as personal or family stock holdings and consulting relationships would drive the numbers of authors with financial interests up. In addition, we were not able to assess financial interest in privately-held biotechnology companies which are not required to report their data to public agencies. Nor did we have access to data that would identify academic consultants to biotechnology companies who are likely to have more technical interactions with client companies than are scientific advisory board members. Furthermore, we were unable to document scientist-authors who received unexercised stock options from biotechnology companies or who purchased company stock in the open markets. Finally, we had no way of determining whether the companies on whose SABs any authors in this study were serving intended to exploit commercially the content of the authors' papers. Thus SAB membership, by itself, may not be a useful indicator of financial interest when examining published manuscripts.

Author Selection

It is possible that some authors listed between the first and last are scientists who hold financial interests of the type we were seeking to document. By limiting our analysis to first- and last-named authors, we could have underestimated the presence of such interests for any article's set of authors.

Applications and Patents

Patents or patent applications of U.S. origin but not filed under the PCT would not show up in our data set, and thus patent inventorships among scientists in our study may be underreported. The degree to which academic scientists are being listed as inventors on patents for biological materials and processes, and becoming eligible for royalties on successfully commercialized products or techniques, introduces an important new source of financial interest among life science and biomedical authors, but this reality must be placed in some context. Under university policies, faculty scientist-inventors themselves receive only a portion (often one-third) of royalty income from patents on their inventions. For most authors, the professional status they receive from publications is far more important than any financial interest they may realize from their research results.

Relevance of Financial Interests to Published Research

Financial interests of some kind may be inescapable to researchers and universities in the late twentieth century, and the mere existence of a financial interest in no way establishes a "conflict of interest" or automatically makes questionable the data and conclusions presented. It is the appearance of a potential conflict that the various guidelines mentioned earlier seek to prevent, and the belief that disclosure of such interests to editors, reviewers or readers will eliminate all potential for such a conflict continues to be a hotly debated topic.

Our results understate the actual financial interests held among members of our study sample. No definitive data are available, but observers in the intellectual property field have speculated that in the early 1980s, more patent applications of U.S. origin were filed under the PCT. With greater financial cost now an issue to both universities and companies, these observers speculate that only the "best" and "most commercially promising" applications are now filed under the PCT.

Disclosure Policy

The notions of what constitutes a "financial interest" and what is considered a "disclosable financial interest" have been discussed at the federal and institutional levels. The broadest interpretation includes any activity that might give the appearance of impropriety or bias in the published or proposed research. The Public Health Service regulations[59] and those of the

National Science Foundation[60] distinguish between what investigators must *consider* and what they must *disclose.* "The investigator must consider all Significant Financial Interests, but need disclose, only those that would reasonably appear to be affected by the research proposed for funding by the PHS.[61] The PHS lists intellectual property rights under "Significant Financial Interest." It excludes income from service on public or nonprofit advisory boards, but not for-profit advisory boards, although it is unclear whether the monetary threshold applies to SAB membership.

Interpreted narrowly, a "disclosable financial interest" might be limited to actual dollar payments above a threshold in areas related to the published or proposed research. Under this interpretation, holding a patent that has not generated income or serving on an SAB where the annual compensation is below $10,000 would not be considered a disclosable interest. We have chosen the broader meaning of "financial interest" but leave it to others to determine whether such interests are disclosable under the prevailing standards.

Of the fourteen journals in the study sample, four currently require disclosure of financial or other potential conflicts of interest: *Science*, *NEJM*, *The Lancet* and *PNAS*. However, *The Lancet* introduced its policy in January 1994, two years after our reference year. In 1992, *NEJM* and *Science* (*Science* instituted its policy July 31, 1992) required some form of disclosure.[62–65] *PNAS*, as of May 1996, requires that all authors "disclose any commercial association that might be a conflict of interest in connection with the manuscript."[66]

Almost all the scientific journals surveyed did not in 1992 and still do not in 1996 require any disclosure to their editors and reviewers of the type of information we have characterized as constituting a financial interest. Since *Science* and *NEJM* carried a relatively small percentage of the in the database (6.78 percent), and since the requirement in *Science* only came into effect in mid-1992, the results of our study do not provide a baseline for mandatory disclosure of financial interest.

One article published in *Science* after the July 31, 1992 date that mandatory disclosure took effect had a patent application matching the manuscript without a disclosure statement; however, *Science* does not *require* disclosure of patent inventorship to the editors and we have no way of knowing whether it was disclosed.

It should be noted that we had no mechanism with which to examine whether any of the authors of the articles we examined submitted information regarding financial interests to editors of the journals involved, whether required to do so or not. The mere absence of such disclosure to readers of the printed articles does not indicate whether in fact specific disclosure occurred in any given case.

CONCLUSION

The degree of financial interest found among authors of published articles in this study is noteworthy. Of particular significance is the finding that one in every three articles in our sample has at least one Massachusetts-based author with a financial interest and that 15 percent of authors in our sample have a financial interest in one of their publications. Moreover, the results provide a minimal estimation of the degree of financial interests held among authors in scientific publications.

Although we found no formal study on the subject, it seems unlikely, however, that the vast majority of researchers realize any significant financial gain from any individual publication. Yet it also seems increasingly evident that the goal of "science for the sake of science," that is research objectively performed for the sake of pure knowledge (if that were ever the reality), is challenged by the perception that someone—the researcher, the institution for whom the researcher works, a biotechnology company with whom the researcher or the institution is affiliated—stands to benefit from the research in a way that could bias the manuscript or its findings. At a time of diminishing research funds, and institutional budgets this is a perceptual problem worthy of consideration.

We can conclude with confidence that for the year 1992 the rate of *published* voluntary disclosures of financial interest (as defined in our study) for fourteen leading journals by Massachusetts-based academic scientists (*i.e*, those who do not give a company affiliation) is virtually zero. Similarly, most life science and biomedical journals do not require that any of these financial interests be disclosed to their editors and reviewers. Further research is needed to determine the effectiveness of mandatory disclosure requirements instituted by several journals, a policy development described by one critic as "the new McCarthyism in science."[67] The newly established Public Health Service and National Science Foundation guidelines on conflict of interest in research will broaden the scope of mandatory financial disclosure within the grant process, leaving open the issue of financial disclosure in published research. The need for a clearly defined and pragmatic set of guidelines governing the relationships between academic researchers and industry has been persuasively articulated.[68]

It is conceivable that the day may come when financial interests of authors of scientific articles will be so ubiquitous that readers will assume automatically the existence of such interests unless there is a specific disclaimer to the contrary. That approach may represent a less burdensome and intrusive one to some scientists, but it may only heighten the anxiety that the reputed independence and disinterested nature of universities, and the accompanying public support, will suffer in the process.

ACKNOWLEDGMENT

This research was supported by a grant from the Greenwall Foundation.

TABLE 4.1: Fourteen Journals Surveyed on Financial Interest of Authors

Journal Title	Total number of original articles on biology and genetics in 1992	Number of original articles with Massachusetts authors in 1992	Percentage of total	Percentage of database	Disclosure of interests required by journal in 1992
Am. J. Human Genetics	271	11	4.05	1.35	No
Cell	357	51	14.3	6.28	No
ENBO Journal	546	28	5.13	3.45	No
Genes & Development	217	27	12.4	3.32	No
J. Biol. Chemistry	3854	226	5.86	27.83	No
J. Cell Biology	516	42	8.13	5.17	No[a]
The Lancet	46	1	0.02	0.12	No
Molecular and Cell Biology	610	51	8.36	6.28	No
Nature	535	61	11.4	7.51	No
Nature Genetics	120	15	12.5	1.85	No
New England J. Medicine	23	5	21.7	0.62	Yes[b]
Plant Cell	117	7	5.98	0.86	No
Proc. Nat'l Acad. Sciences	2454	236	9.62	29.06	No[c]
Science	482	50	10.4	6.16	Yes[d]

a. *Lancet's* conflict of interest policy, introduced to authors on January 1, 1994, expects authors to list all relevant sources of financial support that could potentially embarrass an author if the grant, business interest, or consultancy became known after publication.

b. *The New England Journal of Medicine* first introduced its conflict of interest policy in July 1990. The journal expects that its authors will not have any financial interests in a company or competitor that makes a product discussed in the article.

c. *The Proceedings of the National Academy of Sciences*, as of May 1996, requires that all authors "disclose any commercial association that might be a conflict of interest in connection with the manuscript."

d. *Science* instituted a policy of requiring authors "to reveal to us any relationship that they believe could be construed as causing a conflict of interest, whether or not the individual believes that is actually so" as of July 31, 1992.

TABLE 4.2: Financial Interests and Publications

	A	B	C	A, B, or C
REFERENCE CLASS	PATENTS	SABS	CORPORATE OFFICERS	COMBINED (excludes overlapping members)
AUTHORS n=1,105	112	69	15	169
PERCENT	10.1	6.2	1.4	15.3
ARTICLES n=789	175	160	57	267
PERCENT	22.2	20.2	7.2	33.8

Chapter 5

FINANCIAL INTEREST AND ITS DISCLOSURE IN SCIENTIFIC PUBLICATIONS[6]

SHELDON KRIMSKY
Tufts University, Department of Urban and Environmental Policy

L. S. ROTHENBERG
University of California, Los Angeles,
Division of Medical Genetics, Department of Medicine

Both in the clinical context and in the context of the publication of academic research, there is the potential for a conflict of interest, as defined by Thompson,[1] when a set of conditions exist "in which professional judgment regarding a primary interest (such as a patient's welfare or the validity of research) tends to be unduly influenced by a secondary interest (such as financial gain)." Although the mere existence of a financial interest does not imply a conflict and the potential for financial gain is only one of many factors that can generate such conflicts (including "personal relationships, academic competition, and intellectual passion"), the International Committee of Medical Journal Editors (ICMJE) has identified "financial relationships with industry (for example, employment, consultancies, stock ownership, honoraria, expert testimony), either directly or through immediate family," as the most important conflicts of interest. Moreover, the

6Originally published in *JAMA* 280, no. 3 (1999): 225–226.

ICMJE[2] considers that the manner in which authors, reviewers, and editors deal with such conflicts can affect in part the credibility of published articles in scientific journals.

For readers unfamiliar with the controversies over disclosure of financial interests by researchers and/or authors, a brief review may be useful. Prior to the 1980s, the emphasis of any guidelines or policies regarding financial interests of scientists tended to focus on voluntary disclosure and self-regulation. Beginning in the early 1980s and continuing to the late 1990s, journals, federal agencies, university and medical associations, and the media have issued policies on financial disclosure for authors, reviewers, or grant applicants.

An Institute of Medicine report describes 2 competing models for the management of conflicts of interest: the "prohibition" model, which is "based on a presumption against any relationships that might present a conflict," and the "disclosure and peer review" model, which is "based on a presumption for such relationships with a provision for disclosure and review." A demonstration of "sufficient social benefit" (e.g., improved transfer of medical innovations to the bedside, creation of jobs, furtherance of economic development generally, and facilitation of private support of research programs and public universities) can override the prohibition model and outweigh the risk of bias. The disclosure and peer review model, by contrast, "holds that conflicts of interest are unavoidable and that financial conflicts are only the most visible and perhaps the least scientifically dangerous."[3]

Richard Horton, editor of *The Lancet*, has argued that the case in favor of full disclosure rests on three fallacies: (1) scientific writing can be free from common prejudices; (2) financial conflicts of interest are of greater concern than academic, personal, and political rivalries and beliefs; and (3) disclosure can "heal the wound inflicted by financial conflict."[4] An editorial writer in *Nature* suggests that, barring a demonstrated link between such financial interests and a lack of objectivity or other factors that weaken the credibility of a manuscript, disclosure should only be voluntary.[5]

Arguments favoring disclosure echo the conclusion reached by the American Medical Association, Chicago, Illinois, that "the best mechanism available to assuage public (and professional) doubts about the propriety of a research arrangement is full disclosure" and that such disclosure "should be made to the journals that publish the results of the research."[6]

Since the 1980s, when the commercialization of the biomedical sciences was becoming acutely visible in the American press[7] and the U.S. Congress held hearings on federal research funds and their relationship to conflicts of interest,[8] biomedical journals began adding conflict-of-interest requirements in their instructions to authors.

Even if the information is disclosed to journal editors, however, the question remains of whether it should be shared with journal readers. Some editors view their role as the administrators of such information.[9] We are persuaded by the views of Bernat and colleagues,[10] leaders in the American Academy of Neurology, who argue that the purpose of public disclosure of conflicts of interest is not to remove the conflict but to publicize it "so that all relevant observers become aware of it and can modify their opinions on the credibility of statements of the conflicted person accordingly," which mitigates but does not resolve the conflict.

In a survey of North American medical journal editors published in 1995, Wilkes and Kravitz[11] reported that 26 percent of responding editors required authors to reveal sources of their funding, 28 percent required disclosure of all institutional affiliations, and 13 percent and 10 percent, respectively, required disclosure of consultant positions and of stock ownership in companies that may pose a conflict of interest. This lack of editorial unanimity was revealed in the same year by the nation's two leading funding agencies, the National Institutes of Health[12] and the National Science Foundation,[13] which issued conflict of interest regulations requiring disclosure by researchers to their host institutions of financial interests in connection with grant proposals. It also comes at a time of changing conditions of scientific research funding and of the growth of a more entrepreneurial spirit among academic scientists and research institutions.[14,15]

Thus, although the ICMJE has expressed the majority view that "published articles and letters should include a description of all financial support and any conflict of interest that, in the editors' judgment, readers should know about"[16] the policies of medical and basic science journals vary significantly in their requirements to disclose financial interests to editors.

In our view, journal editors should begin to take seriously the implementation of disclosure policies in response to the escalation of financial interests of authors in their publications.[17,18] Journals should be specific in their instructions to authors on the types of financial associations related to their submission and the form of communication (original research, letters, book reviews, and scientific review articles) that warrant disclosure. We also believe that the scientific community and the public will be best served by the open publication of financial disclosures for readers and reviewers to evaluate. While financial interest in itself does not imply any bias in the results of a paper and should not disqualify it from publication, readers and reviewers are the best judges of whether there is evidence of bias and whether that evidence favors those interests.

Chapter 6

THE PROFIT OF SCIENTIFIC DISCOVERY AND ITS NORMATIVE IMPLICATIONS[7]

SHELDON KRIMSKY

Tufts University, Department of Urban and Environmental Policy

INTRODUCTION

Until the late nineteenth century, the profession of scientist in Western societies was comprised almost exclusively of men from the propertied classes or bourgeoisie who were educated at the elite European universities. It was a calling of sorts, not unlike the ministry for those with means and pedigree who could afford the luxury of investigating the workings of the universe by expanding and challenging their intellect. There was no vast wealth to be made maybe a comfortable living at the peak of one's career.

With the rise of federal land grant colleges in the United States and the expansion of free national universities throughout the world, new scientific career options were created for people of diverse socioeconomic status. Through much of the early twentieth century a career in academic science was much like a monastic order. The pursuit of knowledge, the sharing of

7Originally published in *Chicago Kent Law Review* 75, no. 1 (1999): 15–39.

its fruits, the gratification of self-enlightenment, and mentoring students were all the reward one required to sustain and nurture a career.

The goals of science were already being recast during the Baconian period in the seventeenth and eighteenth centuries when the distinction was made between "experiments of light" that seek to discover the causes of things, and "experiments of fruit" that apply the knowledge to practical ends.[1] The new European nation states began to recognize the practical significance of scientific discovery in areas such as weaponry, mining, and transportation. Building on the European experience with science and technology, the framers of the U.S. Constitution established intellectual property as a fundamental right and conferred to Congress the powers to make laws fostering the "useful arts."

A second transformation in science took place in its purest and most unfettered form during post–World War II American economic expansion. Scientific research was now a matter of public policy. The image of the lone scientist, broadly educated with the grasp of the large picture, working tirelessly in a makeshift laboratory furnished with hand-crafted equipment, and pursuing a path to knowledge according to some ineffable sixth sense, was undergoing a great transformation. The new image was for a strategically planned science consisting of teams of investigators working on large scale projects, competing for limited funds and positioning themselves in a social structure that would ensure the continuity of funding through volatile political times.[2]

In addition to the areas of academic science and engineering that became beneficiaries of state funding, industrial science also expanded significantly during this period. The American industrial system had become fully converted to the need for continuous technological innovation. Chemistry, chemical engineering, electronics, geology, and material sciences were among the academic fields to which industry developed close working ties. By the late 1940s, over 300 U.S. companies funded research in universities through fellowships and direct grants.[3] According to Porter and Malone, "[i]n the decades after World War II, connections between academia and industry slowly weakened, reaching a low point in the early 1970s."[4] As federal government support of basic research increased, the connections between academia and industry declined. In the mid-1950s, the federal government provided about 55 percent of the support for university research, industrial firms supplied 8 percent of the funds and the remaining 37 percent came from foundations and state governments.[5] By the late 1960s, the government share expanded to more than 70 percent while the industry's share fell to under 3 percent.[6] Industry support for universities began to rise again with rising budget deficits and the leveling of science funding in the 1980s.[7]

Until shortly after the World War II, biology was largely a science in the classical pre-Baconian tradition. We studied how things worked rather than how we could improve on nature. The Green Revolution provided the first significant post-war agricultural application of advances in plant biology.[8] The innovations were based on methods of plant breeding and developing hybrid plant varieties that were most efficient when used with chemical inputs.[9] Yields of staple crops were dramatically increased.[10] It is reported that in 1985 the average yield for corn was six times the 1930s figure.[11] Those innovations, however, were not a consequence of a fundamental transformation in the biological sciences. That transformation took place in 1973 with the discovery of recombinant DNA ("rDNA") technology.[12] In that monumental discovery, the biological sciences had made the transition from an analytic to a synthetic science. It was now possible to rearrange the basic architecture of living things by transplanting genes. There were some prior attempts at creating a synthetic biology through discoveries like the hybridization of crops or the cross breeding of animals.[13] But the changes one could make in animals and plants through those procedures were limited by the constraints nature imposed on sexual and asexual reproduction. These constraints were rooted in the genomes of these organisms.

The introduction of rDNA technology established the absolute fungibility of genes, opening up possibilities for synthesizing new organisms and establishing revolutionary methods for mass producing biological products. The commercial opportunities of this discovery were recognized immediately by scientists.[14] Seven years after the discovery of rDNA technology, the journal *Nature* published its cover page with the headline "Setting Up in Biology Business."[15] The issues' editorial raised the conflicts that arise when academic biologists and industry become "partners in progress."[16] There was an uneasiness expressed about the rapid and aggressive commercialization of the biological sciences.

> Problems and acrimony have arisen between the biologists and industry, and among the biologists, particularly over the question of confidentiality. Scientists with one foot in the academic and one in the business world would have felt a conflict between the need to publish fast and first in the former, and to keep secrets and respect patent law in the latter . . . . It is an interesting moment for biologists: they have great power in their hands. Do they let the entrepreneur guide them, willy-nilly, to the fastest return? Or do they, if ever so slightly, change his priorities?[17]

The scientific grounds for this transformation in the biological sciences are well studied and understood. But there is much to learn about the symbiosis between academic biology and industry. How is the field of biology changing as a result of the commercial interests in its discoveries? What has been the response of government to the new technological revolution in molecular genetics? What conflicts have arisen in the engagement of two cultures that have pursued a partnership of mutual interests? This paper discusses the factors responsible for and the manifestations of the intense commercialization of the biological sciences and their impact on the normative structure of science. What are the consequences for science and society of the commodification of scientific knowledge and the growth of entrepreneurship in scientific research? I shall begin with a discussion of the historical background.

THE RAPID COMMERCIALIZATION OF MOLECULAR GENETICS

The excitement in 1973 was palpable. Biology, it appeared, had just come of age. Chemists had been creating new molecules for over 100 years and were responsible for tens of thousands of new compounds that became the signature of our industrial world. Physicists split the atom and unleashed a form of energy that transformed the concept of war and created new forms of mass destruction. In the mid-1970s, developments in the cutting and splicing of genes brought immediate applications in therapeutic drugs, agriculture and food sciences.[18] Savan reports that "industrial funding of academic research and development in the United States quadrupled between 1973 and 1983, increasing from $84 million to $370 million."[19] Unlike other fields where scientists left academia to create their own for-profit business to exploit new discoveries, most of the leading molecular biologists retained their academic positions while pursuing their commercial interests.[20]

In March 1981, the front cover of *Time* reflected the new trend in academic biology.[21] It pictured the disembodied head of scientist entrepreneur Herbert Boyer set within a background of strands of deoxyribonucleic acid ("DNA") bursting from a cell.[22] Boyer, a co-patent holder of the Cohen–Boyer rDNA technique, was described as a new breed of millionaire-scientists, who commercialized their discoveries and helped establish a new industrial sector. While a professor at the University of California, Boyer was a co-founder in 1976 of Genentech, a first-generation biotechnology company formed to apply genetic engineering applications for drug development.[23] Five years after Boyer became involved with Genentech, his personal stock in the company was valued at about $40 million in a volatile biotech securities market. His university salary at the time was $50,000.[24] The lure of rapid financial success ran through the field of biology like an infectious virus.

A report of the Congressional Research Service, released in 1982, described the differences between the academic industrial relations in biology with that of other academic fields. The report indicated that the commercialization of biological techniques occurred at a more accelerated rate than similar instances in physics and chemistry, involved a broader spectrum of disciplines in its participation, and had a much wider range of applications than the commercialization of discoveries in the physical sciences.[25]

By the mid-1980s, hundreds of new venture capital companies had colonized faculty from the leading universities.[26] Following that trend, according to Kenney, "[b]y 1983 every large chemical and pharmaceutical company had made a multi-million dollar investment in biotechnology" and established major funding partnerships with universities.[27]

The climate for research in the biological sciences, particularly molecular genetics, both within the universities and throughout the federal government, changed dramatically. In describing the changes, a 1980 editorial in *Nature* commented that "a new breed of molecular biologist is emerging which concerns itself not so much with understanding basic mechanisms as with manipulating them to reach desired goals."[28] Meanwhile, U.S. government budget deficits began to take their toll on scientific funding.[29] The rapid expansion of universities and federal funding for science that had taken place in the 1960s and early 1970s had appeared to come to an abrupt end before and during the Reagan presidency. Fears had spread throughout academia that the Halcyon years of abundance for scientific research would not continue.[30] To retain their leadership in research, many of the elite research universities began to consider private sources of funding for filling the shortfalls resulting from declining federal budgets.[31]

As the decade of the 1970s came to an end, American business analysts reported declines in productivity and global markets of major industries like steel and microelectronics. They attributed the problem to the failure of American industry to maintain an innovative climate and make use of scientific breakthroughs. The obstacle, they argued, was not in America's scientific leadership, but in the time lapse between discovery and application. It was, in effect, as MIT's President Paul Gray wrote, a problem of technology transfer.[32] Congress responded favorably to this explanation of the United States' declining competitive edge. A series of legislative acts and executive orders were premised on the concept of technology transfer and university-industry partnerships.

Already in gestation during the Carter administration, several pieces of legislation were enacted in 1980 to create more cooperation between industries and universities. The Stevenson–Wydler Technology Transfer Act of 1980[33] encouraged interaction and cooperation among government laboratories, universities, big industries, and small businesses. In the same year, Congress passed the Bayh–Dole Patent and Trademark Laws Amendment,[34] which gave intellectual property rights to research findings to institutions that had received federal grants. Discoveries and inventions from public funds could be patented and licensed, initially to small businesses, with exclusive rights of royalties given to the grantee. The Economic Recovery Tax Act of 1981[35] gave companies a 25 percent tax credit for 65 percent of their direct funding to universities for basic research. In 1983, by executive order, President Reagan extended the Bayh–Dole Act to all industry. To close the circle of research partnerships among industry, universities and government, Congress passed the Federal Technology Transfer Act of 1986,[36] which expanded science-industry collaboration to laboratories run by the federal government. Governmental standards for keeping an arm's length from industry were being turned on their head. Through this act, a government scientist could form a "Cooperative Research and Development Agreement" ("CRADA") with a company as a route to commercializing discoveries made in a federal laboratory.[37] Government scientists could accept royalty income up to a given amount, 15 percent of the National Institutes of Health (the "NIH") share, to supplement their salaries. At the time this policy was enacted, there was virtually no public discussion about the blatant conflicts of interest that this would introduce. The CRADA required government scientists to keep company data confidential and impeded the sharing of information in government laboratories.[38]

The new federal initiatives on technology transfer and academic-industry-government collaborations were responsible for a marked rise in university patents.[39] In 1980, American university patents represented

1 percent of all U.S. origin patents.[40] By 1990, the figure rose to 2.4 percent.[41] Within that decade, the number of applications for patents on NIH-supported inventions increased by nearly 300 percent.[42]

The legislative and executive branches of government have invested in the idea that the ivory tower of academic science and the insulated domains of federal laboratories had to build bridges to the industrial sector. The operative term for these new arrangements was "mutualism." Industry, universities and government had something to offer one another. If carefully planned and consummated, the concerns and impediments to these partnerships could be overcome, and the American society would be the big winner. The engine of innovation was re-ignited, and in its wake universities would have to make adjustments and compromises.

But there was one additional national sector that could connect all the institutional pieces and give the fledgling field of applied genetics the boost it had so aggressively lobbied for. The courts laid the groundwork for establishing intellectual property rights over a broad spectrum of genetic discoveries, creating the new concept of "life patents."[43]

ORGANISMS, ANIMALS, AND GENES AS INTELLECTUAL PROPERTY

Supporters of technology transfer incentives as a means to jump start the lagging U.S. economy and establish the country's global leadership in biotechnology and information technology were faced with a legal obstacle to the fulfillment of the vision. Unless the discoveries and innovations in molecular genetics and software could be protected under patents or copyrights, the new products could easily be replicated by foreign companies and produced more cheaply. Two solutions to this problem were the establishment of trade barriers to prevent foreign products from competing with American goods or extending patent/copyright protection. As the new economic philosophy of free markets took a foothold in United States' policy, trade barriers were out of fashion. Transnational agreements like the North Atlantic Free Trade Agreement were opening up markets and eliminating tariffs. Patent and property right protection became the choice solution for protecting the United States' competitive position in a global economy. Decisions by the Supreme Court and the U.S. Patent Office, and non-decisions by Congress, established the final element of an economic policy that fully incorporated life forms and their parts into the market system. By establishing property rights over discoveries in biology, the courts and the U.S. Patent Office turned scientific knowledge into an invention, thereby creating new opportunities for scientists to acquire wealth, establishing new incentives for product development, and, downstream, adding to the costs of consumer goods and medical care.[44]

In June 1980, a single vote on the Supreme Court transformed the social and legal matrix within which science and the nascent biotechnology industry operated.[45] In a 5–4 decision, the Supreme Court overturned the Patent and Trademark Office's (the "PTO") denial of a patent for a microorganism, thereby paving the way for the use of the patent system for all sorts of life forms and their parts, including human genes.[46] The Court ruled that a human-modified microorganism can be classified as a product of manufacture, and thus falls under patent protection.[47]

At the time the Court reviewed *Diamond v. Chakrabarty,* there was a backlog of 114 patent applications of living organisms, with an estimated 50 applications added per year.[48] The Court was well aware of the commercial interests in this case. Biotechnology trade organizations were growing rapidly and the financial publications were replete with investment opportunities in new capital venture companies.

The Supreme Court saw as its task the determination of whether living organisms fit under U.S. patent law.[49] Without explicit statutory language, the Court sought to interpret the intention of Congress by examining the language and intent of historical documents such as congressional reports and revisions of the patent laws.[50]

In 1973, biologist Ananda Chakrabarty, while working for General Electric Corporation, modified a microorganism by transporting extra-chromosomal elements, called plasmids, from several organisms into one bacterium.[51] Each plasmid was capable of degrading a component of crude oil.[52] The patent application was for both the process of cleaning up oil spills and for the bacterium containing multiple plasmids.[53] The PTO accepted the process application, but rejected the application for a patent on the organism sui generis, claiming that microorganisms were products of nature and therefore not patentable.[54] The decision of the PTO was appealed to the Patent and Trademark Office Board of Appeals, which claimed that Chakrabarty's microorganism was man-made and not found in nature, but upheld the lower court on the claim that living things are not patentable subject matter.[55] The next appeals court, the Court of Customs and Patent Appeals, argued that a key section of the U.S. Patent Act[56] did not preclude man-made living organisms, dismissing the lower court's interpretation of the patent laws.[57] Its decision, upheld by the Supreme Court, extended the concept of intellectual property and provided a liberal interpretation of the terms "manufacture" and "composition of matter" within the meaning of the statute.[58]

On what legal grounds did the five Supreme Court justices base their decision? The justices could find no explicit language in the Act on the issue of whether living things were or were not patentable.[59] They reviewed two statutes on plant patents, one passed in 1930 and the other in 1970.[60] These statutes were used as evidence against the interpretation of blanket congressional intent to include living things as patentable material, because if that were so, Congress would have not chosen to enact two pieces of legislation that extended intellectual property ownership to plants.[61] Moreover, the 1970 plant patent statute explicitly excluded microorganisms from patent protection.[62]

The Court did find in a 1952 recodification of the patent statutes a congressional report that stated that patentable subject matter may include "anything under the sun made by man."[63] It was this phrasing that persuaded the majority of the justices of the congressional intent to include living things as patentable subject matter.[64] How much of the Court's decision was based on the practical considerations of the nascent biotechnology industry is left for interpretation by legal historians.

The dissenting four justices were persuaded by the passage of the two plant protection acts of quite a different congressional intent. For the minority, Justice Brennan wrote: "[B]ecause Congress thought it had to legislate in order to make agricultural "human-made inventions" patentable and because the legislation Congress enacted is limited, it follows that Congress never meant to make items outside the scope of the legislation patentable."[65] Several years later, in 1987, a patent was awarded for a multicellular organism (polyploid oysters), and, in 1988, the first animal patent was approved for a transgenic mouse.

Meanwhile, patents on human gene fragments were issued by the U.S. Patent Office not as living things, but as compositions of matter. Human genes are not manufactured or modified, and therefore could not receive a patent on the criteria of intellectual property. There were two strategies open to patent applicants. They could claim that the isolation of the genes within the genome of the organism was novel and therefore deserving of a patent, or that the form of the gene for which a patent was sought was not derived from nature.

Strictly speaking, genes cannot be patented because, like proteins, they are products of nature. Scientists argued before the PTO that their modification of the genes could qualify for patents because the natural molecular sequence has been altered and a new composition of matter replaced it. To make this argument, scientists used the version of a genetic sequence called copy or complementary DNA ("cDNA"). Typically, a gene that codes for a protein has many redundant or irrelevant nucleotides in the sequence that are not essential for the synthesis of a protein. When the extraneous sequences (called introns) are removed, the version of the gene is called copy DNA. Because this version of the gene is not present in the cell and can be created by using certain enzymes, it was considered patentable under section 101 of the U.S. Patent Act where the subject matter must be novel, useful, and non-obvious. Following the Court's reasoning, the term cDNA is described in books on genetics as "a man-made copy of the coding sequences of a gene."[66]

In extending intellectual property rights to genes, the PTO had ostensibly created a future's market in gene sequences and spurred a competitive frenzy among scientists in molecular biology.[67] The PTO's decision, in effect, meant that natural DNA sequences when modified as cDNA, were considered "artificial products" and therefore patentable.[68] Anyone involved in sequencing genes was encouraged by their colleagues and institution to apply for patents on the grounds that a competitor group will do so and control licensing fees. Patents were awarded to sequences whose functional role in the genome was not yet understood. The patent application usually covered broad uses of the gene sequence. Ostensibly, the human genome was under colonization. Universities, private companies, and, for a while, even government agencies sought intellectual property protection over segments of the genome.[69] Amgen, Inc. of California is credited with holding the most valuable patent of a human gene with estimated earnings of one billion dollars a year.[70] Its patent is for the human erythropoietin gene, which codes for a hormone needed by kidney disease patients.[71] In 1991, the Supreme Court ruled favorably on the validity of Amgen's 1987 patent of this human gene.[72]

Merz et al. note that the PTO has been awarding an increasing number of patents on methods for detecting disease-related genes.[73] The patent holder has exclusive rights to develop and perform diagnostic tests that detect specific genetic mutations.[74] By patenting genes that are responsible for disease, the authors argue that physicians are faced with serious conflicts of interests.[75] "The profit motive may lead to unwarranted promotion of genetic tests which are still in many ways experimental."[76]

The implications of the new financial opportunities in molecular genetics were profound because they contributed to a new set of relations between scientists and their work.[77] First, the time lapse between scientific discovery and commercial use, which had traditionally buffered scientists from the lure of pecuniary affairs, was now very short. This meant that almost any discovery of a new gene had potential commercial value. Second, scientists internalized a new set of values. Added to the traditional value in academia of the "pursuit and dissemination of knowledge" was the responsibility to use that knowledge for the development of marketable products.[78] A new ethos emerged in the biological sciences that meant balancing interests between two independent and potentially conflicting premises.[79] The first states that knowledge is part of the common human heritage, while the second treats knowledge as possessing economic value that should be realized.[80] This change in the culture of the biological sciences brought with it a new set of social relations between academic research and private industry.[81] The question on everyone's mind was how would these new relations affect the practice and integrity of scientific work?

THE GROWTH OF RESEARCHER FINANCIAL INTERESTS IN BIOMEDICAL RESEARCH

By the mid-1980s, genetic technology had spawned hundreds of new companies, many with academic scientists as officers; board members or consultants.[82] Small venture capital companies colonized the faculty of prestigious universities for building their intellectual capital.[83] Major corporations that had sector interests in drugs, therapeutics, and agriculture invested large sums into multi-year contracts with universities.[84] Several of the most notable examples of university-industry partnerships involved Monsanto Corporation. In 1974, Monsanto and Harvard University signed a contract after a year and a half of negotiations. Under the agreement, Monsanto gave Harvard $23 million in research support, laboratory space, construction, and endowment money.[85] In return, Harvard gave Monsanto the patent rights to a substance called tumor angiogenesis factor ("TAF"), which was involved in the growth of cancerous tumors.[86]

Washington University entered into a five-year, $23.5 million agreement with Monsanto in 1982.[87] The agreement involved the support of biomedical research at the university. It was renewed three times, most recently in 1989.[88] Monsanto's total investment with the university came to about $100 million.[89] The agreement was widely publicized and became the subject of a federal subcommittee hearing.[90] According to the agreement, manuscripts and abstracts resulting from Monsanto support would not be delayed for publication for longer than thirty days, during which time Monsanto had to make a decision about whether to patent a result.[91] Professors working under the Monsanto agreement had to give allegiance to the company, avoiding any relationships with other institutions that might create a conflict of interest for Monsanto.[92]

A number of studies published in the early 1990s began to shed some light on the extent to which the burgeoning field of biotechnology had begun to impact universities. One national study investigated faculty linkages to new biotechnology companies in the period 1985 to 1988.[93] At leading American universities, dozens of faculty in departments of medicine and biology developed formal relationships with venture capital companies.[94] At Harvard, the biotechnology faculty had ties to forty-three different firms, at Stanford twenty-five and at MIT twenty-seven.[95] For MIT, at least 31 percent of the faculty in its Department of Biology had commercial affiliations, while for Stanford and Harvard the figure for biotechnology faculty across several departments was 19 percent.[96] The authors of the study reported that their methodology was likely to underestimate the academic-industry connections, and that the trend for faculty to develop a commercial outlet for their work was rising.[97]

During the early stages of the commercialization of academic biology, science writer Philip Hilts wrote: "It is already apparently true that there is no notable biologist in this field anywhere in America who is not working in some way for business. I interviewed some two dozen of the best molecular biologists in the country and found none."[98]

The impact of university-industry relationships on the behavior and values of scientists began to emerge through a series of surveys of biomedical scientists taken by researchers from Harvard's School of Public Health between 1984 and 1994.[99] One of the clear outcomes of these new arrangements that was revealed in the Harvard surveys was that they impeded the "free, rapid, and unbiased dissemination of research results."[100]

Biotechnology faculty with industry support were four times as likely as other biotechnology faculty to report that trade secrets had resulted from their research.[101] The vast majority of the faculty without industry support viewed the commercial relationships as undermining intellectual exchange and cooperation within departments.[102] The surveys also revealed that faculty believed the new relationships were responsible for skewing the research agenda in biology toward applied research.[103]

A 1994 survey of senior executives in 210 life science companies confirmed that the changes taking place in academia were not transitory.[104] Trade secrets and confidentiality for industry-supported faculty were becoming the norm.[105] The majority (58 percent) of the respondents in this survey reported that "their companies typically require academic investigators to keep information confidential for more than six months in order to file a patent application."[106] The authors of the study also reported that 88 percent of the respondents indicated that their funding agreements with universities require students and fellows to keep research confidential.[107]

In another study involving researchers at Harvard University and the University of Minnesota, over 2,000 scientists were surveyed at fifty research intensive universities.[108] About 43 percent of the respondents indicated that they receive private gifts for their research, which their universities do not regulate.[109] Of the 920 scientists who received gifts, one-third reported that their corporate benefactors expected to review their academic papers before publication, and 19 percent indicated that the donors wanted the patent rights to commercialize discoveries arising from the gifts.[110]

Some have argued that the potential liabilities of federal policies designed to create university-industry partnerships and expand intellectual property rights for scientific discoveries were far outweighed by the benefits of the new social contracts in science.[111] They maintain that trade secrecy, at least for a period of time, was not going to disrupt the scientific agenda or impair the quality of research.[112] Universities would recalibrate their guidelines for sponsored research and establish some constraints for the more commercially adventurous faculty. However, two concerns flowing from the intense commercialization of science that could not be resolved by ethical standards established within universities were conflicts of interest and scientific bias. Cases of conflicts of interest were periodically highlighted in investigative journalistic reports.[113] As an example, in 1988, a federally-sponsored study published in the *Journal of the American Medical Association* reported that an anticlotting medication to prevent heart attacks—called tissue plasminogen activator ("TPA"), which is manufactured by Genentech, was significantly more effective than the drug currently in use called streptokinase.[114] Subsequently, it was reported widely in the print media that some of the study's investigators were long-term Genentech stockholders.[115] A journal that prided itself in its ethical standards for its contributors was criticized harshly for not disclosing the authors' conflicts of interests.

In another case, a conflict of interest at a university was connected with violations of federal regulations. It involved a University of Minnesota Medical School surgeon, who started a company in collaboration with his university to develop, manufacture and market an anti-rejection drug for use in organ transplants.[116] Federal investigators discovered procedural violations in how the drug was used, including failure of the clinical faculty to report in proper fashion deaths and other serious adverse reactions. The investigators also determined that the faculty had illegally made profits from the sales.[117] The university was forced to stop its sales of the multimillion dollar drug.[118] Some have argued that universities should not be in a situation where they are producing products, even if they do not violate any laws. According to Porter and Malone, "[B]y 1991, more than 100 American universities had started financing new companies to exploit the research findings of their faculties for commercial advantage."[119] Thus, in addition to conflicts of interests held by individual faculty, institutional conflicts of interest among universities present a special challenge.

Until 1996, there was little quantitative information on the extent to which the authors of scientific publications had a financial interest in the subject matter. of their research. In that year, a study appeared in the journal *Science and Engineering Ethics* that examined leading biomedical journals for the financial interests of authored original papers.[120] The researchers used 1992 as a test year and selected Massachusetts-based authors of articles in fourteen journals selected from *Science Citation Index* for their high impact rating.[121] In the study, authors are said to "possess a financial interest" if they are listed as inventors in a patent or patent application closely related to their published work; serve on a scientific advisory board of a biotechnology company; or are officers, directors or major shareholders in a firm that has commercial interests related to their research.[122] Of the 789 articles and 1,105 Massachusetts authors reviewed in the study, 34 percent of the papers met one or more of the criteria for possessing a financial interest.[123] Furthermore, none of the articles revealed the authors' financial interests. Most of the journals examined did not have disclosure or conflict of interest requirements in their "Instructions to Authors" during 1992, the target year of the study.[124] During that period, scientists seemed less concerned about conflicts of interest than government, journal editors, the media, and the general public.[125]

Some scientists and journal editors scoffed at the idea that such relationships could compromise the integrity of their work.[126] They argued that professional rewards like tenure and promotion or leadership positions in professional societies were more important in influencing the behavior of scientists than consulting relationships or patent applications.[127] The new attention placed on financial conflicts of interest was compared by some to "witch trials," where guilt was determined by association.[128] These critics are correct, however, in pointing out that there has never been a credible link found between cases of scientific misconduct and conflicts of interest, despite the effort by a congressional subcommittee report to draw the connection.[129]

Although many more journals have added financial disclosure and conflicts of interest requirements since 1992, some editors argue that, without evidence of a connection between possessing a financial interest and bias or misconduct in research, they have no need to require their authors to disclose any financial affiliations.[130] This position was highlighted in an editorial published by Nature that acknowledges the pervasiveness of conflicts of interest in science but sees no point in requiring disclosure of financial interests:

> This journal has never required that authors declare such affiliations, because the reasons proposed by others are less than compelling. It would be reasonable to assume, nowadays, that virtually every good paper with a conceivable biotechnological relevance emerging from the west and east coasts of the United States, as well as many European laboratories, has at least one author with a financial interest-but what of it? . . . Such appeals for openness are selective-other pieces of information would be just as (ir) relevant to a paper's content. . . . The work published (Science and Engineering Ethics)[131] makes no claim that the undeclared interests led to any fraud, deception or bias in presentation, and until there is evidence that there are serious risks of such malpractice, this journal will persist in its stubborn belief that research as we publish it is indeed research, not business.[132]

The conjecture that possessing a financial interest increases research bias or misconduct is empirically testable, but few rigorous and convincing studies have been published. In the mid-1980s, a study which appeared in the *Journal of General Internal Medicine* reported: that clinical trials sponsored by pharmaceutical companies were much more likely to favor new drugs (an outcome beneficial to the sponsoring companies in this case) than studies not supported by the companies.[133]

A recent study that brought significant media attention was published in the wake of the controversy over the use of calcium channel blockers to treat obesity.[134] A University of Toronto research team reviewed seventy English language articles, reviews, and letters to the editor on the effectiveness and safety of the new generation of drugs for treating obesity.[135] The investigators classified each author's position (supportive, neutral, or critical) on the use of the drugs and surveyed the authors, inquiring whether they had financial interests with drug manufacturers of calcium channel blockers or competing products.[136] The results of the study, published in the influential *New England Journal of Medicine,* indicated that those authors who were supportive of the obesity drugs were significantly more likely than the authors who were neutral or critical of the drugs to have a financial agreement with a manufacturer of a calcium channel blocker (96 percent, 60 percent and 37 percent. respectively).[137] It was also found that critics of the obesity drugs were not more likely to have financial ties to manufacturers of competing products.[138] Since the interpretation of drug safety and efficacy evidence is hardly an exact science, it is highly plausible that the discretionary factors in one's interpretation can be weighted (whether consciously or unconsciously) toward the interests and values of the institutions that provide the support. Further confirmation of this effect could provide journal editors ample justification for requiring financial disclosure.

The relationship of funding sources to bias in research has been a matter of increasing concern among journal editors and federal funding institutions. A great effort is put into clinical drug trials, and the public trust in their outcome is crucial to rational drug policies. The National Academy of Sciences (the "NAS"), which convenes scientific panels on complex science-policy debates, has been faced with panel members who have private company affiliations. The NAS leadership has acknowledged concerns that a panelist's commercial ties could affect his or her scientific judgment.[139]

THE PRIVATIZATION OF KNOWLEDGE

According to the Hippocratic Oath, physicians have an obligation to share their knowledge with others in the profession in the interest of the patient. The oath includes the phrase:

> [A]nd to teach them this art—if they desire to learn it—without fee and covenant; to give a share of precepts and oral instruction and all other learning to my sons and to the sons of him who has instructed me and to the pupils who have signed the covenant and have taken an oath according to the medical law, and no one else.[140] The code of ethics of the American Medical Association contains the phrase, make available information to patients, colleagues, and the public. . . .[141]

Medical knowledge must serve the common good. This fundamental value which survived through millennia of medical practice, is superseded by the normative changes taking place in biomedical sciences. Because every biomedical discovery has potential monetary value, the new culture of science will seek to protect that discovery from becoming part of the "knowledge commons." Filing patent applications prior to publication establishes a proprietary interest in the discovery. Even after the patent application is filed, it is not in the interest of the applicant to disseminate too much information about the discovery in the event that a competitor will find a way to use the knowledge that avoids patent infringement. Scientists, instead of sharing their discoveries in a timely fashion, are protecting them as trade secrets. This has resulted in wasteful duplication of research, not for the sake of verifying results, but rather for establishing the unpublished data needed to secure intellectual property rights over the discovery. Writing in *Science,* Eliot Marshall noted, "[w]hile some duplication is normal in research, experts say it is getting out of hand in microbe sequencing. Tuberculosis, like *Staph aureus* and *H. Pylori* will be sequenced many times over in part because sequencers aren't sharing data, whether for business reasons or because of interlab rivalries."[142]

The concept of intellectual property in biomedical research has become so inclusive that it embraces genes, plants, animals, microorganisms, including viruses; and even medical procedures. Seth Shulman, in his book *Owning the Future,* tells the story of an eye surgeon who inadvertently discovered a method of making incisions in cataract surgery that can heal without sutures.[143] While eager to share this knowledge with his students and colleagues, to his amazement and dismay, the physician learned that the surgical procedure that he had developed quite independently had been patented.[144] Moreover, by not paying licensing fees to the patent holder when he used the procedure, he was guilty of patent infringement.[145] Shulman disclosed a startling trend: "In 1993 the Patent Office was already awarding scores of patents each month on medical procedures, and by early 1996 the number had reached an unprecedented one hundred per month."[146]

Companies have taken out patents on disease causing bacteria and viruses, sometimes keeping confidential parts of the sequenced genome.[147] This may inhibit two companies competing in the search for a cure or treatment for a disease. Why should anyone own the natural sequence of a natural microorganism? Pharmaceutical companies can now exercise property ownership over both the drug to treat a disease and the microorganism that causes it. The intense privatization of biomedical knowledge that has evolved since the 1980s threatens the entire edifice of public health medicine.

It may be difficult for some to understand how turning federal research funds into discoveries that are privately controlled, how classifying scientific results of therapeutic significance as trade secrets, and how a publicly funded research enterprise in which conflicts of interest are endemic can serve the public interest. The answer may be found in the escalating price of pharmaceuticals and therapeutic tests. An investigative report in the *Boston Globe* noted that forty-eight of the fifty top-selling drugs approved by the FDA received money from the FDA or the NIH in discovery, development, or testing.[148] A drug called Proleukin, used for renal cell cancer, received a federal subsidy of nearly $46 million; the price for a typical course of treatment is $19,900.[149] A second drug called Taxol, used for ovarian and breast cancer, was federally subsidized at nearly $27 million with a treatment cost of $5,500.[150] The effect of escalating pharmaceutical prices is explained by geneticist Jonathan King:

> The patenting process, by granting a monopoly of seventeen to twenty years to the patent holders, allows a company to prevent other efforts to utilize the same information, genes or technology. As a result, it offers the possibilities of superprofits to investors. . . . The profitability of these agents stems in part from the potential lacking competition to charge very high prices for the product.[151]

The cost of genetic screening tests may also be influenced by the private ownership of genetic sequences. The patent for the Tay–Sachs disease is held by the Department of Health and Human Services. A screening test costs about $100.[152] In contrast, the patent for two breast cancer genes (BRCAl and BRCA2) is held by Myriad Genetics and a screening test costs $2,400.[153]

The field of biomedical research has become so infused with profiteering that bioprospecting for cell lines has taken scientists to distant places to negotiate the retrieval of tissue or blood samples.[154] Some indigenous groups have become suspicious of cell line prospecting and consider it a form of Western thievery of Third World resources.[155] The Biodiversity Convention was designed to protect species-rich Third World nations from exploitation by the industrial nations.[156] There is no comparable treaty for the protection of human genes or tissue samples that have commercial value. An unusual antibody discovered in an isolated tribe in South America can be used to develop a drug against a rare disease. A Western company sequences and patents the gene for the antibody, manufactures the drug, and develops a global market that includes the nation in which the tribal group lives. It is probably true that the drug would not have been developed if there were no company willing to invest capital in the tribal antibody. But does the human host of the unusual antibody and the community that supported him have any entitlements to share the profits on his biological material? Should someone have the right to profit from someone else's cell line? A group called the Indigenous Peoples Coalition Against Biopiracy has been organized to address such questions. In a draft of its ethical guidelines, the Human Genome Diversity Project has committed itself to sharing the financial rewards it might receive from cell lines with the communities from whom the cell lines were obtained. These arrangements are beginning to address the equity considerations involving the commercialization of human genetic resources, but do not get at the root issue of the appropriation and privatization of scientific knowledge.

The U.S. courts have thus far ruled against individuals and in favor of surgeons who claimed ownership of cell lines removed during an operation.[157] A case of considerable significance involved the California Supreme Court and a cell line developed from patient John Moore's spleen.[158] The court denied to Moore ownership of the cells taken from his body.[159] The judges did not rule that his cells belonged to the common pool of human knowledge.[160] Instead, they transferred ownership of the cells to his medical care team and their institutions.[161]

It has been over fifty years since sociologist Robert Merton published two classic papers that articulated the normative conditions of scientific practice.[162] Merton wrote that science consisted of an "emotionally toned complex of rules, prescriptions, mores, beliefs, values and presuppositions which are held to be binding upon the scientist."[163] Among the norms Merton cited were communalism—open and free exchange and the shared fruits of knowledge, and disinterestedness—knowledge as the sole interest of research.[164] The commercialization of scientific research has compromised the traditional Mertonian norms on the dubious assumption that the appropriation of knowledge as intellectual property, and in its wake the erosion of communalism and disinterestedness, will yield a greater public good in the long run.

Chapter 7

CONFLICT OF INTEREST AND COST-EFFECTIVENESS ANALYSIS[8]

SHELDON KRIMSKY
Tufts University, Department of Urban and Environmental Policy

More than thirty years ago, Jacob Bronowski, who spoke so poetically about the ethos of science, wrote that "the body of scientists is trained to avoid and organized to resist every form of persuasion but the fact." The values of science, he argued, are "inescapable conditions for its practice."[1] But the practice of science, especially the biomedical sciences, has changed significantly since Bronowski made his observations. Increasingly, academic biomedicine has become commingled with corporate interests. Spurred by the burgeoning commercial opportunities of new discoveries such as those in genetics, the growth in academic-industry collaborations has created uneasiness among some observers who suspect that conditions beyond the pure facts of science can influence its outcome.

During the mid-1980s, several medical journals, including *JAMA*, adopted policies on conflicts of interest.[2,3] According to one survey conducted in 1994–1995, nearly 50 percent of U.S. medical journals with a circulation greater than 1,000 had written policies regarding conflicts of

8Originally published in the *Journal of the American Association (JAMA)* 282, no. 15 (October 20, 1999): 1474–5.

interest.[4] Those who support policies on author disclosure of financial interests in scientific publications generally do not assume that such policies will improve the quality of science. They recognize that transparency of interests is not an antidote to bias or misconduct in science but believe it can foster public trust. Others, skeptical of the emphasis given to financial interests as opposed to other potential sources of conflicts of interest, see no justification for requirements that raise suspicions without contributing to the scientific agenda.[5,6] As noted by Rothman, conflict-of-interest policies are "ethically questionable, because they impugn authors with the implied accusation of wrongdoing without evidence and without recourse."[7] Thus, without an empirically established connection between conflict of interest and scientific outcome, many scientists and journal editors who favor financial disclosure are inclined to view it as sound public relations or as a gesture of moral correctness. The journal *Nature* explained in an editorial that until there is evidence that "undeclared interests led to any fraud, deception or bias in presentation" the journal will continue "in its stubborn belief that research as we publish it is indeed research and not business."[8]

Specific cases of industry-funded science and their relationship to bias and misconduct were investigated by Congress in 1988.[9] The report of the investigating committee, however, brought no additional clarity to the influence of funding sources and conflict of interest on scientific results.[10]

Subsequently, several studies have shown that clinical decisions are affected by physicians' financial incentives or their interactions with drug companies.[11,12] Other studies have explored the effects of industry funding on scientific outcomes. In a retrospective analysis of 107 trials in 5 leading medical journals with regard to outcome and sources of funding, Davidson[13] found that studies sponsored by pharmaceutical companies were much less likely to favor traditional therapy over new drug treatment. Stelfox et al.[14] found that authors who had a financial association with manufacturers were much more likely than those who did not to have a favorable published position on the safety of calcium channel antagonists as a treatment for cardiovascular disorders. That study reported that 96 percent of the authors who were supportive of calcium channel antagonists had financial relationships with manufacturers compared with 60 percent who were neutral and 37 percent who were critical. Only two of the seventy articles included in the study disclosed the authors' potential conflicts of interest. After reviewing these and other results, the editor of *BMJ* wrote, [these studies] "begin to build a solid case that conflict of interest has an impact on the conclusions reached by papers in medical journals."[15]

In this issue of the journal (*JAMA*), Friedberg et a1.[16] have focused the question of conflict of interest on health economics. With the increasing importance of managed care, studies of cost-effectiveness and cost-benefit analyses of pharmaceutical agents have become key factors in health reimbursement decisions. Today, the financial success of a pharmaceutical product depends on meeting not only standards of safety and efficacy but also cost-effectiveness.

Friedberg et al. questioned whether there was an association between industry-favored outcomes of cost-effectiveness studies for high-profile, expensive oncology drugs and corporate funding of the research. The sample of articles used in the study was well balanced between those funded by pharmaceutical companies and those funded by nonprofit organizations. The most noteworthy finding is that studies funded by pharmaceutical companies were nearly eight times less likely to reach unfavorable qualitative conclusions than similar studies funded by nonprofit organizations. These results are consistent with the hypothesis that private funding sources can bias outcomes of pharmacoeconomic studies. However, as the authors point out, there are other hypotheses that can account for the results, including the plausible conjecture that pharmaceutical companies may perform in-house prescreening of cost-effectiveness studies before they are contracted out to independent scientists. If only the drugs that screen favorably on cost-effectiveness are contracted by companies for external analyses, the likelihood of association between the outcome and the funder's interests is increased without the specter of bias.

Another finding of Friedberg et al. that is less favorable to an interpretation other than bias related to conflict of interest is that industry-sponsored studies were more likely to contain qualitative overstatements of quantitative results. However, the statistical power of this result is low and the methods for correlating quantitative and qualitative outcomes are not explained.

The primary challenge raised by this study is to distinguish among several plausible explanations for the apparent biases in cost-effectiveness analyses. This effort would be aided by a comparative study of several pharmacoeconomic assessments of a single drug under different funding arrangements that includes an analysis of how the analytic framework is selected and an examination of the assumptions used in studies funded by for-profit and nonprofit organizations. Such studies could determine whether any differences in outcome can be explained by structural elements in the modeling or other subtle biases related to conflicts of interest.

Most important, the field of pharmacoeconomic analysis must continue to pursue higher levels of professionalization. Standardized methods of analysis should be developed and adopted by health economists through their professional societies. There has been some progress in establishing consensus-based recommendations on cost-effectiveness analyses.[17] However, without a standard set of methods it is not possible to make comparisons across studies to assess the factors that account for varying outcomes. The differences observed between studies funded by industry and nonprofit organizations may be a result of methods chosen, prescreening, or bias due to the source of funding.

By following the traditions of professional societies, such as those of engineering[18] and psychiatry,[19] in setting guidelines of practice, pharmacoeconomists can attain a special role in the health care policy community in developing independent studies that are based on accepted canons that meet the highest standards of their profession. Government agencies that depend on such studies to set health care reimbursements can contribute guidelines that will help in promoting standards of professional practice.

Canada and the United Kingdom have developed national guidelines for cost-effectiveness studies.[20,21]

Biomedical journals should consider developing guidelines for the submission and review of cost-effectiveness studies.[22] Under such guidelines, for instance, authors would be required to clearly describe their assumptions, provide sound justification of the choice of methods, and fully disclose any financial relationships that exist between them and the company that manufactures the product, including whether the sponsor required written approval of the manuscript before submission. Although such an approach does not completely eliminate the potential for bias related to conflict of interest, clearly defined guidelines should foster more transparent reporting of pharmacoeconomic analyses and should enable clinicians and policymakers to better interpret and more appropriately apply the results to patient care decisions.

Chapter 8

CONFLICT OF INTEREST POLICIES IN SCIENCE AND MEDICAL JOURNALS: EDITORIAL PRACTICES AND AUTHOR DISCLOSURES[9]

SHELDON KRIMSKY

Tufts University, Department of Urban and Environmental Policy

L. S. ROTHENBERG

Department of Medicine, University of California, Los Angeles

KEYWORDS

Conflict of Interests, Financial Disclosure, Scientists, Scientific Literature, Editors, Universities and Colleges, Industry and Education

INTRODUCTION

Prompted by a new awareness of the growth of academic-industry collaborations and author financial interests related to their research, scientific journals began introducing conflict of interest (COI) policies in the mid-1980s. Conflict of interest has been defined as a set of conditions in which professional judgment concerning a primary interest (such as patients'

9Originally published in *Science and Engineering Ethics* 7 (2001): 205–218.

welfare or the validity of research) tends to be unduly influenced by a secondary interest (such as financial gain).[1]

Biomedical journals were among the earliest to adopt such requirements in their "Instructions to Authors." The International Committee of Medical Journal Editors, a small non-representative group of medical journal editors, has recommended both to authors and editors that financial associations that "may pose a conflict of interest" should be disclosed.[2,3] Recent media and journal commentaries have raised questions about how journal editors should respond to authors' personal financial disclosures.[4–12]

To date, there are no published studies on the impact of journal disclosure policies. Our 1996 pilot study of fourteen leading biomedical and science journals quantified the personal financial interests of Massachusetts' authors in papers published during 1992.[13] The results showed that 34 percent of 789 articles in the test sample had at least one lead author with a personal financial interest in the results. Since at the time few journals had COI requirements, including but one of the fourteen journals in our study sample, it was not unexpected that the authors that were found to have personal financial interests in the subject of their papers did not disclose that fact in their published articles. With increasing numbers of biomedical and science journals adopting COI requirements for authors since 1992, one might expect to find higher frequencies of disclosure in subsequent years.

This paper reports on an analysis of COI policies and disclosure frequencies of author personal financial interests in biomedical and science journals for 1997, and on a survey taken of 181 editors of peer-reviewed journals on the implementation of their publications' COI policies.

METHODS

To ascertain what percentage of biomedical and science journals had personal financial disclosure policies in place during 1997, we selected a study sample of journals with demonstrated high recognition based on the analysis of journal impact factors by the Institute for Scientific Information

(ISI).10 Two journal impact indicators, published in ISI's Journal Citation Reports (JCR), were used in the selection: the "impact factor" [the average number of times recent articles in a specific journal were cited in the JCR cover year],[14] which is reported annually and the "times cited factor" [the cumulative number of times the article has been cited once or more by all source items],[15] which is measured both annually and cumulatively.

We merged the top thousand journals listed in JCR (1996) both under "Journals Ranked by Times Cited" and "Journals Ranked By Impact Factor,"11 yielding a sample of 1,396 distinct high-impact science and biomedical journals (about 600 journals overlapped the two lists). We then determined which of the 1,396 journals had published COI requirements for authors in effect during 1997 by reviewing the "Instructions to Authors" on the journal web sites or by checking individual volumes of the journal.12 A journal was considered to have a COI requirement for authors if its published instructions to authors states that authors are required to disclose any benefits (financial or otherwise) related to their study that accrue or might accrue to them or their institution, exclusive of the funding they receive for their research; or any relationship they may have had that might be construed as a conflict of interest with respect to the subject matter of their contributed paper. Journal COI policies in our sample varied in their disclosure requirements according to whether they included nonfinancial as well as financial interests and whether they focused exclusively on products. Initially, we

10A review of electronic databases such as Index Medicus and Medline showed that each review several thousand journals. As of July 1997 Index Medicus listed 3,209 journals while Medline indexed 3,854 titles. The National Library of Medicine's Serline data base listed 21,111 current English language periodicals as of November 1997. The American Biological Association's Biological Abstract covers 5,425 current titles while Elsevier's EMBASE indexed 3,500 biomedical journals during the period of our study; the 1996 Science Citation Index ranked 3,700 biomedical publications; and Ulrich's online data base in medicine and biology contained 29,683 publications in all languages in 1997.

11The ISI numerical values of the "times cited" started at 29,926 (first) and ended at 148 (one thousandth), while the numerical values of the ISI "impact factor" went from 51.00 (first) to 1.66 (one thousandth).

12Journals vary in how they disseminate their Instructions to Authors (ITA). Some publish the ITA in each issue of the journal; others publish them in selected issues of a volume. Another group does not publish the ITA, but rather sends them to inquiring contributors.

found that 220 journals (15.8 percent of the study sample) had COI requirements. The list was then reduced to 181 (13.0 percent) journals (see Table 8.1) by eliminating: journals that did not have a COI policy in effect during the entire year; 5 journals that could not be found in any library in the Boston Metropolitan Area or at UCLA;13 and non-peer-reviewed journals. We physically examined all of the original research items in the journals (totaling 61,134) published in the test year 1997, noting all disclosures of private funding sources and personal financial interests of authors.

To determine the types of potential COIs reported in the publications, all author disclosures citing personal financial interests relevant to the published research (including consultancies, scientific advisory board memberships, stock/equity holdings, patents/inventorships, honoraria, expert witness fees, and direct employment/officer positions) were tabulated. Excluded were authors who had company affiliations in their title since that is tantamount to an explicit COI disclosure and not the subject of this study. We also verified whether the COI policy for each journal was in effect for the entire duration of 1997 by comparing the Internet home page listings of "Instructions to Authors" (ITA) with the ITA in published volumes.

To examine the practices of journal editors with regard to COI disclosures, we mailed a survey form to the editors (one per journal) of the 181 peer-reviewed journals that we sampled for our study of journal disclosure rates and received responses from 138 editors (76.2 percent).14 Of the 135 known respondent journals (3 remained anonymous), 91.1 percent (123) are biomedical (including dentistry) and 8.9 percent (12) are classified as other science journals. The survey was designed to learn whether low rates of published disclosures resulted from editorial practices (editors not publishing author COI disclosures), whether journals rejected submissions because of an author COI, and whether the journals' COI policies were broadly applied.

13The journals that were deleted because of unavailability were: *Amyloid: International Journal of Experimental and Clinical Investigation*, *Applied Immunochemistry*, *Cancer Gene Therapy*, *Journal of Cardiovascular Electrophysiology*, and *Journal of Nuclear Cardiology*.

14The survey was certified as exempt from Human Subject Protection Committee review by the Office of Protection of Research Subjects at the University of California, Los Angeles.

RESULTS

The 181 peer-reviewed journals examined by the research team published 61,134 original research items in 1997. Of the original research items, the number that contained at least one positive disclosure of an author's personal financial interests related to the publication was 327 (0.5 percent; denominator 61,134). We distinguish "positive" disclosures of financial interest from "negative" disclosures, which are statements that assert authors have is a disclosure, for purposes of consistency in this study, only the statements that asserted positive financial interests were recorded as primary data. Our final tally of 0.5 percent personal disclosures of financial interest in the peer-reviewed journals excludes cases of such negative disclosures (e.g., no benefits were or will be received from a commercial party related to the subject of this article) because, unlike positive disclosures, they were required only by a small number of journals. Statements of consultancies for and stock/equity holdings in companies that are involved in product development in areas close to the author's research were most frequently cited in disclosures.

One-hundred-nineteen of the 181 journals (65.7 percent), comprising 58.1 percent of the articles examined, published no (0 percent) positive disclosures of authors' personal financial interests. Thirty-seven journals (20.4 percent) had such disclosures in 1 percent or less of the original research items published in their journals. The rates of published positive disclosures of personal financial interests for the remaining journals are given in Table 8.2.

In our review of journal conflict-of-interest policies we discovered that 6 of the 181 journals we examined for this study (3.3 percent) used a standardized template to elicit information on whether the author or the author's institution benefited financially from the research. We use the phrase "template journals" to characterize a group of journals that require authors to choose among a set of standardized statements regarding the financial interests of all coauthors, and that publish the author's response in each article in the form of statements or symbols. The six "template journals" in our sample are: *Journal of Bone and Joint Surgery* (both the American and British volumes), *Journal of Hand Surgery* (American volume), *Journal of*

Refractive Surgery, *Investigative Ophthalmology and Visual Science*, and the *Archives of Physical Medicine and Rehabilitation*. For example, the *Archives of Physical Medicine and Rehabilitation* asks authors to select one of four statements on financial interest: one response indicates that financial interests accrue directly to authors; a second states that the financial interests accrue to organizations with which the authors are affiliated; a third states that the results of the research will not benefit the author(s) or their organization, and the fourth indicates that the author(s) has/have chosen not to select a disclosure statement.

The American and British volumes of the *Journal of Bone and Joint Surgery* both require contributors to check one of five options, the last of which asserts that the authors choose not to furnish disclosure information. Both the *Journal of Refractive Surgery* and *Investigative Ophthalmology and Visual Science* provide templates with specific categories of financial interests, seven for the former and seventeen for the latter.

Five of the "template journals," all except the *Journal of Hand Surgery*, were among the ten journals with the highest rates of positive disclosures; however, the template responses required by them gave no specificity about the authors' financial relationships. The six "template journals" had total rates of disclosure (positive plus negative) that were much higher than any of the other journals. However, five of the six do not require the authors to disclose the names of commercial entities with which they or their institutions have an interest or to disclose the specific nature of that interest. Also, authors in three of the template journals may check a response that states "we choose not to select a disclosure statement." Among the three journals with this option, only seven papers (1 percent) published in 1997 cited this non-disclosure option. This suggests that journals requiring a standardized response to financial disclosure have a very high participation rate among authors. We consider the response category, "we choose not to select a disclosure statement," as a proxy for "we do not wish to participate in financial disclosure."

In the survey of journal editors, 72.4 percent (71, n=98) reported that they rarely or never discuss with authors (or decide jointly with authors) whether a

COI disclosure should be published; and 73.7 percent (84, n=114) of the editors reported that they always or almost always publish such disclosures while 10.5 percent (12, n=114) never do so (see Table 8.3). A majority of responding journal editors (60.2 percent, 77, n=128) have never rejected a submitted manuscript based primarily on COI, but 18.8 percent (24, n=128) have done so primarily for that reason and 19.5 percent (25, n=128) have done so but only in conjunction with other factors (see Table 8.4). Table 8.5 shows the responses of editors to questions concerning who receives and who supplies COI information. About one-third of the journals request conflict-of-interest information from peer reviewers (35.6 percent; 47, n=132), while nearly half the journals (48.8 percent; 62, n=127) require such information from their editors.

CONCLUSION

Based on our sample of 1396 high impact journals, we found that 15.8 percent had a published policy on conflicts of interest during 1997. In our sub-sample of 181 peer-reviewed journals with COI policies, 87 percent were medical journals (n=157) and 13 percent (n=24) were science journals. In contrast, medical journals made up only 34 percent (n=474) of our original sample of 1,396 high impact journals. We believe the reason for the higher percentage of medical journals with COI requirements is that medical research receives greater public scrutiny and media attention compared to basic science. As a result, medical journal editors have become more responsive to disclosing in publications even the appearance of a conflict of interest held by a contributor. By the mid-1990s one survey found that as many as 34 percent of all medical journals and 46 percent of U.S. medical journals with circulation over 1,000 reported they had written COI policies for contributing authors.15

15Glass, R. M. "A Survey of Journal Conflict of Interest Policies." A talk given at the Third International Congress on Peer Review in Biomedical Publication, Prague, The Czech Republic, September 18, 1997. The survey was sent to 1,251 medical journals, excluding dentistry, nursing, and medical sciences, with circulation over 1,000 as listed in Ulrich's 1994 guide; 648 responses were reported.

In our sample of peer reviewed journals that had published COI policies in effect throughout 1997 we found that 0.5 percent (n=327) of the original research items (n=61,134) had, at least, one positive disclosure of personal financial interests by an author, while 65.7 percent of the journals had zero positive disclosures of author personal financial interests during that year. Our survey of journal editors reveals that the vast majority usually publish author disclosure statements suggesting that low rates of personal financial disclosures are either a result of low rates of financial interest (nothing to disclose) or poor compliance among authors to the journals' COI policies. Based on the previously mentioned pilot study,[16] higher disclosure rates in the template journals, and the growth of commercialization in the biomedical sciences,[17–20] we believe that poor compliance is the more likely explanation for low disclosure rates in most journals with COI policies. The approach taken by the "template journals," while sacrificing detail on the specific nature of the financial interests involved in exchange for the more generalized alert to readers, deserves greater consideration by editors.

ACKNOWLEDGMENTS

This research was supported in part by a grant from the Greenwall Foundation. Some preliminary results of this paper were reported on January 25, 1999 at the annual meeting of the American Association for the Advancement of Science in Anaheim, California. The authors wish to acknowledge with gratitude the research assistance of Sharon Wolfson and Darcy Byrne. While this paper was in preparation, Dr. Krimsky served as an expert witness in a legal case involving scientific integrity and conflict of interest. None of the journals discussed in this paper or their editors are parties to the case.

TABLE 8.1: Journals Analyzed and Surveyed

- Accounts of Chemical Research AIDS
- American Heart Journal
- American Journal of Clinical Pathology
American Journal of Medicine
- American Journal of Obstetrics and Gynecology
- American Journal of Ophthalmology
- American Journal of Orthodontics and Dentofacial Orthopedics
- American Journal of Physiology American Journal of Psychiatry
- American Journal of Public Health
- American Journal of Respiratory and Critical Care Medicine
- American Journal of Respiratory Cell and Molecular Biology
- American Journal of Surgery
- American Journal of Tropical Medicine & Hygiene
- Analytical Chemistry
- Anesthesia & Analgesia Anesthesiology
- Annals of Emergency Medicine
- Annals of Internal Medicine Annals of Neurology
- Annals of Thoracic Surgery Antiviral Research
- Archives of General Psychiatry
- Archives of Internal Medicine
- Archives of Neurology
- Archives of Ophthalmology
- Archives of Otolaryngology
- Archives of Pathology & Laboratory Medicine
- Archives of Physical Medicine and Rehabilitation Archives of Surgery
- Arteriosclerosis Thrombosis and Vascular Biology
- Arthritis & Rheumatism
- Behavioural Pharmacology Biochemistry
- Bioconjugate Chemistry
- Biological Psychiatry Bone
- British Journal of Haematology
- British Journal of Obstetrics and Gynaecology
- British Journal of Radiology British Journal of Surgery
- British Medical Journal
- Canadian Journal of Anaesthesiology
- Canadian Journal of Microbiology Cancer
- Cancer Epidemiology: Biomarkers and Prevention
- Cancer Research
- Cell Growth & Differentiation
- Chemical Research in Toxicology
- Chemistry of Materials
- Chest
- Circulation
- Circulation Research
- Clinical Cancer Research Clinical Chemistry
- Clinical Pharmacokinetics
- CMAJ-Canadian Medical Association Journal
- Critical Care Medicine
- Current Eye Research Diabetes
- Diabetes Care
- Drug Development Research
- Drug Metabolism & Disposition Drugs
- Environmental Science and Technology
- Epidemiology and Infection Epilepsia

- European Heart Journal
- European Journal of Endocrinology
- European Journal of Vascular Surgery
- European Respiratory Journal FASEB Journal
- Fertility and Sterility Gastroenterology
- Gastrointestinal Endoscopy Gynecologic Oncology
- Heart (formerly British Heart Journal)
- Hepatology
- Human Reproduction
- Hypertension
- Infection Control and Hospital Epidemiology Inorganic Chemistry
- Investigative Ophthalmology & Visual Science
- JAMA–Journal of the American Medical Association
- JMRI–Journal of Magnetic Resonance Imaging Journal of Agricultural and Food Chemistry
- Journal of Allergy and Clinical Immunology
- Journal of Applied Physiology
- Journal of Bone and Joint Surgery-American Volume
- Journal of Bone and Joint Surgery-British Volume
- Journal of Chemical and Engineering Data
- Journal of Chemical Information and Computer Sciences
- Journal of Clinical Investigation
- Journal of Clinical Neurophysiology Journal of Clinical Oncology
- Journal of Clinical Psychiatry
- Journal of General Physiology Journal of Gerontology
- Journal of Hand Surgery Journal of Hepatology
- Journal of Hypertension
- Journal of Internal Medicine
- Journal of Investigative Dermatology Journal of Investigative Medicine Journal of Medicinal Chemistry Journal of Natural Products
- Journal of Neurology, Neurosurgery & Psychiatry
- Journal of Neuropathology and Experimental Neurology
- Journal of Neurophysiology Journal of Neurosurgery
- Journal of Nuclear Medicine
- Journal of Nutrition
- Journal of Oral and Maxillofacial Surgery
- Journal of Organic Chemistry
- Journal of Pediatric Gastroenterology and Nutrition
- Journal of Pediatrics
- Journal of Periodontology
- Journal of Pharmaceutical Sciences
- Journal of Physical and Chemical Reference Data
- Journal of Physical Chemistry
- Journal of Prosthetic Dentistry
- Journal of Refractive Surgery
- Journal of Rheumatology

TABLE 8.1: (continued)

- Journal of the American Academy of Dermatology
- Journal of the American Chemical Society
- Journal of the American College of Cardiology
- Journal of the American College of Surgeons
- Journal of the American Dental Association
- Journal of the American Dietetic Association
- Journal of the American Geriatrics Society
- Journal of the National Cancer Institute
- Journal of Thoracic and Cardiovascular Surgery
- Journal of Trauma-Injury, Infection & Critical Care
- Journal of Urology
- Journal of Vascular Surgery
- JPEN–Journal of Parenteral and Enteral Nutrition
- Langmuir
- Laryngoscope
- Lasers in Surgery and Medicine
- Lipids
- Macromolecules Medical Care
- Medical Journal of Australia
- Medicine
- Medicine and Science in Sports and Exercise
- Modern Pathology
- Molecular Pharmacology
- Muscle & Nerve Neurology
- Neuropsychopharmacology Neuroreport
- New England Journal of Medicine
- Obstetrics and Gynecology Ophthalmology
- Organometallics
- Otolaryngology-Head and Neck Surgery
- PACE–Pacing and Clinical Electrophysiology Pain
- Pediatric Infectious Disease Journal
- Pediatric Research
- Pediatrics
- Plastic and Reconstructive Surgery Preventive Medicine
- Proceedings of the National Academy of Sciences of the USA
- Proceedings of the Society for Experimental Biology and Medicine Psychiatric Services
- Psychopharmacology Psychosomatics Radiology
- Science
- Seminars in Hematology Seminars in Oncology
- South African Medical Journal
- Southern Medical Journal
- Spine
- Stroke
- Surgery
- The Lancet
- Transfusion
- Transplantation

TABLE 8.2: Rates of Author Personal Financial Disclosure in 1997 Journals (n=181)

Personal Financial Disclosures (percent interval)[a]	Number of Journals (n=181)	Percent of Total	Number of Articles (n=61,134)	Percent of Total
0	119	65.7	35,498	58.1
>0–1	37	20.4	19,199	31.4
>1–2	12	6.6	2,931	4.8
>2–3	3	1.7	1,088	1.8
>3–4	2	1.1	516	0.8
>4–5	2	1.1	471	0.8
>5–6	0	0.0	0	0.0
>6–7	1	0.6	298	0.5
>7–8	1	0.6	349	0.6
>8–9	1	0.6	314	0.5
>9–10	0	0.0	0	0.0
>10	3	1.7	470	0.8

a. Indicates the number of journals and articles whose rates of positive personal financial disclosures fall within designated intervals. The percentage for each journal is calculated by dividing the number of articles that have at least one published personal financial disclosure by the total number of original research articles published by the journal during 1997.

TABLE 8.3: Editorial Practice on Receipt of Personal Financial Interest Disclosures

3–A: Discuss with author and decide jointly		
1 = Sometimes (but rarely)	46	(46.9%)
2 = Usually (almost always)	18	(18.4%)
3 = Always (can't think of an exception)	9	(09.2%)
4 = Never	25	(25.5%) n=98
3–B: Make an independent determination with/without additional author information		
1 = Sometimes (but rarely)	25	(24.3%)
2 = Usually (almost always)	43	(41.7%)
3 = Always (can't think of an exception)	17	(16.5%)
4 = Never	18	(17.5%) n=103
3–C: Publish author disclosure statements of private financial interests		
1 = Sometimes (but rarely)	18	(15.8%)
2 = Usually (almost always)	20	(17.5%)
3 = Always (can't think of an exception)	64	(56.1%)
4 = Never	12	(10.5%) n=114

TABLE 8.4: Rejection of Manuscripts Based on Conflicts of Interest

Have you or your journal ever rejected a submitted manuscript based primarily on potential financial conflict of interest?

		n=128
Yes, primarily for that reason	24	18.8%
Yes, in conjunction with other factors	25	19.5%
No	77	60.2%
Options 1 & 2	1	0.8%
Options 1, 2, & 3	1	0.8%

TABLE 8.5: Sharing of Financial Interest Disclosure*

	YES	NO
Do you share with reviewers financial interest disclosures:		
contained only in cover letters? (n=102)	37 (36.3%)	65 (63.7%)
contained in the submitted manuscript? (n=121)	93 (76.9%)	28 (23.1%)
Do you expect the peer reviewers to use financial interest information in their evaluation of manuscripts? (n=103)	56 (54.4%)	47 (45.6%)
Is it the practice of your journal to request financial conflict of interest information from:		
peer reviewers? (n=132)	47 (35.6%)	85 (64.4%)
editors? (n=127)**	62 (48.8%)	65 (51.2%)
Do you ask reviewers or editors to disqualify themselves if they perceive a potential conflict of interest? (n=104)	77(70.0%)	27 (26.0%)

*The survey was carried out in December 1998.

**"n" designates the number of editors responding to the specific question.

Chapter 9

AUTONOMY, DISINTEREST, AND ENTREPRENEURIAL SCIENCE[16]

SHELDON KRIMSKY
Tufts University, Department of Urban and Environmental Policy

There is general agreement among a wide spectrum of informed observers that, beginning in the late 1970s and early 1980s, American research universities underwent changes that brought them greater commercialization.[1] The changes involved:

- Greater emphasis on intellectual property
- Stronger ties between academia and the private sector, including research and development partnerships
- Greater emphasis on research over teaching and service
- Less concern about protecting traditional academic values such as free and open exchange of ideas, and sharing of biological materials such as cell lines
- More emphasis in new income streams, whatever their source
- Institutional involvement in business ventures, equity partnerships, and new faculty start-up companies

16Adapted from a published paper in *Society* 43, no. 4 (May 2006), 22–29.

There is no consensus about the meaning of these changes. Are they simply an accentuation of trends that already existed? Do they represent changes in degree rather than in kind? Do they foreshadow a new ethos in science and/or a new breed of professoriate? Is there a change in the norms of "disinterestedness" and "autonomy" within academic science? Will closer ties between academia and business portend less independence for the university? Is this a permanent and irreversible change or can some traditional values be restored?

Several of these questions require a careful historical study of American research universities. Because the United States does not have a homogeneous system of universities, the effects are not being felt uniformly. The American Association of University Professors (AAUP) collects and organizes data based on a five-part classification system of universities. Category I comprises the elite universities that award doctoral degrees. To be classified a category I institution in the AAUP database, a university must grant a minimum of thirty doctoral-level degrees annually, awarded by three or more unrelated disciplines. Category Il institutions either do not grant doctoral degrees or they grant too few to qualify as Category I. The lowest rung of the classification consists of two-year colleges whose faculty has no ranks. Between the first and second ranked schools (Categories I and Il) the average full professor salaries differ by $25,000.[2] Out of 3,773 institutions AAUP surveyed in 2004–2005, about 8 percent (299) were classified as Category I. However, Category I institutions have 48 percent of the total faculty in the entire system of universities: 176,000 out of 368,000. Moreover, the top 200 research institutions accounted for 95 percent of all the Research and Development (R&D) expenditures in 1997.[3] It is certainly true that universities aspire to become Category I because that is where the prestige lies, and where there are better salaries, lower teaching schedules, and more amenities.

In recent years, American universities have become more commercialized and more managerially (and therefore less democratically) administered. These observations are supported by indicators such as patents, a proxy measure of the commercial orientation of faculty, and reported declines in faculty governance.[4] The number of patents obtained by universities has risen dramatically from 96 granted in 1965 to 3,200 in 2000. Similarly, private sector contributions to university R&D budgets have grown. There has also been a trend in university investment into faculty start-up companies and Offices of Technology Transfer on campuses.

University R&D funding from industry was 2.5 percent in 1966 and rose to about 4.1 percent in 1980, 6.1 percent in 1992, and then to 7.7 percent in 2000.[5] According to the *National Science Foundation's Science and Technology Indicators,* "The funds provided for academic R&D by the industrial sector grew faster than funding from any other source during the past three decades although industrial support still accounts for one of the smallest shares of funding."[6] Expressing the changing R&D budgets of universities on average does not tell the whole story. Averages can fluctuate and obfuscate what is occurring at any given institution. The number of universities whose industry contribution to its R&D funding exceeds twice the average or about 15 percent appears to be rising. In 2000, Duke University received 35 percent of its R&D budget from industry, followed by Georgia Tech with 24 percent, Penn State with 17 percent, and University of Texas with 15 percent. Smaller schools also followed the trend: Alfred University, University of Tulsa, Eastern Virginia Medical School, and Lehigh University had 48, 32, 24, and 22 percent, respectively, of industry contributions to their R&D budgets.[7]

Given the changes in university culture and funding, we may question whether the norms of autonomy and disinterestedness have changed among academic scientists. Before we can address these claims, we should not assume that the meanings of these terms are generally understood. If the autonomy of scientists has declined, then against what standard is it measured? And with respect to disinterestedness, what does that mean and how can we assess or interpret a change in it?

I shall address these questions by examining the concepts of autonomy and disinterestedness in relationship to science. I shall explore how and under what conditions these terms are relevant to scientific practice in the context of what some observers have referred to as the new era of academic capitalism.

AUTONOMY

Absolute autonomy in the university is a myth. It would involve a set of conditions that allow a scientist to investigate any problem he or she chooses, regardless of the cost, and without any external accountability—such as peer review or human subjects committees. This is not a functional view of autonomy. Surely, it is not reasonable to assume that a scientist has the right to pursue any investigation (other than pencil and paper operations that do not require external support) regardless of cost, ethical impact, merit, social values, or environmental consequences. But relative autonomy is not unrealistic and distinguishes university professors from scientists in other sectors. It operates within the constraints of peer review, the funding sources, human and animal subject requirements, and accountability to other legal and ethical norms.

Relative autonomy requires the following minimum conditions:

1. The investigator selects the subject matter of the study. A biologist trained in genetics can choose from an array of research programs from human to bacterial genetics, analysis of existing life to synthesis of new life, or from genetics to behavioral genetics. A frontier area of research—such as endocrine disruptors or epigenetics—might attract new investigators. The new area of research might be given primacy for federal funding, which signals the researcher to apply—but does not dictate it. Or a new technology like micro-arrays in genomics might also attract new users among scientists in the light of new grant opportunities. Although external funding opportunities play a large role in

shaping what people decide to study, it is still their choice on how they want to frame a study, which grants to apply for, what methodology to use or, in the context of declining support, whether to switch areas of research or to stop doing sponsored work.

2. Methods and protocols developed to undertake the research are the choice of the investigator and not imposed on the investigator by a sponsor.

3. Data that comes from the investigation is controlled by the investigator and his/her research team and not by the funding source.

4. Interpretation of results is largely the responsibility of the investigative team and not subject to external control by a funding agency or sponsor.

5. Publication of the results is also controlled by the researchers involved in the study. Because of concerns that research sponsors impose covenants on scientists restricting publication of certain findings, the International Committee of Medical Journal Editors has issued guidelines to participating journals. Authors must sign an affidavit that they and not the sponsors are in control of the publication.

Autonomy in the university extends beyond the investigator's choice of research project, control of the methodology, ownership of data, and the right to interpret results. It also means the right of faculty to speak freely and independently on subjects of their expertise or for that matter on any subject that they choose. This is what we generally mean by academic freedom.

The legal concept of academic freedom has been traced back to Germany around 1850 when the Prussian Constitution asserted the freedom of scientific research *(Freiheit der Wissenschaft)*.[8] Harvard University, chartered in 1650, appointed Edward Wigglesworth its first professorship (of Divinity) in 1722 without limit of time, ushering in the tenure system into North America.

At the turn of the twentieth century, Edward A. Ross, Stanford University economist and secretary of the American Economics Association, spoke out in favor of municipal ownership of utilities and supported the socialist Eugene V. Debs for his presidential bid. Ross was fired from Stanford University in 1900 because of the influence of a member of the university Board of Trustees whose wealth was tied to the private ownership of utilities. Two philosophers, Arthur Lovejoy and John Dewey acted in response to this abuse of board power by establishing the American Association of University Professors (AAUP). AAUP first issued the doctrine on tenure in 1915. It changed the balance of power at American universities and established the idea that professors with tenure had a measure of autonomy, at least in principle, to speak freely and undertake research in areas of their choosing without controls placed on them by their institution—administrators or Board of Trustees—and without the fear that they could lose their position because of what they believed or wrote. The privilege of tenure, which included protected speech and employment, was adopted by most American universities by the mid-twentieth century as a result of the continued leadership and influence of the American Association of University Professors. Without job security and academic freedom, it can be argued, there would be no autonomy in the professoriate.

During the period of the 1950s, a number of universities fell prey to the cult of Senator Joseph McCarthy, who exploited the public's fears about the spread of communism. A number of faculty who were known to be Marxists or socialists or simply open-minded enough to discuss these ideas were fired for being a national security risk and, therefore, untrustworthy to hold academic positions.

Of course, tenure does not guarantee a fair wage, salary increases, or a good working environment. There remain many ways that outspoken faculty with minority views can be punished and ostracized. There is no absolute protection against academic tyranny, but job security and the protection of "academic freedom" of speech remain a strong protection in universities that are not replicated widely in other sectors of society.

My own case illustrates the point. While a young untenured professor, I supervised a class research project where students investigated a controversy over the contamination of an aquifer in a small New England town. Under my direction, the students prepared a report on contested hydrogeology of contamination, the role of public agencies, and the behavior of a multinational corporation at the center of the controversy. Before the report was issued, a vice president of the company tried to block publication and dissemination of the students' study by lobbying the president of my university. The president had not seen the report and could not comment on its veracity. But he did understand the concept of academic freedom in the university. The quality of the report was my responsibility—not the president's nor the university. I could be sued but my university would not be held accountable in this case because it was not a party to a contract or grant.

When Robert Merton wrote about the autonomy of science it was in the context of the rise of fascism and the totalitarian state. He questioned how science, which is embedded in the social order, can remain independent from state ideology, while it is largely dependent on state resources. Merton saw a growing tension between the scientific and political cultures. "The conflict between the totalitarian state and the scientist derives in part, then, from an incompatibility between the ethic of science and the new political code which is imposed upon all, irrespective of occupational creed."[9] Scientific autonomy means that scientists can operate according to their professionally derived norms rather than the dicta of state ideology.[10] Currently, scientific autonomy is threatened by a president who introduces religious and political beliefs as a litmus test for responsible science.[11] "Science must not suffer itself to become the handmaiden of theology or economy or state. The function of this sentiment is likewise to preserve the autonomy of science."[12] In 1938, Merton addressed the autonomy of science, as an institution, when it faced threats by authoritarian state interests. He wrote, "the social stability of science can be ensured only if adequate defenses are set up against changes imposed from outside the scientific fraternity itself."[13]

Autonomy in science is more complex, however, than protection of the scientific subculture from a totalitarian state. To understand this we have to recognize that decline in autonomy may arise externally or internally. External causes of decline result from factors outside the control of the scientist, where power and authority are taken from him or her. As an example, suppose a scientist requires funding for research but can only get the funds from a sponsor who requires control over how the data shall be interpreted. Similarly, a scientist might give up control to the sponsor of designing a research protocol in exchange for receiving funding for the study.

An internal decline in autonomy results from the choices made by the individual scientist, who chooses to conform to a new social structure within the university. For example, scientists who start their own companies and continue as full-time professors will embody a new corporate persona. The norms and values of the corporate scientist are different from those of an academic scientist in as much as the latter will more likely feel freer to criticize products and activities that appear in conflict with the public interest. Once the norms of the commercial culture are superimposed onto the university, the autonomy of the academic scientist is destined to decline. Professors (like other workers) will eventually internalize the values of their institutions.

Also, a person who was hired at the university but who does not have the security of tenure, loses some autonomy—or the protection against dismissal resulting from his or her alignment with unpopular views including political and intellectual deviance. Autonomy, academic freedom, and the tenure system are interlocking and mutually reinforcing systems. It is possible to have the benefits of academic freedom and tenure but yet decide not to use either in the ways that a fully autonomous professor might. In other words, there is something we can call autonomous self-repression resulting from the adaptation of an academic to an emerging corporate culture that has colonized the university.

A faculty member who is also involved with a start-up company is straddling two cultures, one that has a tradition of independence of thought and critical self-expression and the other that is premised on withholding public criticism that might malign the corporate name or create enemies in an industry that works out its differences though backroom negotiations and quiet litigation rather than public discourse.

Many scientists require research funding in order to do their work. Often, they will follow the fashions of the funding sources by necessity. There are some fields like philosophy, classical studies, mathematics, literature and the like that produce scholarship without much funding, if any at all. Because their choices are less constrained by funding opportunities, I would argue that they have more autonomy to select their areas of study than other scientists who must work within the constraints of sponsored research. Nevertheless, there is still an advantage to autonomy in university science over that of industry or government, particularly when sponsored research does not determine whether the individual has a job, in other words where "soft money" is survival. Many medical schools operate on a system where faculty may have tenure but must raise their own salary from sponsored research.

The case of Tyrone Hayes, an endocrinologist at the University of California at Berkeley, shows what happens when funding imperatives meet the core values of an investigator. Hayes, who studied the effects of low concentrations of the herbicide atrazine on frogs, found that minute amounts of the chemical inhibit the development of the frog's larynx and feminize males. His research initially received funding from the private sector. When his sponsors learned that he was getting positive results they tried to delay publication because the studies were not in their financial interest. The contract he signed stated that the company retained the final say over what could be published. Hayes eventually disassociated from the company and reasserted his right to publish his findings.[14]

There is no absolute autonomy in science. Those scientists who do not work in a hierarchical institution, who are not beholden to a single source of external funding, or possess no external funding at all, who have academic freedom and are not self-constrained or externally constrained from exercising it, whose livelihoods are not wholly dependent on external funding (soft money), who are not financially or professionally vulnerable to pressure for suppressing or distorting results or for remaining silent while others violate the ethical norms of science, score higher on the scale of autonomy, ceteris paribus, than those who cannot claim to be working under these condition.

Government scientists report greater constraints on their scientific autonomy than academic scientists. Those government scientists who speak freely end up either taking on the role of "whistleblower" or becoming marginalized by their federal agency.

University administrators also can exert pressure on scientists whom they view as a liability or embarrassment to the institution. For example, Theodore Postol, a professor of security studies at MIT issued a critique of the effectiveness of the Patriot missile system. His institution, MIT, had an interest in a positive evaluation of the missile shield since the developer of the Patriot missile, Raytheon, sponsored research at MIT. Because the university system has traditionally been non-hierarchical and protective of basic job security for its faculty, the opportunity still exists for professors to exercise a higher degree of autonomy even under the risk of marginalization and recriminations in salary and space by their institution.[15] It is testimony to the importance of autonomy that some academics embrace it—in the public interest—despite the penalties they incur.

In 1937, Robert Merton delivered a talk to the American Sociological Society Conference titled "Science and the Social Order," subsequently reprinted in the journal, *Philosophy of Science*, in which he stated: "The sentiments embodied in the ethos of science—characterized by such terms as intellectual honesty, integrity, organized skepticism, disinterestedness, impersonality—are outraged by the set of new sentiments which the State would impose in the sphere of scientific research."[16] As previously noted, Merton was responding largely to the growth of fascism and its challenge to the autonomy of science. He continued in his later writings to emphasize "disinterestedness" as a norm of scientific activity without defining it outright. His writings indicate that by "disinterestedness" he meant "of no concern for the utility of science" including any fame or fortune it can bring the scientist. Consider the following passage from his 1938 paper. "One sentiment which is assimilated by the scientist from the very outset of his training pertains to the purity of science. Science must not suffer itself to become the handmaiden of theology or economy or state."[17]

DISINTERESTEDNESS

One of Merton's interpreters and colleagues Norman Storer wrote: "In Merton's original delineation of disinterestedness he relates it almost exclusively to prohibiting the scientist from making the search for professional recognition his explicit goal."[18] The currency of science, according to Storer, can only be its contribution to knowledge, as an end in itself. Any other commodities consciously sought from the work such as fame or money would denigrate the integrity of the enterprise. He states that "the norm [disinterestedness] acts primarily to prohibit the scientist from planning his work so that he will personally benefit from it in terms of money, influence, public esteem, or even professional reputation."[19] Storer also includes in the norm of "disinterestedness" an emotional detachment from any theory or hypothesis in which the scientist may be investigating. Such an attachment can result in inflexibility and recalcitrance toward accepting negative results.

> Emotional neutrality, hand in hand with rationality and universalism, enjoin the Scientist to avoid so much emotional involvement in the work that he cannot adopt a new approach or reject an old answer, when his

> findings suggest that this is necessary, or that he unintentionally distorts his findings in order to support a particular hypothesis.[20]

Notwithstanding these rather idealized views of the emotionally detached and disinterested scientist, there is the popular image of the multi-vested scientist who holds many competing personal, professional, institutional, and fiduciary interests. How do we reconcile the norm of "disinterestedness" with the new profile of the entrepreneurial scientist? Let me proceed with a typology of interests held by contemporary academic scientists:

1. Predilection (intellectual, ideological and/or emotional attachment) to a hypothesis or theory.

2. Applied applications of inquiry.

3. Patent or intellectual property derived from a research discovery.

4. Financial interests in the sponsor of the research beyond funding.

5. Propensity toward and interest in a positive outcome (for publication).

6. Academic promotion.

7. Securing a grant.

8. Implications of the science on public policy.

9. Ethical significance of the science.

10. Professional standing among scientific peer group.

11. Equity interest in a company with a commercial interest in one's work.

12. Paid consultant in one's area of expertise.

Some have argued that these interests are indistinguishable with respect to their potential influencing effects on a scientist's judgment and therefore emphasizing financial interests is a form of political correctness.[21] However, the popular culture, professional organizations, legislators, and scientific/medical journals have selected "financial interests" as a special category. The requirement of transparency of any real or perceived financial conflicts of interest speaks to the special consideration placed on monetary relationships between scientists and other parties related to the subject matter of their research. But is it justified?

DISTINGUISHING FINANCIAL CONFLICTS OF INTEREST

It is true that all scientists have multiple interests in their work, whether financial or otherwise. But is there a rational basis for distinguishing a scientist's financial interest from other interests such as seeking fame or promotion or desiring a positive outcome from a study to maximize the chance of publication?

The argument that all interests are fungible and that financial interests cannot and should not justifiably be set apart for ethical considerations is disingenuous on several counts. First, our legal and regulatory institutions have long distinguished financial interests from other types of relationships and interests. The Ethics in Government Act, for example, places special emphasis on government employees who have financial interests associated with their public responsibilities in policy or regulatory matters. Section

208(a) in the rule issued by the Office of Government Ethics prohibits employees of the executive branch from participating in an official capacity in matters in which they, or persons or entities in which they have a relationship, have financial interests. Senior government officials must set aside their stock holdings in a blind trust, but they are not required to set aside their friendships or disqualify themselves from votes because of friendships or campaign support.[22] Judges, on the other hand, are expected to recuse themselves in cases where they have a financial interest or personal association. The judiciary has a higher ethical standard than other federal and state government employees.

Second, some interests cannot be disassociated from the scientific enterprise. For example, the desire for fame, professional advancement, or to win the race to scientific discovery are inextricably part of the enterprise of science. While it is plausible that such interests can bias the outcome of one's scientific work, they cannot be distilled from the practice of science. The Korean stem cell scientist Hwang Woo Suk appeared to be on the path to international fame as the first scientist allegedly to clone a dog and human embryonic stem cells. The drive for international honor and prestige, both for Dr. Suk and his country, blinded him to the standards of honesty and integrity in science. In an investigation of his claims, Seoul National University administrators charged Dr. Suk with fabricating scientific data. He subsequently resigned from his institution admitting to having submitted false data in a major publication.[23]

Third, a scientist's predilection for a certain theory or hypothesis is generally part of his or her published record. Nothing is hidden about a scientist's preferred approach toward understanding or explaining physical phenomenon. One scientist may accept the linear dose-response relationship for chemicals at very low doses, while another might support the threshold hypothesis. Because the correct hypothesis cannot currently be decided empirically, it is understood by the scientific community that a preference for one hypothesis over another can influence the interpretation of data. In another example, scientists are beginning to question the somatic mutation theory of cancer (cancer begins from a mutated cell) and in its place have proposed a field theory of cancer etiology whereby neoplasms arise from the interaction between bodily cells and tissues. No one doubts that a scientist's association with one theory or hypothesis over a competitor will shape his or her interpretation of experimental evidence.

We expect scientists to become passionate about their theories. It is part of the social and psychological enterprise of science that scientists advocate for a preferred explanatory framework. The eminent twentieth-century philosopher of science Karl Popper believed that scientists should seek to falsify their theories because theory verification, unlike theory falsification, was logically impossible. He wrote:

> . . . what characterizes the empirical method is its manner of exposing to falsification, in every conceivable way, the system to be tested. Its aim is not to save the lives of untenable systems but, on the contrary, to select the one which is by comparison the fittest, by exposing them all to the fiercest struggle for survival.[24]

While it is true that most scientists do not seek to falsify their hypotheses, the norm of organized skepticism obligates the scientific community to subject a theory or hypothesis to the most rigorous examination possible to determine if it stands up against the evidence at least as well as other explanations.

Thus, it can be argued that it is part of the enterprise of science that scientists align themselves with one or another explanatory framework. These interests have become part of their intellectual legacy, which is open to public view in their scholarly writings. Any bias associated with a scientist's propensity toward one theory or another can be discussed and addressed in open communication with the scientific community.

In contrast, a scientist's financial interest in the subject matter of his or her research is typically not in the public record and cannot be helpful in judging epistemological bias. In other words, financial bias is outside the discourse of rational debate. It should also be recognized that good science can be produced without financial interests in the outcome of research, but cannot be done without a scientist's epistemological and psychological interests in the outcome.

If one were to examine the effect of conflicts of interests on the bias of research, it would be possible to design a study that investigated whether financial interests were more heavily weighted toward one outcome over another. A group of such studies was conducted in drug research and the reports will be summarized in this article. Investigators divided published papers into two groups. One group consisted of authors who were funded by government or nonprofit organizations. In another group of matched studies, authors received their funding from for-profit companies. Using selected outcomes such as potential risks associated with the drug or efficacy comparing new versus old versions of a drug, the investigators were able to determine whether financial interests correlated with interests of the for-profit sponsor.

Such studies can be undertaken because "holding a financial interest" or "receiving funding from a for-profit sponsor" is a contingent part of any study, and thus can be compared to a control group where these relationships do not appear. The same is not true about having a passion for one's hypothesis or seeking recognition for one's work. All scientists share these interests and as a result their influences are ubiquitous, mitigating the possibility of a controlled study for ascertaining the effects of such interests.

For the above reasons, the financial interest of scientists as a source of potential bias remains an area of concern among journal editors, members of the print media (where conflict of interest among journalists is a serious matter), medical societies, and government ethics committees.

FUNDING EFFECT AND OBJECTIVITY

Beginning around the mid-1980s, medical sociologists and biomedical ethicists began to study whether corporate-funded research was more biased toward the sponsor as compared to similar research funded by nonprofit institutions or government agencies. Within a decade, a body of work had evolved that confirmed the "funding effect" in studies on the safety and efficacy of drugs. Some examples follow. One report found that authors with financial interests in drug studies were ten to twenty times less likely to present negative findings compared to those without such interests.[25] In another study, the author reported that "in no case was a therapeutic agent manufactured by the sponsoring company found to be inferior to an alternative product manufactured by another company."[26] A third study design took the form of a meta-analysis where the authors reviewed numerous published papers to determine whether the source of funding had an effect on the outcome. The investigators began by screening 1,664 original research articles. These were culled to 144 that were potentially eligible for their analysis. They ended up with 37 studies that met their criteria and found 11 that showed industry-sponsored research confirmed a funding effect:

> Although only 37 articles met [our] inclusion criteria, evidence suggests that the financial ties that intertwine industry, investigators and academic institutions can influence the research process. Strong and consistent evidence shows that industry-sponsored research tends to draw pro-industry conclusions.[27]

The fact that "possessing a financial interest" in one's research has a biasing effect is a troubling finding for science. Simply disclosing the effect does not minimize or bring to light the particular bias. The funding effect raises the question of whether scientists who hold such interests can be "disinterested," whether as a norm "disinterestedness" is linked to objectivity, and whether "financial conflict of interest" is a good predictor of bias.

The distinguished British physicist John Ziman, who contributed important scholarship to the sociology of science, tackled these issues in his book *Real Science*. He observed that science was undergoing a transformation from its traditional role in universities pursuing basic research to a new variant he called post-academic science. According to Ziman, "post-academic scientists are expected to be continually conscious of the potential applications of their work."[28] Moreover, he believes that the norm of disinterestedness no longer applies in most contemporary science. "What cannot be denied is that the academic norm of disinterestedness no longer operates. . . . Even the genteel pages of the official scientific literature . . . are being bypassed by the self-promoting press releases."[29]

However, Ziman does not view the loss of disinterestedness as epistemologically connected to the demise of objectivity. Rather, he argues, objectivity can be preserved in post-academic science even with the loss of disinterestedness. An individual scientist's interests will be sorted out, selected, and adjusted for bias by the community of scientists who will operate with the other critical norms including organized skepticism (rigorously critiquing every result and testing it for its validity); communalism (sharing information); and universalism (operating under a shared set of criteria and scientific methodology). Ziman writes:

> The production of objective knowledge thus depends less on genuine persona] "distinterestedness" than on the effective operation of the other norms, especially the norms of communalism, universalism and skepticism. So long as post-academic science abides by -these norms its long term cognitive objectivity is not in serious doubt . . . but provided that organized skepticism continues to be practiced conscientiously, we need not revise our belief—or otherwise—in the "objective reality" of the scientific world view.[30]

What becomes lost in the age of post-academic science is short-term "social objectivity" or what Ziman describes as the public's trust in science. Ziman trusts the self-correcting power of science as evidenced by cases of scientific misconduct that come to light even after generations. This is why, I believe, Ziman considers that the loss of social trust in science will be restored once science corrects itself. But not all published knowledge is rechecked. Few experiments are or even can be replicated. It could take many years before mistakes are corrected or fraud, misconduct, and bias are discovered. During this period there can be damage to human lives, particularly in drug research. Manufactured tobacco science is such an example. If the funding effect proves to be pervasive, then it tells us that conflicts of interest distort objectivity. Whether the biasing effect will be discovered and how long it will take is another question. But should we build a system of scientific integrity on the hope that we can in due time reverse the distortions of biased, commercially vested science rather than disinterestedness serving as the norm?

FUTURE OF AUTONOMY AND DISINTERESTEDNESS

Both concepts make sense when applied to universities and independent research centers. The term "autonomy" describes the scientist's total control over his or her research from planning and executing a study to the interpretation and publication of results. The tendency of some observers to give up disinterestedness as a norm fails to distinguish financial interests external to the practice and epistemology of science from those interests that cannot be distilled from scientific work. They also neglect to fully acknowledge the pervasiveness of the funding effect in science and its implications for objectivity. Moreover, scientific pluralism—a marketplace of stakeholder interests—is not a substitute for protecting the (financial) disinterestedness of academic science.[31] Transformation to a new stage of science that is blended seamlessly into commerce is neither desirable nor inevitable. Craig Calhoun's contribution provides an excellent overview and diagnosis of the problem, but I take issue with his comment that "it is important to realize there is no easy return to some earlier, better version of the university."[32] The pendulum in academia has swung out of balance on more than one occasion—two examples being the McCarthy period and the Vietnam War. Subsequently, most universities reinstituted their traditional values of academic freedom and unclassified research. Efforts are currently underway to protect academia from egregious commercialization and we may yet see the pendulum return to a place where external influences, such as Big Pharma and tobacco money, do not distort research policy and damage scientific integrity. Universities and independent research institutes are beginning to understand the importance of reestablishing public trust in science while rewarding autonomy and disinterestedness.

Chapter 10

FINANCIAL TIES BETWEEN DSM-IV PANEL MEMBERS AND THE PHARMACEUTICAL INDUSTRY[17]

LISA COSGROVE
University of Massachusetts

SHELDON KRIMSKY
Tufts University, Department of Urban and Environmental Policy

MANISHA VIJAYARAGHAVAN
University of Massachusetts

LISA SCHNEIDER
University of Massachusetts

KEYWORDS
Conflicts of interest, Ethics, Financial Interests, Psychopharmacologics

Medical journals began introducing conflict-of-interest (COI) disclosure requirements for authors two decades ago beginning with articles of original research. A growing awareness of the importance of author disclosure in biomedical publications is reflected in the rising number of medical

17Originally published in *Psychotherapy and Psychosomatics* 75 (2006): 154–160.

journals that have adopted COI policies over the past decade and the support for such policies among professional societies. If financial COI among medical researchers can bias the outcome of a study (as recent research shows),[1,2] there is as much reason to believe it can also bias the recommendations made by members of advisory panels.[3–5] The importance of protecting the integrity and public trust in scientific and medical advisory committees has been widely discussed.[6,7] Yet, there remain areas that lack the transparency of financial COIs that have become standard procedures in many medical publications.[8,9]

To date there has been no study examining the relationship between the pharmaceutical industry and the scientists comprising the advisory panels that recommended changes in the *Diagnostic and Statistical Manual of Mental Disorders* (DSM), a leading medical manual used for the diagnosis of psychiatric disorders. Pharmaceutical companies provide substantial funding for conventions, journals, and research related to what is included in the DSM, because what is considered diagnosable directly impacts the sale of their drugs.[10] This "uneasy alliance"[11] was evidenced when a prominent journal reported that it was difficult to find research psychiatrists to write an editorial about the treatment of depression who did not have financial ties to the pharmaceutical companies that manufacture antidepressant medications.[12] Recently some members of the American Psychiatric Association (APA) have expressed concern about the potential for COI that arises with the increase in industry support.[13] For example, at the annual meeting of the APA in 2004 there were fifty-four industry-supported symposia. Also, pharmaceutical advertising revenue in APA journals, totaling $7.5 million in 2003, increased 22 percent in one year.[14]

OBJECTIVES OF THE STUDY

Continuously produced since 1952 by the APA, the DSM, currently in its sixth revision,[15] is the official manual for psychiatric diagnosis in the United States. Its classification system is used by government agencies and for all mental health professionals who seek third-party reimbursements. The manual provides the standard psychiatric taxonomy found in psychiatry and psychology textbooks.[16] Nearly 400,000 mental health professionals, including psychiatric nurses, social workers, psychologists and psychiatrists practice in the United States, most of whom have taken instruction on the DSM.[17]

In this study, we investigated the financial relationships that members of the advisory boards to the DSM-IV and the DSM-IV-TR have had with the pharmaceutical industry, which manufactures drugs used by clinicians for the treatment of mental disorders. We assessed the extent and types of financial relationships for each of the diagnostic panels.

METHODS

The *Diagnostic and Statistical Manual for Mental Disorders (*DSM*)* is organized around working groups or panels. Most of the panels address a specific category such as "Mood Disorders." The members of each panel have significant influence in determining whether a new diagnosis should be added or an older diagnosis revised for the next edition of the manual. From the latest edition of the DSM (DSM-IV) and the edition with text revision (DSM-IVTR), we identified 228 individuals associated with the development of the volume. After deleting clerical staff, we were left with 170 expert members who comprised a total of 18 distinct panels.

Each panel member was put through a series of "screens" to determine whether he or she has had any financial associations with one or more pharmaceutical companies whose business is potentially affected by decisions or recommendations made by the panel. The "screens" involved tracking publications of the panel member for disclosures of potential COIs (in journals that have disclosure policies), and submitting the panel member's name into the news media database Lexis-Nexis and Internet search engines. We also submitted panel members' names into databases of the U.S. Patent and Trademark Office to screen for holdings of intellectual property in a drug whose sales could be affected by recommendations of the DSM. For example, the FDA's approval of Sarafem (fluoxetine hydrochloride) for the treatment of premenstrual dysphoric disorder was contingent upon expert testimony that concluded that premenstrual dysphoric disorder was a distinct clinical entity that should be included as a mental disorder in the DSM.[18]

Panel members were screened for any financial affiliations they had with the drug industry between the years 1989 (the DSM-IV was published in 1994) through 2004. By using multiple screening techniques to gather published or Internet data on financial affiliations, we were able to avoid a methodology that relied solely on self-reporting (e.g., surveying panel members).

Financial associations of interest for this study include: honoraria, equity holdings in a drug company; principal in a start-up company, member of a scientific advisory board or speakers' bureau of a drug company; expert witness for a company in litigation; patent or copyright holder; consultancy; gifts from drug companies including travel, grants, contracts, and research materials. We use the term "financial interest" in describing the relationship between panel members and the pharmaceutical industry rather than the term "conflict of interest" (COI) because the latter term implies an interpretation of the interest. Thus, we choose not to define COI. Rather, we identify categories of financial interest and reserve judgment on whether they represent a real, perceived, or potential COI.

Our approach is congruent with other publications that make a distinction between the finding of "financial interests" and the judgment of "conflicts of interest."[19–21] Specifically, there is less disagreement about what constitutes a "financial interest" than there is about what makes a "conflict of interest."

The specific screening methods we applied included Lexis-Nexis, Internet search engines (such as Google and Yahoo), the U.S. Patent and Trademark Office Internet Site on patents pending or awarded, and Medline. Each finding of a panel member's financial connection to a drug company was coded: H = honorarium; RF = research funding; RM = research materials (equipment, drugs, cell cultures, etc.); EQ = equity in a company; CB = member of a corporate board (advisory board or board of directors); SB = speakers' bureau; CON = consultant; ET = expert testimony; DINP = drug industry nonprofit (a nonprofit organization funded primarily by the pharmaceutical industry such as the Novartis Foundation); P = holds a patent, patent application or royalties on a product relevant to the treatment of mental disorders, and CIFS = collaborator in industry-funded study. Individuals classified under CIFS are DSM panel members who participated in a study as a co-investigator or collaborator, where the principal investigator (rather than the panel member) was described as funded by the pharmaceutical company.

A DSM panel member who received a fee from a drug company for speaking at a session of a symposium sponsored by a pharmaceutical company was coded as an H (received an honorarium). However, when a university or medical school receives unrestricted educational grants from a drug company, the company usually does not have a role in selecting the speakers, setting the honoraria, or signing the speaker's honoraria check. For our study, a panel member who spoke at a symposium but did not receive *direct* payment from a drug company (e.g., he/she spoke at a professional meeting or Grand Rounds of a medical school sponsored wholly or in part by a pharmaceutical company) was not coded as H. In other words, lecture honoraria paid by a university were considered as non-disclosable financial interests for the purpose of this study. Although not included in our analysis, we found twenty-three panel members who gave talks sponsored, in part, by drug companies but who were compensated by universities under unrestricted grants.

While our methodology allowed us to ascertain varied financial relationships that existed between DSM panel members and the pharmaceutical industry during the period of analysis, it did not allow us to make causal inferences about those relationships. We could not determine whether, or to what extent, an individual's association with the pharmaceutical industry influenced his/her behavior on a DSM panel or, conversely, whether participation on a DSM panel influenced his/her subsequent involvement with the pharmaceutical industry. For the most part, the data on panel member associations with the pharmaceutical industry are atemporal. Some of the financial relationships might have occurred before, during, or after publication of the DSM volumes. Ethical considerations are relevant whether the panel member's involvement in the drug industry occurred prior to or after DSM publication.

Three investigators independently conducted screens on the panel members. Any questions about coding were resolved by a fourth investigator. No panel member was coded as having a financial connection unless there was unambiguous information confirming the relationship.

Interrater Reliability

We chose 20 percent of the panel members (every fifth name) for an interrater reliability test. One investigator, who was not involved in coding, reviewed the data for nineteen names representing forty-four coding decisions. This investigator missed two coding decisions; all other coding decisions matched the results of the coding team. Our test demonstrated that the most likely error of using another coder was missing a financial interest. This error was minimized by having three independent coders who compared findings and reached consensus. Our methodology tends to err in understating rather than overstating the financial interests of panel members.

RESULTS

Of the 170 DSM panel members 95 (56 percent) had one or more of the eleven financial links to a company in the pharmaceutical industry. Figure 10.1 shows the percentages of the panel members listed in the DSM-IV and DSM-IV-TR with financial linkages to drug companies. Unless otherwise noted, the percentages given for each DSM category include the members from both the 1994 and 2000 editions. In six out of eighteen panels more than 80 percent of panel members were found to have financial ties to the pharmaceutical industry. These include 100 percent of the panels for the "Mood Disorders Work Group" (n=8) and the "Schizophrenia and Other Psychotic Disorders Work Group" (n=7), 81 percent for "Anxiety Disorders" (n=16), 83 percent for

"Eating Disorders" (n=6), 88 percent for "Medication-Induced Movement Disorders" (n=8) and 83 percent for "Premenstrual Dysphoric Disorders" (n=6) (see Table 10.1). The mental illness categories denoted by "Mood Disorders" and "Schizophrenia and Other Psychotic Disorders" are the two main categories for which psychopharmacological treatment is standard practice, whereas it is far less likely for individuals diagnosed with "Substance-Related Disorders" (17 percent; n=6) to receive such treatment (unless there is a coexisting mental disorder such as a mood disorder).

The most frequent financial link we found between the DSM expert panels and the drug industry is "research funding." Among the 170 panel members, 42 percent received research funding from pharmaceutical companies; 22 percent were consultants and 16 percent served as members of a drug company's speakers' bureau (see Figure 10.2).

Of those panel members who had financial links with the pharmaceutical industry (n= 95) 76 percent had research funding, 40 percent had consulting income, 29 percent served on a speakers' bureau, and 25 percent received honoraria other than from serving on a speakers' bureau (see Figure 10.3). More than half of the panel members with financial ties were found to have more than one category of financial interests with the pharmaceutical industry.

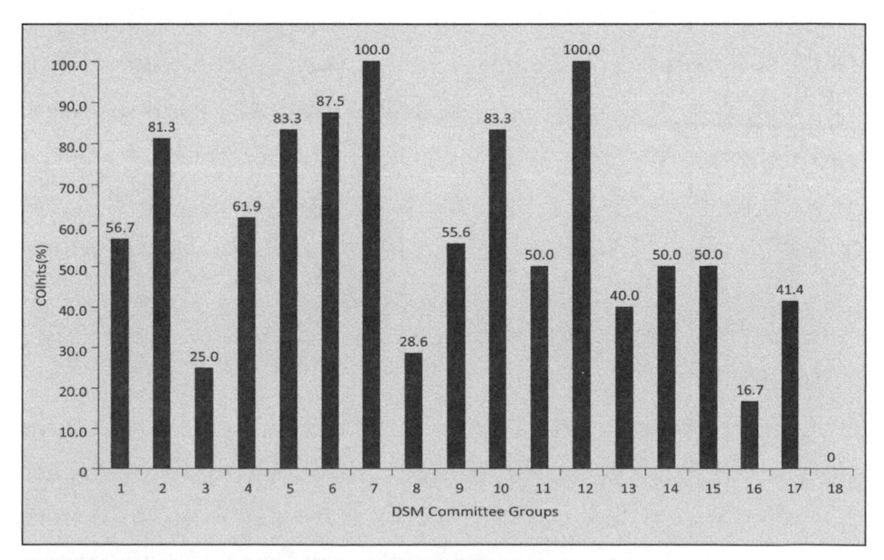

FIGURE 10.1: Percentage of panel members of DSM with financial ties.

TABLE 10.1:

	Percentage
1. TaskForce-IV	56.7
2. Anxiety Disorders	81.3
3. Delirium, Dementia, Amnestic, and Other Cognitive Disorders	25.0
4. Disorders Usually First Diagnosed during Infancy, Childhood, and Adolescence	61.9
5. Eating Disorders	83.3
6. Medication-Induced Movement Disorders (TR)	87.5
7. Mood Disorders	100.0
8. Multiaxial Issues	28.6
9. Personality Disorders	55.6
10. Premenstrual Dysphoric Disorder	83.3
11. Psychiatric Systems Interface Disorders	50.0
12. Schizophrenia and Other Psychotic Disorders	100.0
13. Sexual Disorders (IV)	40.0
14. Sexual and Gender Identity Disorders (TR)	50.0
15. Sleep Disorders	50.0
16. Substance-Related Disorders	16.7
17. Committee on Psychiatric Diagnosis and Assessment	41.4
18. Joint Committee of the Board of Trustees and Assembly of District Branches on Issues to DSM-IV	0.00

Eleven panel members had five different financial ties (see Figure 10.4). When there were competing coding categories, only the most precise descriptor was coded. For example, a panel member listed on a speakers' bureau, where he/she received honoraria, was coded only as "SB." Hence, the percentages in Figure 10.4 represent the most conservative estimate of DSM panel members with multiple financial associations to the pharmaceutical industry.

LIMITATIONS

The results of this study need to be interpreted in light of several limitations. First, it is reasonable to expect that some types of financial relationships were not detected by our screening methods. For example, expert witnesses serving in litigation are difficult to detect with standard screening tools. Second, our screening methods fell short of allowing us to quantify or to set a temporal sequence for the association. In most instances information about the amount of money received from pharmaceutical companies was not disclosed. Also, disclosures were reported strictly in terms of whether a person was a current or past recipient of industry support.

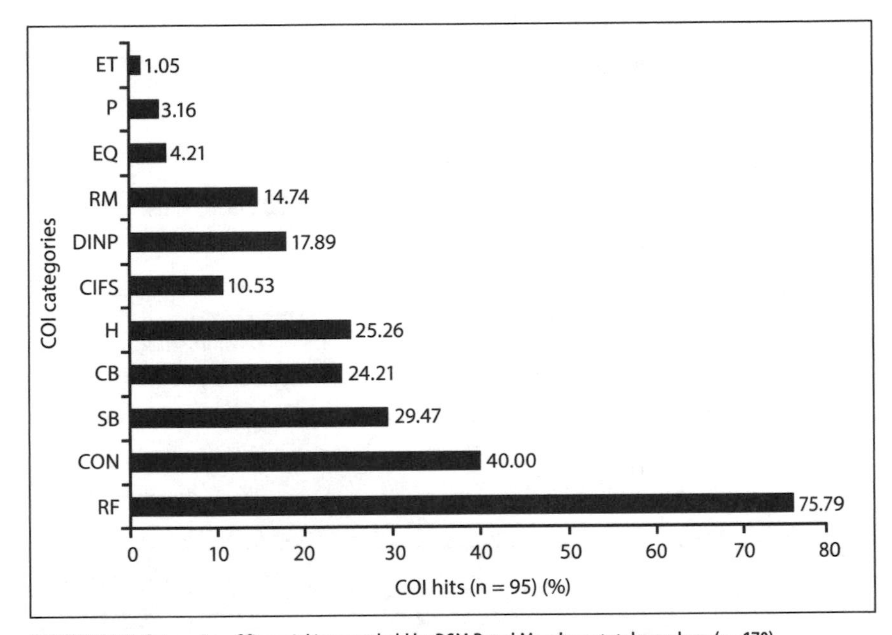

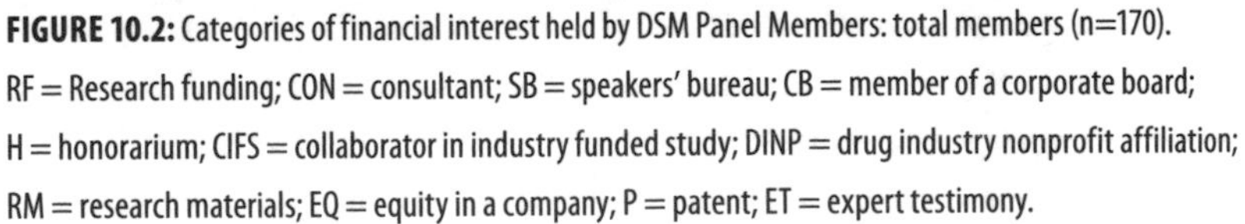

FIGURE 10.2: Categories of financial interest held by DSM Panel Members: total members (n=170). RF = Research funding; CON = consultant; SB = speakers' bureau; CB = member of a corporate board; H = honorarium; CIFS = collaborator in industry funded study; DINP = drug industry nonprofit affiliation; RM = research materials; EQ = equity in a company; P = patent; ET = expert testimony.

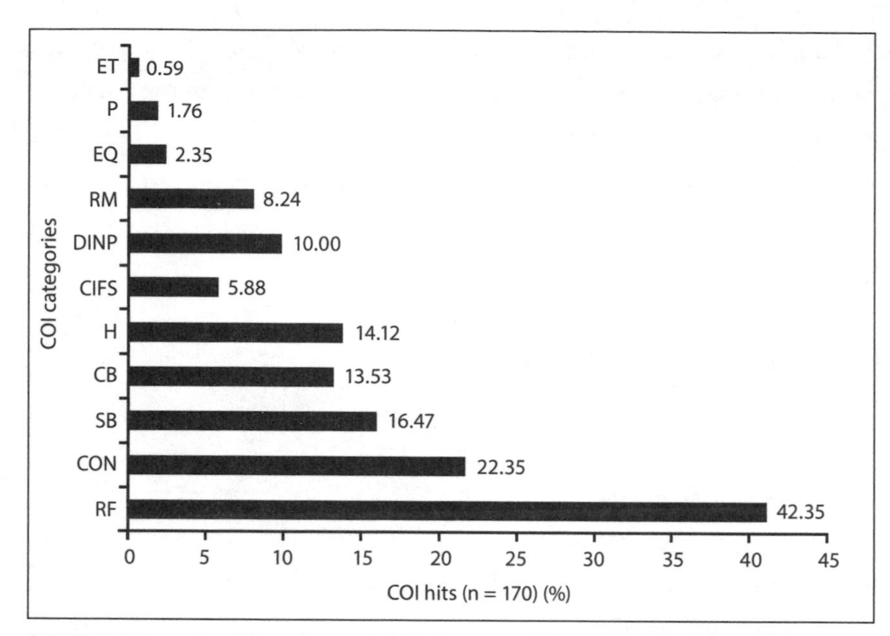

FIGURE 10.3: Categories of financial interest held by DSM Panel Members: members with financial interests (n=95). RF = Research funding; CON = consultant; SB = speakers' bureau; CB = member of a corporate board; H = honorarium; CIFS = collaborator in industry-funded study; DINP = drug industry nonprofit affiliation; RM = research materials; EQ = equity in a company; P = patent; ET = expert testimony

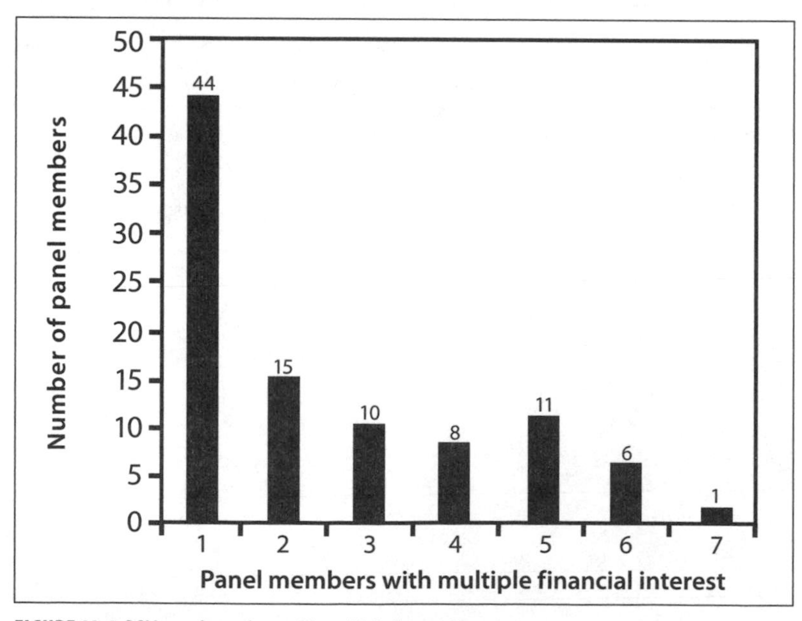

FIGURE 10.4: DSM panel members with multiple financial interests.

Even among journals that have set high standards on author disclosure, no specific dates were given. For example, the following is a typical disclosure statement: "Dr. X has received honoraria and research support from Company Y and Company Z and he also serves on their speakers' bureau." Because it was not possible to quantify these relationships or locate the precise time they occurred, only qualitative conclusions can be drawn about the extent of the financial relationships of DSM panel members with the pharmaceutical industry. Also, no conclusions can be drawn about the impact these relationships may have had on a panel member's behavior. However, there are ethical concerns in these relationships regardless of the amount of money given or the timing of the financial association. A financial relationship held during (or just before) participation as an expert panel member might influence or be perceived to influence the outcome of a DSM work group.

On the other hand, if their association with the pharmaceutical industry occurred after work on the panel was completed, panel members might be perceived as using their prestige to leverage lucrative consulting arrangements with the pharmaceutical industry. In the latter case public trust in the independence of medical science is eroded when former panel members, who are valued as "thought leaders,"[22] exert their influence on prescription practices through consulting, public speaking, and participation in industry-sponsored educational workshops.

Finally, we relied on self-reported disclosure data that was available in the open literature. Author noncompliance with journal COI policies has been cited as a problem.[23] Therefore, the percentages of DSM panel members identified as having financial ties and the variety of their ties to the pharmaceutical industry should be considered de minimis figures.

COMMENTS

Receiving financial support from a pharmaceutical company should not automatically disqualify an individual from serving on a DSM panel.

However, the public and mental health professionals have a right to know about these financial ties, because pharmaceutical companies have a vested interest in what mental disorders are included in the DSM. Transparency is especially important when there are multiple and continuous financial relationships between panel members and the pharmaceutical industry, because of the greater likelihood that the drug industry may be exerting an undue influence on the DSM. As previously noted, the DSM working groups that had the highest percentage of financial ties to the pharmaceutical industry were those groups working in diagnostic areas (e.g., mood disorders and psychotic disorders) where pharmacological interventions are standard treatment. In light of the extreme profitability of the psychotropic drug market, the connections found in this study between the DSM and the pharmaceutical industry are cause for concern. For example, antidepressants and antipsychotics were the fourth and fifth leading therapy classes of drugs in 2004, with annual global sales totaling $20.3 and $14.1 billion dollars, respectively.[24] One antidepressant alone, venlafaxine (Effexor, Wyeth) achieved $3.7 billion in sales in 2004.[25] The antipsychotic drug market has been identified as one of main therapeutic areas for global market growth with sales of $8.5 billion in 2002 and projected sales of $18.2 billion by 2007.[26] Therefore, we recommend that the APA institute a disclosure policy for panel members of the DSM who have financial ties to the drug industry. This is consistent with the trend for greater transparency in the membership of federal advisory panels. Raising awareness about the real or perceived COI of panel members is an important public health issue.[27–32]

Transparency should also apply to journal publication. We undertook a search of Ulrich's Periodicals Directory using the terms "psychiatry and drugs," "psychopharmacology," "drugs and mental illness" and "psychiatry and medication." When restricted to the descriptors "active," "academic/scholarly," "English language' and "refereed journals," the search identified forty-five journals of psychiatry. Of those, nineteen had COI disclosure policies (42 percent). Many of the financial ties that were found in this study were discovered because of these disclosure policies. Thus, we also recommend that all psychiatry journals adopt COI policies following the recommendations of the International Committee of Medical Journal Editors.

ACKNOWLEDGMENTS

We are most grateful for the assistance of two UMass graduate students, Michelle Levinson and Greg Noga, and one Tufts University undergraduate student, Naomi Mower, all of whom conducted preliminary searches. Ms. Mower participated through a summer scholars grant from Tufts University provided through the university's endowment. Dr. Krimsky was her mentor under the program. He received a small stipend for supplies. No other participants of the study were supported by grants.

Chapter 11

THE ETHICAL AND LEGAL FOUNDATIONS OF SCIENTIFIC "CONFLICT OF INTEREST"[18]

SHELDON KRIMSKY

Tufts University, Department of Urban and Environmental Policy

"Conflict of interest" is embedded in many areas of public ethics. Certain enactments named for their ethical content, such as the U.S. Ethics in Government Act, have sections devoted to "conflict of interest," and the legal community, government officials, financial organizations, and many news organizations have strict guidelines on such conflict. Yet the term is rather new to the scientific and medical research communities. Prior to 1980 little public attention was given to scientists with competing interests in their research. The first medical journal to introduce a conflict of interest disclosure requirement was the *New England Journal of Medicine* in 1984, followed a year later by the *Journal of the American Medical Association*.[1]

One might ask whether conflicts of interest among scientists should be treated differently than they are in other professions. Why, moreover, did the concern about conflicts of interest arise so much later among scientists, compared to public policy and law? This chapter explores the ethical and legal foundations of conflict of interest (COI) in the sciences and asks

18Originally published in *Law and Ethics Biomedical Research: Regulation, Conflict of Interest and Liability*, Trudo Lemmens and Duff R. Waring, eds. (Toronto: University of Toronto Press, 2006).

whether COI among scientists, in contrast to other professions, represents an ethical problem.

SCIENCE AND ETHICS

Conflict of interest in science and medicine has been defined as a set of conditions in which professional judgment concerning a primary interest (i.e., integrity of research) tends to be adversely influenced by a secondary interest (i.e., financial gain).[2] There are two possible explanations for why the issue of conflict of interest arose late among the scientific professions: (1) scientists were believed to operate within a normative system that mitigates any concerns about such conflict, and (2) considerable public trust afforded to scientists, including clinical investigators, eclipsed any potential societal concerns about competing interests.

Science is a self-governing system, subdivided into professional societies, journals, and communication networks, referred to as the "invisible colleges" that define the shared areas of study, outlets of publication, and collaboration of similarly trained individuals.[3] Its normative structure, emphasizing the importance of skepticism, replication, and empirical verifiability, according to some observers, makes any other interests scientists may have irrelevant to the mission of science. Only one set of interests can lead to success within the profession—the unfettered commitment to methodological rigor and the pursuit of verifiable knowledge.

Throughout the nineteenth and much of the twentieth centuries, with the exception of Nazi science, which is generally viewed as an aberration, the ethics of science was uniquely tied to its epistemology, insulating it from public oversight. It was not until the 1970s that the bubble of normative insularity of science was burst, specifically for clinical trials and human experiments generally. Scientists using human subjects are in a fundamental conflict of interest that is inherent to the process. As researcher, the primary concern of the scientist is to determine the truth about the effectiveness and efficacy of a treatment. The focus has to be on the observance of sound methodology, honest data gathering, and statistical rigor. If too many subjects are dropped from a trial, the results may not be publishable in the most competitive journals.

As a clinician, however, the researcher has a responsibility (as expressed by the Hippocratic Oath) to do no harm and to try to help a sick patient get better. In their effort to balance these goals, clinical researchers sometimes fail to disclose all the risks facing the subject, or fail to stop the trial for a subject who is having adverse reactions. Alternatively, they may make a premature leap from animal studies to human trials in their enthusiasm to reach a positive outcome for a drug before their competitors do. In the wake of highly publicized cases where the concerns of human subjects were discounted in favor of a researcher's professional interests, legislation or regulation emanating from funding agencies focusing on the protection of human subjects was adopted in the United States, Canada, and other countries with advanced centers of biomedical research.

The long tradition of trust in science was rooted in the myth of the scientist as a selfless investigator of universal truths. As scientists proved their utility to civilization in the eighteenth and nineteenth centuries, some philosophers and scientists alike began to think of science as also providing insights into the moral order of the universe—an idea with roots in Greek philosophy. As J. H. Randall noted in his classic work, *The Making of the Modern Mind*, "The Order of Nature" contained an order of natural moral law as well, to be discovered and followed like any other rational principles of the Newtonian world machine.[4] Philosophers of science who believed that there was a parallel between the formal structure of science and that of ethics proposed a theory of ethics based upon a deductive nomological system comparable to mathematical physics.[5] The aspiration of developing a system of ethics derived from natural law or modelled on the mathematical sciences met its demise concurrently with the refutation of logical positivism as the foundation of philosophy.

Another view held by some sociologists of science and natural scientists writing qua humanist is that the culture of science has its own ethical system, which serves as a model for other sectors of society. Jacob Bronowski popularized the view that "science has humanized our values,"[6] while Robert Merton introduced the normative structure of the social system of scientific organizations, which he observed in the early twentieth century.[7]

The conditions under which scientists and government officials, including judges, carry out their fiduciary responsibilities may be quite different. In addition, the normative constraints on science and government and the lines of accountability are distinct. Conflicts of interest in science and government are not necessarily rooted in a similar ethical matrix or based on a comparable legal foundation. In fact, one might justifiably question whether COI in science can be grounded in any ethical matrix. The discourse over COI in science might just be about political correctness. We can, however, object to conflicts of interest in science and medicine on grounds other than ethical ones. If there is an ethical basis for addressing conflicts of interest among academic scientists, we need to consider the ethical principle (or principles) on which it rests. I shall begin this inquiry by asking the following questions: What factors establish COI as an ethical concern in public affairs? Do those factors apply to scientific COI? If not, do other considerations apply to scientists?

COI IN GOVERNMENT

Government employees are stewards of the public's policies, its land, laws, and regulations. Federal officials who use their positions to gain personal financial benefit are in conflict with their fiduciary role as stewards of public resources. In this sense they are trustees of the public's properties, regulations, and its legal traditions. COI *behavior* (financially self-serving decision making) is a violation of the ethical principle that government employees should not use their positions for personal gain. Generally, we cannot know whether a decision of a public official was made out of self-interest or whether that self-interest and public interest happen to coincide.

Why not regulate or punish only the behavior that violates the ethical principle?

Andrew Stark, in his book *Conflict of Interest in American Public Life*, provides a three-stage anatomy of conflict of interest.[8] The *antecedent* acts (stage 1) are factors that condition the state of mind of an individual towards partiality, thereby compromising the potential of that individual from exercising his or her responsibility to foster public rather than private or personal interests. Examples are government employees accepting gifts, paid dinners, and the like. The *states of mind* (stage 2) represents the affected sentiments, dispositions, proclivities, or affinities conditioned by the antecedent acts. Thus, a politician who accepts a substantial campaign contribution from an individual may be more inclined to favor that individual's special business needs in legislative decisions than if no contribution were given.

The final stage represents the outcome behavior or *behavior of partiality* (stage 3) of the public official or those actions taken by that individual (decision behavior) arising from a state of mind affected by the antecedent acts. The outcome behavior could result in self-aggrandizement or in rewarding friends at the expense of the general public interest.

If public conflict-of-interest law were directed only at stage 3, the behavior of partiality, this would have several implications. First, a person could be found guilty of conflict of interest only if it could be proved that his or her behavior resulted from gifts, favors, or mutually self-serving relationships. We cannot infer the disposition or intention of the public employee from the outcome of a policy or regulatory decision. It is difficult to characterize a person's state of mind. Consider the case where a U.S. president issued a pardon to a person living outside the United States who was charged with a felony and who never stood trial. Funds contributed to the president's campaign can be traced to the alleged felon's immediate family. How would one show that there is a link between the gifts, the President's state of mind, and the decision to issue a pardon?

The third implication of focusing exclusively on outcome behavior in COI law is that it would have little prophylactic effect. Most of the damage is already done by the time the legal processes kick off. Only a small number of cases would be prosecuted, since the burden of demonstrating violations would be high.

As Stark notes, "because we cannot prevent officials from mentally taking notice of their own interests, we prohibit the act of holding certain kinds of interests in the first place."[9] Therefore, the law operates on the public health model of "primary prevention." Public employees are required to be free of any conditions that may dispose them to act in a way that elevates self-interest (particularly financial self-interest) over public interest.

In public health, "primary prevention" means eliminating the exposure. In COI terminology, "primary prevention" means "avoiding the appearance of conflict of interest." The best of our journalists operate on a preventative principle by not accepting lunch, gifts, or drinks from a person they interview. The ethical principle may be stated as follows: To protect the public's confidence in a free and independent press, journalists must comport themselves in such a way that avoids even the appearance that they could gain a financial benefit from the slant or context of a story or the way in which they present an individual.

Does the stewardship frame apply to scientists? In fact, scientists do have some stewardship functions. A great majority of the grants scientists receive in academic research are from public funds. Scientists are obligated to use the funds according to the provisions of the grant. They are expected to publish the results of their research in the open literature. If an American scientist makes a commercially useful discovery from his/her publicly funded grant, then under the Bayh-Dole Act (1980), the U.S. government transfers all intellectual property rights to the discovery to the researcher and his/her institution. This is a case where public investment is turned into private wealth, indicating a limited stewardship role of the scientist over the grant income.

The stewardship frame does not fit well with the self-image of university scientists, who place a high premium on academic freedom and independence. It is also not consistent with federal policies, which have created incentives for faculty to partner with for-profit companies and to start their own businesses. In other words, the U.S. government provides incentives for academic scientists to hold conflicts of interest. The government has reconciled these tensions by requiring disclosure and COI management of federal grant recipients at their institutions.

Another reason the U.S. government does not embrace the stewardship frame for COI is that it would place a high burden on federal agencies for waiving a COI. Moreover, because the university is the legal recipient of the federal grant, it would have to address institutional conflicts of interest, a decision that the U.S. government has deferred.

Under the 1972 Federal Advisory Committee Act (FACA), agency advisory committees are explicitly forbidden to be inappropriately influenced by special interests, and its members must comply with federal conflict of interest laws designed to protect the government process from "actual or apparent conflicts of interest."

Two rules guide the U.S. federal advisory committee structure on conflicts of interest.[10] The first states that no person with a substantial conflict of interest can serve on a federal advisory committee. A federal employee may not "participat[e] personally and substantially in an official capacity in any particular matter in which, to his knowledge, he or any other person . . . has a financial interest if the particular matter will have a direct and predictable effect on that interest."[11] However, the second rule holds that the first rule can be waived.

In a study of Food and Drug Administration (FDA) advisory committees covering more than a year and a half, *USA Today*'s investigative journalists found that there were 803 waivers for conflicts of interests in 1,620 member appearances, or about 50 percent.[12]

Scientists are not stewards of public law or natural resources, certainly not in the way public employees or elected officials are. As recipients of public grants, it might be argued that academic scientists have stewardship of public funds and thus their relationship to those funds must be clear of conflicts of interest. This is not a popular argument, and it was not used to justify the Guidelines on Conflict of Interest issued by the National Science Foundation and the Public Health Service. The title of the Public Health Service Guidelines on conflict of interest is "Objectivity in Research."[13] Thus, managing COIs among scientists was viewed as promoting scientific integrity, not protecting public law, regulations, or property from being compromised by personal interests.

DISCLOSURE OF STEWARDSHIP

While "stewardship ethics" does not seem applicable to academic science, another ethical response to scientific COI, one which has gained moderate acceptance in recent years, is transparency. The argument for scientists to disclose their conflicts of interest might be framed as follows. Scientists are expected to abide by the canons of their discipline even as they hold other interests, such as financial interests, in the subject matter of their research. The disclosure of one's financial interests (patents, equity holdings, honoraria) is deemed a responsibility because it allows peer reviewers, editors, and readers to look at published studies with additional skepticism.

Organized skepticism, one of the four Mertonian norms,[14] plays a central part in the scientific culture. A good scientific paper will discuss possible methodological limitations of a study and sources of bias. In many fields, it is considered the responsibility of the author to invoke a self-referential skepticism. Disclosure of one's financial interest in the subject matter of a paper falls into that tradition of barring all reasonable biases.

An author's financial disclosure might suggest to reviewers or readers that they consider how hidden biases related to the revealed interest might have entered the study. Also, disclosure allows editors to decide whether the conflicts are so egregious that the paper should not be published in their journal.

Disclosure also provides another social value. When an author's commercial affiliations are not cited in the publication of a paper but are learned after a controversy erupts, it makes it appear that the scientist has something to hide, even if he/she does not. In other words, with the lack of transparency of affiliation, public trust in science is diminished.[15]

Is disclosure a sufficient ethical response to scientific COI? Disclosures considered under COI policies or guidelines when scientists submit a paper for publication, testify before Congress, are recipients of a federal grant, or serve in an advisory capacity include whether the scientist:

- is a stockholder in a company that may benefit from research, a review, or an editorial;
- is a paid expert witness in litigation;
- receives honoraria from companies;
- is a patent holder;
- is a principal in a company that funds his/her research;
- serves as a paid member of a scientific advisory board or board of directors of a company.

The application to clinical trials of the informed consent ethical framework has recently come under debate. There are two streams of thinking here. One view is that COI is inherently unethical in clinical trials because it breaks the trust relationship between patient and physician. The second view holds that COI is not inherently unethical but must be part of the well-established informed consent process. Thus far, informed consent has focused on the nature of the medical intervention, including risks and benefits to the subject. Introducing COI into the informed consent process is viewed by some as a marked departure from the ethics of patient care.

In the case of the tragic death of Jesse Gelsinger in September 1999, the young man was not fully advised of the conflicts of interest involved in his experimental gene therapy treatment. During the investigations following Gelsinger's death, it was learned that the director of the University of Pennsylvania's Institute for Human Gene Therapy, James Wilson, founded a biotechnology company called Genovo, Inc. Both he and the University of Pennsylvania (Penn) had equity stakes in the company, which had invested in the genetically altered virus used in the gene therapy experiment. Wilson and one of his colleagues had also been awarded patents on certain aspects of the procedure. Genovo at the time contributed a fifth of the $25 million annual budget of Penn's gene therapy institute and, in retum, had exclusive rights over any commercial products.

The informed consent documents made no mention of the specific financial relationships involving the clinical investigator, the university, and the company. The eleven-page consent form Gelsinger signed had one sentence that stated that the investigators and the University of Pennsylvania had a financial interest in a successful outcome. When Genovo was sold to a larger company, James Wilson had stock options reported to be worth $13.5 million; the university's stock was valued at $1.4 million.[16] According to the report in the *Washington Post*, "numerous internal U. Penn. documents reveal that university officials had extensive discussions about the possible dangers of such financial entanglements."[17]

The Gelsinger family filed a wrongful death lawsuit against the university, which was eventually settled out of court for an undisclosed sum of money. One of the plaintiff's allegations in the suit was that the clinical investigator overseeing his trial had a conflict of interest that was not adequately disclosed prior to Jesse Gelsinger's involvement. They argued that the financial interests in conjunction with other undisclosed or downplayed risks might have altered the family's risk benefit estimate before entering the trial and saved young Gelsinger's life. After Penn settled with the Gelsinger family, the university administration announced new restrictions on faculty involved in drug studies when they have equity in companies sponsoring the research.

In the aftermath of the Gelsinger case, the Department of Health and Human Services (DHHS), under the leadership of Secretary Donna Shalala, held hearings on whether the financial interests of clinical investigators should be listed on informed consent documents given to prospective candidates for clinical trials. In a draft guidance document DHHS suggested that researchers involved in clinical trials disclose any financial interests they have to Institutional Review Boards that monitor other ethical issues and possibly to the patients deciding whether to participate as human subjects. Leading scientific and medical associations, including the Federation of American Societies of Experimental Biology (FASEB) and the American Association of Medical Colleges (AAMC), opposed the idea of a guidance document for clinical trials, arguing that it over-regulates medical research without contributing to the safety of patients.

Millions of Americans participated in more than 40,000 clinical trials in 2002, about 4,000 of which were supported by the National Institutes of Health (NIH). Research scientists and the companies sponsoring those trials were concerned that the additional disclosure requirements with no direct bearing on the safety or benefits of the trials would create unnecessary impediments to attracting human volunteers. On the other hand, the decision to become a human volunteer in a medical experiment can be one of the most important choices a person can make. Why should prospective volunteers not know everything of relevance to the trust relationship they are asked to develop with the clinical investigator?

Conflict of interest in clinical trials has become an ethical issue because of the perceived fiduciary responsibility of the clinical investigator to disclose all relevant information to the human subject. This legal responsibility was upheld by the California Supreme Court in the case of the MO-Cells, cells taken from John Moore during his surgery without his informed consent.[18]

CONSEQUENTIALIST ETHICS OF COI

Does anything intrinsically unethical occur when a scientist engages in research in which he or she has a commercial interest? To answer affirmatively we would have to demonstrate that such a condition would violate the scientist's fiduciary responsibility to some person or persons, or that there is an inherent conflict between any of those relationships and the scientists' goal or mission qua scientist. With the exception of human subjects' research, there is no compelling argument here. I can find no inherent reason why scientists cannot pursue the truth and still participate in the commercialization of that knowledge. The two activities do not appear to be logically or conceptually in conflict. But the context and consequences of scientific COI may be ethically significant. Does possessing a commercial interest in the subject matter of one's research have other, unintended effects?

In his book *Real Science*, John Ziman addresses the question of the significance of "disinterestedness" in ensuring the objectivity of science.[19] He observes that in the current climate of commercial science, "what cannot be denied is that the academic norm of disinterestedness no longer operates."[20] While Ziman asserts that we can no longer assume "disinterestedness" as a norm in this period of "post-academic science," "the real question is whether their [scientists'] interests are so influential and systematic that they turn science into their unwitting tool."[21] In other words, will the loss of "disinterestedness" result in the demise of objectivity?

Ziman distinguishes between two concepts of objectivity. He defines cognitive objectivity as an epistemic concept that refers to the existence of physical entities and their properties and that is independent of what we may know about them. Cognitive objectivity is attained when we tap into the properties of the "objective world," that segment of the physical universe that exists independent of our thought processes.

Social objectivity is defined by Ziman as the perception that the knowledge process is not biased by the personal self-interest of the knower. Despite the loss of "disinterestedness" in science, Ziman believes that cognitive objectivity can be protected. "The production of objective knowledge thus depends less on genuine personal "disinterestedness" than on the effective operation of other norms, especially the norms of communalism, universalism and skepticism. So long as post-academic science abides by these norms, its long term cognitive objectivity is not in serious doubt."[22]

I dwell on Ziman's work because he provides an important context for understanding society's ethical and legal response to scientific COI.

Cognitive objectivity is the verifiable and dependable knowledge science seeks. If that knowledge is not threatened by the loss of disinterestedness, then society's response to COI may be decidedly different than if it were. And while social objectivity (the public's perception of objectivity) may be important, its loss does not, in itself, affect the quality of certifiable knowledge—the published research in our peer-reviewed journals.

How can we determine whether cognitive objectivity is preserved in post-academic science? In contrast to other methods of fixing belief, science is considered to be self-correcting. It is generally understood that, in the long run, systematic bias and errors in science will eventually be disclosed and corrected. However, the time period for self-corrections in science to take place can be quite protracted. It took about 1800 years before Galileo corrected Aristotle's laws of motion. While errors or bias in modern science may not have to wait that long to be discovered, they can be very damaging even for short periods. Witness the work of Sir Cyril Burt on twin studies and IQ: Burt's results influenced cognitive psychologists and educational theorists for decades before it was discovered to be a fraud.[23] The faith we have in the self-corrective nature of science must be viewed against the effects of biased studies in fields like biomedicine, toxicology, and material science. Within this context we may ask whether multi-vested science is distorted by a conflict of interest effect. In Ziman's words, will the loss of disinterestedness affect cognitive objectivity? A relatively new body of research can help us answer this question.

In the consequentialist framework, the ethics of COI is viewed in terms of whether holding a conflicting interest correlates with one of the transgressions in science. The burden is to demonstrate a link between possessing a COI and some level of scientific misconduct or bias. The generally accepted transgressions in science include the following:

- scientific fraud;
- failure to give informed consent;
- wanton endangerment of human or animal subjects;

- plagiarism; and
- systematic bias.

Borderline ethical issues include:

- unwillingness to share scientific data/information; and
- participation in ghost writing.

A COI can be said to be an ethical issue in science if it disposes a scientist to commit an ethical transgression—that is, if it increases the probability that the scientist will violate his/her professional responsibility. If COI does not affect the professional responsibility of scientists, then perhaps efforts taken towards managing COI have, as suggested above, more to do with political correctness than with righting an ethical wrong.

In 1996 Les Rothenberg and I published a study which showed that lead authors of articles published in fourteen highly rated journals had a 34 percent likelihood of having a financial interest in the subject matter of the publication. *Nature Magazine* wrote an editorial stating that:

> It comes at no surprise to find that about one third of a group of life scientists working in the biotechnology rich state of Massachusetts had financial interests in work they published in academic journals in 1992. The work published makes no claim that the undeclared interests led to any fraud, deception or bias in presentation, and until there is evidence that there are serious risks of such malpractice, this journal will persist in its stubborn belief that research as we publish it is indeed research, not business.[24]

Five years later, *Nature* reversed itself and decided it would introduce conflict of interest requirements for authors.[25] In its editorial announcing the change of policy *Nature* wrote, "there is suggestive evidence in the literature that publication practice in biomedical research has been influenced by the commercial interests of authors."[26]

What do we know about the relationship between possessing a financial interest and bias? Is there a funding effect in science? If there is evidence that the private funding of science produces conclusions biased towards the interests of the sponsor, then we have a genuine cause for treating COI as an ethical problem. The first set of systematic studies that looked at whether there was an association between the source of funding and the outcome of a study was centered on the drug industry.

One of the most elegant and influential studies demonstrating an association between funding source and outcome was published in 1998 in the *New England Journal of Medicine* by a Canadian research team at the University of Toronto.[27]

This study began with the question of whether there was an association between authors' published positions on the safety of certain drugs and their financial relationships with the pharmaceutical industry. The authors focused their study on a class of drugs called calcium channel antagonists, which are used to treat hypertension. Their choice was based on the fact that the medical community debated the safety of these drugs. The researchers performed a natural experiment to investigate whether the existing divisions among researchers over the drug's safety could be accounted for by funding sources, whether, that is, medical researchers were financially connected to the pharmaceutical industry, and whether those affiliations explained their conclusions.

First, the authors identified medical journal articles on calcium channel blockers (CCBs, also known as channel antagonists) published between March 10, 1995 and September 30, 1996. Each article (and its author) was classified as being supportive, neutral, or critical with respect to these drugs. Second, the authors were sent questionnaires which queried whether they had received funding in the past five years from companies that manufacture either CCBs or products that compete with them. The investigators ended up with seventy articles (five reports of original research, thirty-two review articles, and thirty-three letters to the editor). From the seventy articles, eighty-nine authors were assigned a classification (supportive, neutral, or critical). Completed questionnaires about author financial interests were received from sixty-nine authors. The study results showed that an overwhelming number of the supportive authors (96 per cent) had financial relationships with manufacturers of CCBs, while only 37 per cent of the critical authors and 60 per cent of the neutral authors had such relationships. The authors of the *New England Journal of Medicine* study wrote that "our results demonstrate a strong association between authors' published positions on the safety of calcium-channel antagonists and their financial relationships with pharmaceutical manufacturers."[28]

Other studies confirm a funding effect for randomized drug trials,[29] economic analyses of new drugs used in oncology,[30] and research on nicotine's effect on human cognitive performance.[31] Marcia Angell, former editor of the *New England Journal of Medicine*, commented that it was her impression that "papers submitted by authors with financial conflicts of interest were far more likely to be biased in both design and interpretation."[32] Angell's impression was validated by findings that appeared in the *Journal of the American Medical Association* from a meta-type analysis on the "extent, impact, and management of financial conflicts of interest in biomedical research."[33] Beginning with a screening of 1,664 original research articles, the authors culled 144 that were potentially eligible for their analysis and ended up with 37 studies that met their criteria. One of the questions the authors pursued in their study was whether there was a funding effect in biomedical research. Eleven of the studies they reviewed found that industry-sponsored research yielded pro-industry outcomes. The authors concluded:

> Although only 37 articles met [our] inclusion criteria, evidence suggests that the financial ties that intertwine industry, investigators, and academic institutions can influence the research process. Strong and consistent evidence shows that industry-sponsored research tends to draw pro-industry conclusions. By combining data from articles examining 1140 studies, we found that industry-sponsored studies were significantly

more likely to reach conclusions that were favorable to the sponsor than were non-industry studies.[34]

There is perhaps a dozen or so studies that confirm the funding effect in science for clinical drug trials. The effect has also been confirmed for tobacco research[35] and postulated but not rigorously analyzed for toxicological studies of industrial chemicals,[36] nutrition research,[37] and policy studies.[38] Notwithstanding these results, there is no evidence that COI is correlated with scientific fraud or other serious ethical violations. Moreover, it is difficult to assess how generalized or pervasive the funding effect is. To reach the conclusion that research studies authored by scientists with commercial interests in the subject matter is inherently unethical (on the basis of a few dozen selected studies) because of a potential funding effect is neither defensible nor practical. There is, after all, over $2 billion in private research and development (R&D) funding going to U.S. universities (about 7 per cent of the total R&D budget in academia). To make the case that privately funded research fails the objectivity test and therefore is unethical would require a vast study of studies in a variety of disciplines as well as replication of results.

A recent survey reported in the journal *Nature* of several thousand U.S. scientists begins to provide some of the answers. Early and mid-career scientists were asked to respond anonymously to sixteen questions on their research behavior. One of the questions scientists were asked was whether they have changed the design, methodology, or results of a study in response to pressure from a funding source; 20.6 per cent of the mid-career scientists and 9.5 per cent of the early career scientists answered affirmatively.[39]

Several sectors, such as privately funded tobacco research, have been targeted as untrustworthy. As a consequence, some universities have refused to accept tobacco money for research or other purposes. In areas where the funding effect in science has not been confirmed, a consequentialist ethic for managing conflicts of interest (where assessing moral significance is predicated on their consequences to science) may not apply. There is, however, another ethical framework which has been incorporated into legal doctrine and applied to other sectors to prevent or minimize COI.

INTEGRITY OF SCIENCE AS AN ETHICAL NORM

Protecting the integrity of scientific enterprise is embedded in the scientific ethos. Organized skepticism, objectivity, disinterestedness, correcting mistakes, punishing scientific misconduct, peer review, and institutional review boards are all part of the system the scientific community has established to protect the integrity of the scientific enterprise. One can argue that a scientific discipline replete with conflicts of interest is likely to lose its integrity in the eyes of the general public because it appears to be accountable to interests other than the pursuit of truth. In Ziman's terms, even if science's cognitive objectivity is protected, the social objectivity of science will be threatened.

In the eyes of the public, the major virtue of academic scientists and their institutions is that, even when they do disagree, they can be trusted to present what they know "without fear or favor." Whether or not this high level of credibility is really justified, it is what gives science its authority in society at large. Without it, not only would the scientific enterprise lose much of its public support: many of the established conventions of a pluralistic, democratic society would be seriously threatened.[40]

There is a quantity-quality relationship. As a field of science becomes increasingly commercialized, the quality of the science and the public's confidence in it suffers. Just think of cigarette science, or the studies funded by the lead or the chemical industry. The goal behind these industry-funded research agendas is to manufacture uncertainty for the purpose of derailing or postponing regulation. If the protection of scientific integrity is a societal goal and COI is an obstacle to reaching that goal, then COI should be viewed as an ethical issue. Moreover, preventing or minimizing COI becomes an ethical imperative.

We try to prevent COI in legal procedures because it erodes the goal of a fair trial. Federal judges cannot own a single stock in a company that is a litigant in their courtroom. It would be inconceivable for society to accept a judge's declaration that, in deference to transparency, he would disclose that he was sentencing a convicted felon to serve his sentence in a for-profit prison in which he, the judge, has equity interests. The courts are exclusively funded by public sources; universities and professors receive funding from public and private sources. We cannot apply the same standards. But there are certain conditions where the integrity of research is so critical to public trust that a response is warranted.

What can be done to restore the integrity of academic science and medicine at a time when turning corporate and blurring the boundaries between nonprofit and for-profit are in such favor? We should perhaps begin by harkening back to the principles on which universities are founded. We should consider the importance of protecting those principles from erosion and compromise for the sake of amassing larger institutional budgets and providing more earning potential for select faculty members. I have proposed several principles:

- the roles of those who produce knowledge in academia and those stakeholders who have a financial interest in that knowledge should be kept separate and distinct;
- the roles of those who have a fiduciary responsibility to care for patients while enlisting them as research subjects and those who have a financial stake in the specific pharmaceuticals, therapies, or other products, clinical trials, or facilities contributing to patient care should be kept separate and distinct; and
- the roles of those who assess therapies, drugs, toxic substances, or consumer products and those who have a financial stake in the success or failure of those products should be kept separate and distinct.[41]

The ethical foundations needed for protecting the integrity of science demand measures that go beyond the mere disclosure of interests.[42] If disclosure were the only solution, scientists would be viewed as simply other stakeholders in an arena of private interests vying for epistemological hegemony. The ethical principles—as ideals—would require that certain relationships in academia be prohibited. The legal foundations, however, remain uncertain. Currently, the law has little to offer on the question of preventing a clinical investigator from having a financial conflict of interest in therapies while caring for patients or supervising clinical trials. Universities have become the self-managers of COI both among their own faculty and for their own institution.

There are no legal sanctions for transgressing a norm, because there are no established legal norms. In other areas of public ethics, the laws are more explicit. In the United States, the roles of financial auditors and accountants have been under more scrutiny since the Enron affair. New rules have separated auditing from other financial dealings. Legal separation of conflicting roles, however, has not reached the scientific community, perhaps because scientists, unlike lawyers, politicians, and accountants, are still viewed as adhering to a standard of virtue that renders them immune from compromise by their involvement with commercial interests. Recent scientific evidence reveals a quite different picture.

Chapter 12

WHEN SPONSORED RESEARCH FAILS THE ADMISSIONS TEST: A NORMATIVE FRAMEWORK[19]

SHELDON KRIMSKY

Tufts University, Department of Urban and Environmental Policy and Planning

INTRODUCTION

Whenever the topic of dubious sources of external funding was raised in conversation, a former president at my university was known to have replied, "The only problem with tainted research funding is there t'aint enough of it." It is a curious statement by the president of a distinguished university, who, among other things, is the steward of the university's endowment, the torch bearer of its mission, and the moral placeholder of its values. This was also the same president who defended the university's policy against accepting grants or contracts for weapons research.

Like many other academic institutions, Tufts University has, on occasion, debated issues related to the ethics of sponsored research. Controversies

19Originally published in *Universities at Risk: How Politics, Special Interests, and Corporatization Threaten Academic Integrity*, James L. Turk, ed. (Toronto: James Lorimer & Co. 2008).

have erupted over whether individual faculty or institutional policy should prevail in deciding whether sponsored research was acceptable to the institution. The individual researcher and his/her own institution are in a symbiotic relationship with respect to research funding. Sponsored research is awarded to the institution with the understanding that an individual faculty member, who is usually called the principal investigator (PI), is the responsible party to whom the research award is designated. If the PI changes institutional affiliation, frequently arrangements are made to transfer the sponsored funding to the PI's new institutional home. The PI works within a set of norms that have been incorporated into the university's bylaws and policies. Some of these norms are unique to the institution, while others are based on federal mandates and are, therefore, uniform across all institutions that receive federal support. Within the context of university sponsored research, I shall examine the following normative questions:

- Should universities be selective in approving grants or contracts applied for by individual faculty on the basis of ethical considerations?
- Is it acceptable, and if so, on what grounds, for university administrators to restrict grants or contracts sought by individual faculty based on a particular type of research activity?
- Is it acceptable, and if so, on what grounds, for university administrators to restrict grants or contracts sought by individual faculty based solely on the types of products manufactured by, or the business sector classification of, the sponsoring organization?
- Is it acceptable, and if so, on what grounds, for university administrators to restrict grants or contracts sought by individual faculty merely on the relationship of the investigator to the sponsoring organization?
- What rules of governance are appropriate in deciding which funding is or is not appropriate?

- What, if any, litmus tests can be used by university administrators for setting standards of acceptable research sponsorship without undermining the idea of a free and open university?

In this chapter I shall address these questions by first developing a framework that possesses generalizable elements, yet is capable of individuating the answers to specific cases by adapting them to variations in university micro-culture policies on sponsored research. I shall focus my framework on contracts and grants and set aside the issue of the standards for acceptable gifts to the university, which generally do not involve research. Second, I shall apply the framework to several cases where there has been controversy over the ethics of sponsored research.

THE UNIVERSITY'S MULTIPLE PERSONALITIES

Universities are unique institutions in the American landscape. The members of the professoriate have considerable autonomy relative to other professions, as characterized by what we teach, the types of research we do, and in our freedom to write and speak without having to meet a litmus test of the institution. And while all universities strive for the three major goals of education, research, and service, there is little homogeneity in the balance given to these goals. Within the research mission, universities express their goals quite differently. Even within the same institution, the concept of research takes different forms. I have previously described this as the university's multiple personalities.[1]

The four archetypal personalities or models that characterize university research are the classical, the Baconian, the defense, and the public interest. While universities have research portfolios representing each of these personalities or models, the weight of funding in any of these categories can affect a university's normative policies on acceptable sponsored research. According to the classical personality, identified with the phrase "knowledge is virtue," research is organized around the attainment of knowledge for its own sake. Inquiry is internally driven by faculty, as captured by the expression "investigator-initiated research." Scientists/scholars adhere to the norms of universal cooperation, free and open communication, and the knowledge commons, where the results of research are freely available for everyone to use.

The Baconian model is described by the expression "knowledge is productivity." (Francis Bacon used the expression "knowledge is power.") The university's role is to provide personnel and intellectual resources to foster economic development. Professional education and corporate-related research is the key to this dimension of the academic mission. The work of the scientist begins with discovery, continues through application, and ends with intellectual property. Universities that emphasize the Baconian personality are typically more receptive to industrial contracts and corporate partnerships or faculty-corporate liaison programs. The agreements permit compromises on such issues as sharing of data, sequestered or confidential research, and single-party restrictive licensing of patents. The Baconian model implies that the pursuit of knowledge is not fully realized unless and until it contributes to productivity. As one study notes: "In recent decades universities have added a component of economic development to their missions, accomplished largely through transfer of university technology to existing or new businesses."[2] Therefore, the responsibility of the investigator includes both discovery and technology transfer. The term "translational research" is au courant for describing this process.

The "defense" model of the research university is guided by the dictum "knowledge is security." Universities devote their resources to capturing grants and contracts from the defense industry. The micro-norms of the research community adapt to weapons research, anti-terrorism and counter-insurgency studies, spy satellites, and code breaking, to name a few. Often, these projects involve research contracts that are fully or partially classified.

Finally, the public-interest model of science is framed around the aphorism "knowledge is human welfare." The university organizes its research facilities to address major societal problems such as the cure of dread disease, environmental pollution, global climate change, and poverty.

Public-interest science is generally more favorable to the idea of the "knowledge commons" on the principle that when knowledge is publicly funded, it should be publicly available. This is the spirit of a bill introduced into Congress titled the Federal Research Public Access Act of 2006, which requires open access to research results funded by the federal government within six months of their first publication.[3] The norms attached to public-interest science can be expected to be different from Baconian and defense-oriented science. Traditionally, corporate science has had many more constraints and covenants embedded in its university contracts. As noted by Resnik, "Companies may suppress results, refuse to apply for patents, or keep research under a cloak of secrecy for many years. They may also refuse to share useful tools, resources or techniques."[4]

Because academic institutions are not homogeneous in their research personality, they choose a different balance in their research portfolios, and as a result the norms that define their research programs will vary. The Baconian personality will tolerate more privacy and confidentiality than the classical personality. Within the balance of these multiple personalities, individual researchers are accorded a degree of autonomy. Their autonomy is set against the norms of the institution, which may limit some of their choices. It should also be recognized that institutional factors can also impede faculty autonomy. Faculty can exercise self-censorship on what they study or how they study their field of interest, if they believe it will have a negative effect on tenure or promotion. In the next section I discuss how faculty autonomy can be in conflict with a university's policies on acceptable sponsored research.

RELATIVE AUTONOMY OF FACULTY

Traditionally, faculties in universities decide the research questions they pursue and the grants and contracts they apply for as well as what results are published and when. These choices are what we refer to as faculty autonomy. Other rights of the faculty include academic freedom, or the right to speak and write on controversial or unpopular subjects, and the right to define the content of a course.

The autonomy of faculty is a privilege associated with tenure, but it is not an absolute privilege. It is largely modulated by two factors: the local university rules and those of government agencies that apply to all federally funded institutions. The federal rules are uniform, although many are merely guidelines for which individual institutions provide local content. In the case of conflict of interest, for example, universities have significant latitude to meet federal compliance standards. Whether strict regulations or flexible guidelines, federal authority sets constraints on scientific autonomy, where autonomy is viewed as "academic libertarianism."

As an example, U.S. federal guidelines on human subjects research has extended informed consent requirements to scholars interviewing other scholars, hardly the intent of protecting vulnerable populations. Many universities set limits on faculty consulting. If not universal, it must be nearly so that a faculty member cannot simultaneously hold tenure in two separate institutions. Some universities prohibit faculty from using their names, titles, and university affiliation on product advertisements—but it is certainly not a university-wide norm. No one claims that university faculty as teachers or scholars have absolute autonomy when it comes to academic-business relationships. When Nobel laureate and Harvard professor Walter Gilbert became the CEO of the BioGen Corporation, the university asked that he give up his professorship while he was in the corporate role.[5]

For some faculty, keeping the funding flowing is the lifeblood of their academic existence. Without this funding, they cannot support post-docs and technicians in their laboratory. On the other hand, it is the post-docs and technicians that are needed to maintain the flow of external funding and to produce publishable results. Increasingly, medical school faculty depend on external funding for their own salaries. As such, research faculty become small entrepreneurs who sell their research services in the form of grants and contracts to foundations, government agencies, or corporations. It is in the interest of the research faculty to be able to leverage their expertise and research skills to attract funds from any source that is willing and able to fund them. People with abundant sources of funding can be selective; those with fewer choices want the maximum latitude to bring in a grant or contract.

Universities benefit from the overhead that accompanies a grant or contract, regardless of its source. Some university administrators even discourage faculty from applying for funding from sponsors that do not honor the government indirect cost standard. From purely an economic standpoint, universities can benefit by having an open admission policy for research funding, allowing their faculty to define the parameters of their research, with little or no restrictions on the substance of the grant or contract or the moral standing of the sponsor. But economics, while perhaps a central driving force behind university policies on sponsored research, is not the only consideration. Sponsored research programs are bound by norms other than maximizing cash flows to the university. For example, universities may decide that they will not accept research contracts that require sequestered student dissertations. When such contracts are opposed, the norm of open science communications trumps the interest in adding more research dollars to the institution. Similarly, universities are increasingly opposed to accepting sponsor control over publication.

Secret covenants that give the sponsor the final approval on publication was the issue behind the Betty Dong case. Professor Dong of the University of California at San Francisco signed a contract that gave the sponsor the ultimate authority for publishing results from the data collected by her research. Left without sufficient resources to take on the company in litigation, Professor Dong withdrew the paper at the eleventh hour from the galleys of the *Journal of the American Medical Association.*[6]

Under a libertarian view of faculty, the individual scientist alone decides what the contract conditions are between the faculty and the sponsoring organization. Where libertarian principles prevail in academia, the university is viewed as an enabler that fosters the interests of the autonomous researcher-entrepreneur while providing infrastructure for the research in exchange for overhead charges.

THE COMMUNITARIAN-LIBERTARIAN DIVIDE IN ACADEMIA

Within most universities there is a tenuous balance between communitarian and libertarian tendencies of the faculty. Under communitarian values, a faculty governance process decides on the rules and norms under which research takes place. At best, academic communitarianism functions out of a democratic process prescribed in the formal rules of the faculty governance structure consistent with the powers of the administration and board of trustees. Libertarianism affords faculty all rights not limited by communitarian norms. Thus, if a university has no constraints on faculty outside employment, then it is assumed that faculty may work as many hours as they wish outside the university while meeting their campus obligation.

The balance between communitarianism and libertarianism is calibrated for each institution. There is no reason to believe the balance should be identical across all institutional cultures. We may refer to this as the communitarian-libertarian balance, which speaks directly to the normative structure of research. The institution may decide that, in the overall interests of the university, some restraints are warranted on the type of grant money accepted. But the story doesn't end there. There are norms which are so deeply at the core of the university's mission and raison d'être that their retraction would bring into question whether the institution is still functioning like a university.

META-LEVEL NORMS

Universities share a family resemblance in their educational mission. However, not all institutions that claim to offer teaching and education would be considered universities. Thus, McDonald's has created "Hamburger University," which is only a university through false analogy and appropriated nomenclature. Other entities that use the term "university" are nothing more than diploma mills.

Not all universities can claim to focus on research. But of those that do, the communitarian-libertarian balance at universities helps to explain the diversity of norms guiding sponsored research. While acknowledging variations in the moral yardstick at different research institutions, I propose a second level of norms whose function it is to protect and preserve the unity of core values that capture the family resemblance across the micro-cultures we call universities. I refer to these as meta-level norms for research integrity. These norms should not be contingent on the calibration set for an institution between communitarian and libertarian values, which we may consider ground-level norms.

I shall argue that "meta-level" norms should be invariant across all universities. Insofar as the norms dictate acceptable criteria for accepting sponsored research, they shall not be traded away by rebalancing communitarian and libertarian interests (ground-level norms). To justify their invariance across the diverse university cultures, meta-level norms must stand up to critical scrutiny. When a meta-level norm is rejected, we must be prepared to argue that the institution falls short of meeting one of the essential qualities we associate with universities. Because they are invariant, meta-level norms are not subject to tradeoffs. Sponsored research contracts that violate meta-level norms should fail the admissions test, however else the university calibrates its balance between communitarian and libertarian interests of its faculty. Ideological preferences should not affect the choice of meta-level norms. For example, the social and regulatory conservative Henry Miller of the Hoover Institute writes, "Universities must also ensure that he who pays the academic piper doesn't get to call the tune by influencing the results of research or by suppressing undesirable findings."[7]

- As a start, I propose the following meta-level norms for externally funded research. The autonomy of the researcher must be protected. The researcher and his/her co-investigators must be in full control of the protocols, the data, the interpretation of results and the decision, venue, and time of publication.
- External research grants or contracts shall not place confidentiality requirements on the research outcome with the sole exception of a

brief period to file a patent. Secrecy has no place in the open university.

- The purpose of external research funding should be to contribute to knowledge and not to produce public relations, promote products, or defend litigants in court.
- Transparency of the sponsor is essential to establishing trust in the university's role of a knowledge generator.
- No contractual constraints shall be imposed on the researchers for reporting results, however inconvenient or objectionable they may be to the sponsor.
- The principal investigators in a sponsored research project should be solely responsible for writing up the results; the sponsor should not play a role as co-investigator or contract with a ghost-writing company that is not cited in the publication.
- No external research grants or contracts should discriminate by age, race, religion, gender, political beliefs, or disability.

A few examples will illustrate these points. Let us suppose a sponsor issues a contract for research that takes away from the principal investigator his or her autonomy for publication of the results. This provision in a sponsored research contract is inconsistent with how we understand a research university is supposed to function. University research is not a product of public relations or advocacy but an independent pursuit of knowledge guided by the researchers' disciplinary canons and reviewed by peers.

While contracts that conflict with the autonomy of the academic researcher are inconsistent with the idea of the university, they are quite prevalent across academia. As noted by Resnik:

> Corporations that sponsor research frequently require scientists and engineers to sign contracts granting the company control over proprietary information. These agreements typically allow companies to review all publications or public presentations of results, to delay or suppress publications,

> or to prevent researchers from sharing equipment or techniques. The agreements are legally binding and have been upheld by the courts.[8]

In a second case, the corporate sponsor issues a contract in which an academic investigator permits his name to be used in a ghost-written article. The sponsor hires a firm that specializes in writing and placing medical articles in the literature. The firm locates a highly regarded scientist whom they consider a "thought leader" and offers him a contract to sign his name as author to an article written for the firm, which would then be placed in a medical journal. To the company, there is an advantage in having a prominent medical researcher write about, for example, off-label uses of a drug, which have not been approved by the LIS Food and Drug Administration (FDA). Companies are prohibited from promoting or advertising the use of a drug for some purpose that has not received FDA approval. Physicians, however, may prescribe off-label uses of drugs based on their own judgments, even as drug companies are prohibited from lobbying them in support of such uses. Through ghost writing, drug companies get around the prohibition of off-label lobbying by having doctors speak to one another about their "best" practices and observations. Journals are beginning to respond to ghost writing as a form of plagiarism and therefore a violation of scientific integrity.[9]

In both the previous cases, the focus of the sponsored research is on the nature of the contract language and not on the company sponsor or the type of research. Meta-norms against plagiarism or sponsor control of research outcome speak to the core principles of a research university. However, the boundary between ground-level and meta-level norms is dynamic and not invariant. As professional associations and accrediting groups reach unanimity over ground-level norms, those norms can rise to the meta-level.

Starting after World War II and continuing through the Vietnam War period, a number of universities in the United States accepted classified research from government agencies, most prominently the Department of Defense. Faculty and graduate students participating in a classified research project had to be investigated and vetted by the Department of Justice to determine whether they represented a security risk—a term that was broadly interpreted to include people who protested the war. Classified research divided the academic community. Some view it favorably as contributing to the "defense" model of the university, where academic scientists contribute to national security. Others, however, saw the system of classified research as an anachronism in the modern research university where free and open exchange of ideas is the hallmark of higher learning.

During the 1960s, student anti-war activists at the University of Pennsylvania (U Penn) learned that their institution hosted classified research on chemical and biological warfare, under the program names Spicerack and Summit, which had direct relevance to the LIS actions in Vietnam.[10] While students criticized the research for being immoral, most faculty criticism focused on the inappropriateness of secrecy at the U Penn campus. Eventually, after a split between the administration and faculty, the U Penn Board of Trustees voted to divest the university of its classified war-related research and refused to transfer the projects to its off-campus sites.[11]

By the Vietnam War's end there was a broad consensus among major research universities that classified or secret research was incompatible with the values of academic science. The pursuit of certifiable knowledge requires transparency to actualize the self-correcting function of science. Classified science was a recipe for perpetuating errors. Science could easily fall victim to ideology in a closed system built on secrecy, sequestered science, and loyalty oaths.

A number of research universities that had once accepted classified research spun off separate research centers that were not part of the core universities. For example, in response to student protests against secret research on campus in the 1960s and early 1970s, MIT turned Lincoln Laboratory into a semi-autonomous, off-campus entity where it located its classified research. Thus, the ground-level norm of "no classified research" adopted by some universities became the meta-level norm adopted by the vast majority of universities by the mid-1970s. However, secret research entered the university in another form: namely, business contracts that contained clauses about protecting confidential business information, including intellectual property. The campus political climate and public criticism that pressured universities to abandon classified research was not recreated to address business secrecy on campus when "academic capitalism" was ignited in the 1980s.

Another meta-level norm grew out of federal mandates that required universities to adopt ethical guidelines for human experiments, especially informed consent requirements for all persons participating as human subjects. By the mid-1970s it was no longer a matter of discretion for universities to protect human subjects in clinical trials. Human subjects' research, whether from public or private funding sources, had to comply with federal guidelines on informed consent.

CASUISTRY AND QUESTIONABLE SPONSORED RESEARCH

The term casuistry refers to the use of cases to draw conclusions in ethics and law. I shall use cases to draw out and test the normative principles developed in previous sections. The framework I described has two levels for building a normative structure of sponsored research. Meta-level norms provide unity, continuity, and invariance across research universities, while ground-level norms offer a degree of flexibility, enabling localized balances of communitarian and libertarian values. The boundaries are not impermeable but respond to consensus-building processes of professional and university associations. In the United States, government regulations and congressional oversight committees can also create a broad consensus for negotiating ethical standards in conducting research.

TOBACCO INDUSTRY-SPONSORED RESEARCH

Let us begin with a case of prohibiting research from a particular industry. The University of California Board of Regents debated whether any of its faculty should be allowed to conduct research financed by the tobacco industry.[12] This proposal was not a restriction on the content of the research but rather on the funding source. Presumably, the same project supported for funding by Philip Morris, Inc. could be funded by another corporation. This proposal to ban any research contract or grant from an entire industrial sector, namely tobacco, from sponsoring research at any of the universities in the California state system grew out of the findings about the tobacco industry's unsavory and rogue activities that were revealed in the state tobacco litigation.

Arguments in favor of "banning tobacco money" cite the health effects of tobacco, the industry's record in distorting research findings that were unfavorable to the sale of tobacco products, and the industry's lack of respect for scientific integrity. Critics of the tobacco industry characterize this sector as flagrantly dishonest and untrustworthy in claims about its products. For years the industry manipulated the amount of habit-forming nicotine in cigarettes and advertised directly to children. Universities, some believe, have an obligation to draw a scarlet letter on rogue industries who wish to gain credibility by funding research at a university. Bans on tobacco money are mostly found in certain schools, such as public health and medicine, because the mission of these schools cannot be reconciled with the reckless disregard of human health shown by the tobacco industry. The pretension that the tobacco industry had a true interest in science was nothing more than a ruse; all the time, it was seeking to buy itself fabricated knowledge. The misdeeds of the industry are well documented. The World Health Organization reports that, "the tobacco companies planned an ambitious series of studies, literature reviews and scientific conferences, to be conducted largely by front organizations or consultants, to demonstrate the weaknesses of the IARC [International Agency for Research on Cancer] study and of epidemiology, to challenge ETS [environmental tobacco smoke] toxicity and to offer alternatives to smoking restrictions."[13]

For over fifty years tobacco companies have placed articles in the medical literature, without revealing their support for the research. They financed a large number of studies intended to show that the research conducted by IARC was flawed, and they created an independent coalition of scientists to manufacture uncertainty on the link between tobacco and disease. The tobacco industry also funded international seminars to develop "good" epidemiological standards of scientific proof that would serve cigarette manufacturers by raising the standards of proof.[14] As noted in the journal *Science*, "The [tobacco] companies frequently killed their own research when it came to unfavorable conclusions, funded biased studies designed to undermine reports critical of smoking, and used the names of respected scientists and institutions to bolster their public image."[15]

Increasingly, studies have shown that health research funded by the tobacco industry is biased in favor of the financial interests of the sponsor. A report published in the journal *Addiction* found that "scientists acknowledging tobacco industry support reported typically that nicotine or smoking improved cognitive performance while researchers not reporting the financial support of the tobacco industry were more nearly split on their conclusions."[16]

Even with its legacy of deceit and malfeasance, there are reasons not to ban research funding from an entire industrial sector. First, some of the grants or contracts funded by the tobacco industry may lead to positive outcomes, particularly post-litigation, as the industry is under the social microscope. It can no longer get away with its unsavory practices. If there is any redemption from past misdeeds, industry money could benefit society. Most of today's mega-foundations obtained their wealth by human exploitation and deceit. By placing a blanket ban on such funding, the university could be foregoing socially valuable research.

A second consideration is how to circumscribe the so-called tobacco sector. Tobacco companies are parts of conglomerates. If there is a contaminated branch of a corporation or industrial sector, does it implicate all other branches from the corporate trunk? If a university bans funding from Philip Morris, should it also ban funding from a food corporation that sits under the same corporate umbrella? To circumvent such restrictions, tobacco companies can provide support to a foundation which then doles out money for research. As an example, a cancer researcher at the Weill Cornell Medical College published a study stating that 80 percent of lung cancer deaths could be prevented through the use of scans. The study had been financed in part by a little-known nonprofit called the Foundation for Lung Cancer. Investigative reporters at the *New York Times* learned that this foundation was largely supported by the parent company of a tobacco group.[17]

Third, what makes tobacco's behavior unique? Consider, for example, drug companies that have withheld important safety information from the FDA, resulting in preventable deaths. Or what shall we say about the asbestos, lead, and chemical industries that knowingly compromise workers' health in favor of maximizing corporate profits? Is there such a thing as tobacco exceptionalism? Failure to distinguish among rogue corporate behavior is one of the reasons the American Association of University Professors issued a policy on sponsored research that opposed the idea of singling out the tobacco sector:

> An institution, which seeks to distinguish between and among different kinds of offensive corporate behavior, presumes that it is competent to distinguish impermissible corporate wrongdoing from wrongful behavior that is acceptable. A university which starts down this path will find it difficult to resist demands that research bans should be imposed on other funding agencies that are seen as reckless or supportive of repellent programs. If the initiative in calling for these bans on the funding of faculty research comes from the faculty itself, our concerns about the restraints on academic freedom are not thereby lessened. A university at which the research is conducted should not be identified with the views and behavior of the tobacco industry because faculty members accept its funding, just as the university should not be identified as necessarily endorsing the content of the researcher's work.[18]

Some critics of tobacco exceptionalism question whether government sources of funding stand on higher moral ground than that of the cigarette industry. For example, funding from the U.S. Homeland Security Agency or from the Department of Defense is said to be tarnished by an illegitimate war effort based on presidential malfeasance in claiming unsubstantiated weapons of mass destruction as justification to make a preemptive strike against an independent state. Once moral criteria enter the decision for determining which funding is worthy of entering the university, research libertarians argue that no clear line can be drawn.

Another opponent against banning tobacco research in universities was quoted in *Science* as opposing the use of moral criteria to evaluate sponsored research. "How do you avoid infringing on academic freedom, and what sort of slippery slope do you create by denying grants on moral grounds?"[19]

Where does the case of the tobacco companies sit with respect to the framework developed in this paper? The meta-norms of the framework are not intended to apply to the historical or current misdeeds of a company. Similarly, the use of guilt by association, which can blemish an entire industrial sector for the malfeasance of a few companies, is not justified for establishing a meta-norm. Neither a company's history nor its current market behavior tells us anything about the quality of the research it could sponsor and its respect for independent and autonomous academic scientists who would plan and execute that research. Meta-norms are specific to the conditions under which research is conceptualized and executed under the sponsor's contract. In this framework, the meta-norms cannot be used to prohibit research dollars from an entire industrial sector, unless a specific action adopted by that sector violates one of the core principles of independent research: for example, if an industrial sector never allows a sponsored researcher to have autonomy over publication.

The framework leaves open the possibility that ground-level norms would bar tobacco companies from sponsoring research. The university community can decide that, regardless of the social value of the proposed research project, the commercial goals of the tobacco industry are in conflict with the values of the institution. Moreover, the university does not wish to lend its honorable name to a dishonorable industry that preys on people prone to addiction and whose product is responsible for untold deaths and illnesses. Forbidding any research sponsor should not be taken lightly, because it is restricting individuals from potentially funding their work. For example, in 1990 the University of Delaware refused to receive grants from the Pioneer Fund, which one faculty member described as an organization with "a long and continuous history of supporting racism, anti-Semitism and other discriminatory practices."[20] University president E. A. Trabant initially defended the ban on Pioneer Fund "so long as the fund remains committed to the interest of its original charter and to a pattern of activities incompatible with the University's mission."[21] After an independent arbiter ruled in favor of two professors who wished to apply for grants from the fund, the University of Delaware reversed its policy.[22] The Harvard School of Public Health and the University of Glasgow prohibit their researchers from applying for tobacco funding. Some foundations like the Wellcome Trust (UK), the American Legacy Foundation, and the American Cancer Society will not fund researchers who have been awarded tobacco money.[23]

While such policies may conflict with an individual faculty member's funding opportunities, and in some cases their ability to maintain their laboratories, they do not rise to the level of infringement of academic freedom (meta-norm) as long as the policies follow appropriate university governance procedures and they do not constrain a faculty member's right to speak or write about a subject.

Those protesting any constraints on sponsored research cite the academic freedom of individual researchers to pursue areas of investigation of their choice. A resolution approved by the University of California's (UC) Academic Senate stated that "no special encumbrances should be placed on faculty members' ability to solicit or accept awards based on the source of funds."[24] Others correctly note that academic freedom refers to speaking, writing, and pencil-and-paper research. No professor has an unbridled right to engage in laboratory research or any research that requires sponsored funding independent of institutional or government norms. The investigator and the university administration are partners in the sponsored research. If an institution refuses to accept sponsored funding from a tobacco company, there are other options open to investigators who wish to pursue a research program. They could find other funding, pursue the study without funding, if possible, or collaborate with someone whose institution will accept the funding. There are no universal norms among universities which state that, "because I can get funding from company X for work Y in institution Z, then I *should* get approval from Z for such funding." Universities, however, should be able and willing to provide a justification, within the tradition of faculty governance, to prohibit a particular funding source. The burden for denial should be on the shoulders of the university.

In September 2007 the Board of Regents of the University of California took a middle-of-the road position between outlawing tobacco funding and giving total authority to individual faculty to negotiate research contracts with tobacco companies. The regents created a scientific review committee whose mandate it is to certify that a tobacco-industry funding proposal "uses sound methodology and appears designed to allow the research to reach objective and scientifically valid conclusions." Once the proposal is vetted and approved, the investigator will be allowed to apply for funds from the tobacco industry at any of the University of California colleges and universities.

WEAPONS RESEARCH

A second case that can be tested against the normative framework is the opposition to a class of research, namely, *weapons research*. Let us assume that a faculty senate is in agreement with the administration on proscribing any sponsored research involving weapons, including building or testing weapons or weapons systems, analysis of weapons, or protecting citizens or the military against weapons, as in the cases of developing vaccines for biological weapons or anti-missile systems. Let us also assume that the research contract does not violate the academic freedom or autonomy of the investigator in conducting or publishing the research, which usually implies that it is not classified. In this example, the contract language is not in conflict with other meta-norms. Because no meta-norms are violated, the decision on the suitability of the research would be made at the ground level. Can the university apply reasonable ethical grounds, based on its mission and core values, sufficient to gain support from the academic community for proscribing weapons research? Can the university establish a sufficiently clear demarcation between weapons and non-weapons research to avoid even the appearance that the decision is whimsical rather than being grounded on an accepted ethical norm?

It is likely that most, if not all, federally funded weapons-related contracts would have some degree of secrecy and, therefore, violate a meta-norm in the normative framework of this chapter. If the research is unclassified, it could be argued that it is fundamental in nature and does not have specific weapons application, such as a system of parallel computing that could be useful for radar tracking of a high-speed projectile in the atmosphere. Alternatively, a novel method of vaccination against Rift Valley fever or anthrax could raise questions about indirect weapons research. If a country has a vaccine, then the biological agent for which citizens or soldiers can be immunized becomes a weapon. Although at various times in history biologists have signed pledges stating they will not work on biological weapons, the distinction between defensive and offensive weapons within the fields of bacteriology and virology can easily be blurred.[25] As a communitarian decision, a university can, on moral grounds, proscribe sponsored research on weapons. But the framework I have introduced does not imply that response as long as meta-level norms are not breached. In my view, the concept of the "weapon" itself does not elevate the research concern to a violation of a meta-norm, therefore making it inherently unfit for a university.

COMMERCIAL TESTING IN ACADEMIA

Throughout much of the twentieth century, during the growth of academic entrepreneurship and government-sponsored research, universities reexamined the standards for tenure at their institutions and debated the criteria for evaluating the quality of faculty productivity. In fields such as chemistry and chemical engineering, faculty were doing extensive consulting and participating in what Karl Taylor Compton, former president of the Massachusetts Institute of Technology (MIT), called "pot-boiling research." According to John Servos's account of industrial relations at MIT, "excessive outside work," "pot-boiling" as Compton called it, would militate against "[academic] advancement."[26] Compton and others warned faculty that they would not get promoted if their work involved routine testing programs, typically handled by consulting companies, rather than the engagement in fundamental advances in science. Those who opposed pot-boiling research considered it outside of the university's educational and research mission to be turned into corporate testing centers that are likely to be accompanied by contracts with confidential business information requirements, potentially violating the meta-norms of openness and publishing rights. However, even if the meta-norms are not violated, an institution is correct in exercising its fiduciary responsibility when it evaluates whether the sponsored project offers any educational value or contributes to new knowledge.

Today, a number of universities make income by selling their testing services to corporations. As an example, Clemson University hosts the Clemson University Packing Service, which "provides contract package/product testing and material evaluation for both food and nonfood industries."[27]

Because the standards for what contributes to educational value and new knowledge may vary widely across disciplines and institutions, decisions about the proper place of testing programs in universities is best left to local governing systems. Nevertheless, extreme cases can easily be identified. For example, if the sponsored activity is not likely to yield published papers in refereed journals, the sponsored contract would fail the test of advancing knowledge. University public health departments have toxicology sections that accept industry contracts to test chemicals by in vitro or in vivo studies, applying standardized protocols that meet the criteria of regulatory bodies. There exist many refereed journals for publishing such studies.

I would include a meta-norm in my framework that sponsored testing activities that have no prospect for advancing knowledge or educational benefits for the university should be proscribed. This meta-norm protects universities, during times of financial exigency, from becoming contract research outposts for corporations. When the interpretations are ambiguous, the decision making should be left at the ground level, where local standards are applied.

During the past quarter-century, universities have found new lucrative income streams in running clinical trials for drug companies. Most of the new drug testing in the United States and Canada is supported by the pharmaceutical industry, Medical faculty benefit by acquiring publications from such trials, when such publications are approved by the sponsor. Published trial data can contribute to the applied knowledge of drug safety and efficacy, but rarely contribute to basic medical knowledge. From the public-interest perspective, universities may offer more quality control and moral accountability in managing clinical trials than one finds among contract research organizations, who hire private institutional review boards and have no accountability outside of their corporate structure.

MEGA-RESEARCH CONTRACTS

In 1980 the U.S. Congress enacted the Bayh–Dole Act, which stimulated aggressive corporate research investment in universities. Under the act, the government gave up to the universities and their business partners all intellectual property rights assigned to discoveries made under federal grants. A host of new multi-year, multi-million dollar corporate grants and contracts were awarded to universities, targeted to academic units as opposed to individuals. These included British Petroleum's $15 million to Princeton,[28]

Chevron's $25 million to the University of California at Davis,[29] and Exxon-Mobil's $100 million to Stanford.[30]

Recently, the University of California at Berkeley (UCB) has been at the epicenter of a controversy over corporate-academic partnerships involving sectors of the university. In the first of two partnership agreements, the dean of UCB's College of Natural Resources sent out letters of inquiry to sixteen agricultural biotechnology and life sciences companies, ostensibly to auction off a research collaboration with the Department of Plant and Microbial Biology. The dean selected Novartis, a $20 billion food and pharmaceutical company, as UCB's corporate partner. Under the contract Novartis provided UCB with $25 million in research dollars over five years. Among the benefits to Novartis were patenting and licensing rights as well as seats on UCB's internal research committee, which decided on the allocation of funds. All faculty members who signed on to the agreement (which it turned out was the vast majority of the department) were subject to restrictions. "Once a faculty member signed the confidentiality agreement, he or she could not publish results that involved data without approval from Novartis."[31] Novartis could request publication delays of up to 120 days and could obtain exclusive licensing rights of UCB patents—which has been argued by legal scholars is not in the public's interest. A second mega-contract with UCB came to fruition in 2007. British Petroleum (BP) signed an agreement worth $500 million in research funds to UCB, Lawrence Berkeley National Laboratory, and the University of Illinois to develop new sources of energy, with a primary interest in biofuel crops.[32] BP's funding supports a major expansion of UCB's clean energy research. The company gains the opportunity of assigning fifty of its researchers to the partnering institutions. Faculty at UCB raised questions about the impact of the agreement on researchers' academic freedom and the external control over the university's research agenda.[33]

Mega-contracts awarded to universities can compromise the autonomy of the institution or its faculty. Corporate partnerships typically involve joint corporate-academic committees that decide on the research agenda for use of the funds. With corporate funds amounting to hundreds of millions of dollars, there is a risk that it could create a monoculture of research in a department or even an entire school that is financially linked to one industrial sector or a single multinational corporation. The grass roots group Stop BP-Berkeley expressed a similar view in their protest literature: "We believe the proportion of corporate funding in public research must be carefully limited, to prevent the over-development of specific areas of research at the expense of others."[34] The academic unit in partnership and under contract with the company begins to take on the appearance of a research satellite of the sponsor. If the partnership lasts long enough, the size and influence of the sponsor's contract can violate the meta-norms that should be common to all universities. The prospect that mega-contracts can override the university's core values by violating meta-norms is, within the framework I have outlined, a reason to oppose them. The scale of the contract, not its specific content or the reputation of its sponsor, is at the root of the conflict. The quantitative changes arising from the size of the contract can result in qualitative changes that can impair the university's autonomy and diminish its role as a broker within the academic marketplace of ideas.

RACIAL OR ETHNIC DISCRIMINATION AND SPONSORED FUNDING

Universities are obligated to abide by national anti-discrimination laws. Let us imagine that a U.S. university is offered a grant to become research partners with a university in a Middle Eastern country. The national science agency in the country is similar to the U.S. National Science Foundation in that it funds basic science and operates under a system of peer review. There is one difference between the two agencies. Our hypothetical Middle Eastern science agency, following its national laws, prohibits anyone who is Jewish from working on the grant. The U.S. university must decide whether it will adopt the standards of another country with regard to personnel on its grants when it considers sponsored research funded by that country. Would accepting such a grant violate U.S. anti-discrimination laws? And if it were possible to get around those laws, would it be ethical to accept such funding? Both the U.S. Constitution and the federal civil rights enactments are sufficient grounds for treating anti-discrimination as a meta-norm in the proposed ethical framework for sponsored research. An external grant that requires a university to violate a constitutional principle—equal treatment under the law, cannot be permitted, whatever national government is footing the bill. It should not be left to the discretion of a university to sign a contract for sponsored research that would prevent members of the university community from fully participating in the research project because of their race or ethnicity. Even the lure of healthy profits from oil-rich countries cannot be an excuse for accepting such a contract or considering it under a fully deliberative communitarian process.

CONCLUSION

This chapter has explored the question: Are there ethical grounds for prohibiting university faculty from applying for certain types or sources of external funding? I propose a two-level normative framework, which I term the ground level and meta-level. The latter consists of a set of norms directed at the core epistemic values of independent and autonomous research institutions. The normative conditions outlined in the meta-level should be invariant across all research universities. Examples include norms such as that the investigators of a study are fully responsible for the data, the contents of published work, and the timing and venue of publication. The National Institute of Environmental Health Sciences journal *Environmental Health Perspectives* emphasizes such a norm in its instructions to authors: "all authors are required to certify that their freedom to design, conduct, interpret, and publish research is not compromised by any controlling sponsor as a condition of review or publication."[35] Another meta-norm should be that, under the conditions of external research, there shall be no discrimination of personnel with regard to race or ethnicity.

Ground-level norms include any factors of social and moral relevance to the institution that, in conjunction with faculty governance standards, allow the institution to calibrate a balance between communitarian and libertarian interests. Under my proposed framework, sponsored funding from tobacco companies, pro-Nazi organizations, or radical animal rights groups could meet meta-level conditions if the research contracts protect core values of the concept of the university. However, to account for the diversity of interests and values across American universities, the ground-level norms are set by proper governance functions at the individual institutions. The burden must be on the university to provide transparency and deliberative justification for taking from individual investigators the prima facie right to apply to funding organizations for sponsorship of their research. In developing this framework, I recognize that I depart from the policy adopted by the American Association of University Professors (AAUP), with whom I agree on many other issues. The AAUP has stated: "Denying a faculty member the opportunity to receive research funding for such reasons would curtail that individual's academic freedom no less than if the university acted directly to halt research that it considers unpalatable."[36]

In my view, as long as faculty members are neither suppressed from nor penalized for writing, teaching, investigating, or, speaking about an issue, they retain their academic freedom. That freedom is not extinguished in the case that a university community takes responsible and transparent collective action, following accepted governance procedures, that prohibits certain funding from entering the university.

ACKNOWLEDGMENTS

The author wishes to thank Carlos Sonnenschein, Charles Weiner, L. S. Rothenberg, and Kristin-Shrader Frechette for their comments on earlier drafts of this chapter.

Chapter 13

CONFLICTS OF INTEREST AND DISCLOSURE IN THE AMERICAN PSYCHIATRIC ASSOCIATION'S CLINICAL PRACTICE GUIDELINES[20]

LISA COSGROVE

University of Massachusetts, Boston,
Department of Counseling and School Psychology

HAROLD J. BURSZTAJN

Harvard Medical School, Beth Israel Deaconess
Medical Center, Department of Psychiatry

SHELDON KRIMSKY

Tufts University, Department of Urban and Environmental Policy

MARIA ANAYA

University of Massachusetts, Boston,
Department of Counseling and School Psychology

JUSTIN WALKER

University of Massachusetts, Boston,
Department of Counseling and School Psychology

20Originally published in *Psychotherapy and Psychosomatics* 78 (2009): 228–232.

KEYWORDS

Clinical practice guidelines, Conflict of interest, Financial associations

H J. B. has served as a consultant to physicians and institutions seeking to craft COI policies and as a plaintiff- or defendant-retained expert in competency to consent to neuropsychoharmaceutical treatment as well as product liability cases. All of the authors had full access to the data in the study; L. C. takes responsibility for the integrity of the data and the accuracy of the data analysis.

INTRODUCTION

Clinical practice guidelines (CPG) have become increasingly influential on health care providers because they are a "primary mechanism for communicating clinical aspects of emerging therapies and "standard of care" expectations to practicing physicians." Given their significance and impact, concerns have been raised about the potential for bias when treatment recommendations are developed by industry-funded researchers.[1–5] In the United States, medical specialties are now grappling with external criticism from the media, Congress,[6] journal editors, and medical writers.[7] The field of psychiatry, in particular, has been at the epicenter of extensive media coverage on conflicts of interest (COI). The reasons for the focus on psychiatry range from special considerations regarding the rights of highly vulnerable patient populations[8,9] to ongoing public concerns regarding the conduct of academic, institutional, and organized psychiatry[10–13] including that of the American Psychiatric Association (APA). For example, it was discovered in 2006 that all of the experts on two DSM-IV panels (schizophrenia and psychotic disorders and mood disorders) were pharmaceutical industry fundees,[14] a finding that raises concerns because psychopharmacology is recommended in the CPG as being the standard of care for the treatment of these 2 categories of disorders. In an effort to protect clients' welfare and help restore the integrity of psychiatry, the APA pledged that there would be greater transparency of any potential COI in its diagnostic and treatment guidelines.[15,16] This commitment is vital because the APA's published CPG influence the care given to millions of patients by both primary health care and mental health care providers. The present study provides data and recommendations that may be used to improve the process for developing CPG that are endorsed by the APA or other medical specialties.

METHODS

The APA issued eleven clinical practice guidelines between 1998 and 2007 that correspond to specific disorders identified in the DSM. This study reports data on the extent and type of financial relationships that authors of the three major CPG (for schizophrenia; bipolar; and major depressive disorder, MDD) had with the pharmaceutical industry. The schizophrenia CPG was published in 2004, MDD in 2000, and bipolar in 2002. Updates for MDD and bipolar, referred to by the APA as "guideline watches," were published in 2005 and were also included in the present study. Following a similar methodology to a previous study, the name of each CPG author was put through a series of "screens" to determine whether he or she has had financial relationships with 1 or more pharmaceutical companies whose business is potentially affected by decisions or recommendations made by the practice guidelines.

Rather than relying solely on self-reporting (e.g., using surveys) these multiple screening techniques allowed for a more thorough assessment of industry relationships. The screening databases we applied included Medline, Lexis-Nexis Academic, and the U.S. Patent and Trademark Office internet site on patents pending or awarded (to determine whether authors had any intellectual property in a drug or medical device whose sales could be affected by practice guideline recommendations). Internet search engines were used to access other reliable disclosures (e.g., author disclosures provided at peer-reviewed conferences). CPG authors were screened for any financial affiliations they had with the drug industry from up to five years before the publication of the 2008 guidelines.

The following categories were used to describe financial associations: honoraria; equity holdings in a drug company; principal in a start-up company; member of a scientific advisory board or speakers' bureau of a drug company; expert witness for a company in litigation; patent or copyright holder; consultancy; collaborator in an industry-funded study; gifts from drug companies including travel, grants, contracts, and research materials. Each finding of a CPG author's financial connection to a medical device manufacturer or drug company was then coded by its respective category. In addition to gathering data on the type of financial relationships, we identified the names and number of pharmaceutical companies with whom members had relationships.

Three investigators independently conducted screens on the work group members. Any questions about coding were resolved by a fourth investigator, who also conducted a random audit of the coding. No member was coded as having a financial connection unless there was unambiguous information confirming the relationship and a consensus reached by the reviewers.

RESULTS

Out of twenty CPG authors comprising the three practice guidelines, eighteen (90 percent) had at least one financial relationship with the pharmaceutical industry. None of the financial associations of the authors were disclosed in the CPG. One hundred percent of the work group members for both the schizophrenia and bipolar practice guidelines were found to have industry relationships, and 60 percent of the work group members for MDD had financial relationships.

Of the eighteen work group members who had industry relationships, 77.7 percent received research funding, 72.2 percent were consultants, 44.4 percent were on corporate or advisory boards to companies, 38.8 percent received honoraria, 33.3 percent served on company speaker's bureaus, and 16.6 percent held equity in a drug company that manufactured the drugs identified in the practice guidelines. The majority of work group members who had ties had multiple ones; 88.8 percent had more than one category and 66.6 percent were found to have three or more categories of financial interest with the pharmaceutical industry (e.g., ties such as being a consultant, sitting on a corporate board, and receiving research funding). Over half (55.5 percent) had four or more categories of financial interest.

One hundred percent of the working group members who had industry relationships had financial relationships with companies whose products were specifically considered or included in the guideline they authored. For example, eleven drugs were identified as meriting "substantial or moderate clinical confidence" for bipolar disorder. All of the authors of the CPG for bipolar disorder had financial relationships with companies whose drugs were identified as "first-line pharmacological treatment" (e.g., had equity in a company that made the medication, was a consultant or corporate board member, received honoraria). Nine drugs were identified as "likely to be optimal medications" for MDD. All of the companies whose drugs were listed as "optimal" provided funding to the authors of the CPG for MDD. Sixteen medications were identified as "commonly used" in the CPG for schizophrenia. All of the authors of the CPG had financial relationships with companies whose drugs were identified as "commonly used."

DISCUSSION

The three CPG examined in this study were selected because of their influence in medical practice and because of the potential revenue generated from sales of the recommended agents. Almost 23 million people are diagnosed with MDD, bipolar disorder, and schizophrenia,[17,18] and the revenue generated from sales of antidepressants and antipsychotics was approximately $25 billion in 2007.[19] The influence of the three CPG examined in the present study makes it incumbent on the APA to insure that authors'

COI are not only made transparent, but also are well managed. As Choudhry et al. note: "Financial conflicts of interest for authors of CPGs are of particular importance since they may not only influence the specific practice of these authors but also those of physicians following the recommendations contained within these guidelines. The APA has made an important first step in advocating for greater transparency of potential COI, and has made available author financial disclosure statements in its diagnostic guidelines[16] as well as in some of its recently published CPG (e.g., obsessive-compulsive disorder, Alzheimer's dementia). However, there has been no information given about how potential COI are managed, and author disclosures of the current MDD, bipolar, and schizophrenia CPG have not been provided. Current policies do not address indirect conflicts (e.g., when a family member owns pharmaceutical company stock) or indirect sources of funding (e.g., having one's salary drawn from pooled industry funds given to a department).[20] They also do not require that individuals specify the amount of money received, making it impossible to gather data on the extent of authors' financial relationships with industry.

The present study has several limitations. Some types of financial relationships were likely to be missed because they are difficult to detect. Our results should be considered de minimis figures. Another limitation is that our study cannot offer generalizability with respect to the incidence of industry financial relationships of authors of other CPG because the present study did not examine all of the APA's CPG. However, other researchers have shown that there is a strong association between authors' published opinions of products and their financial relationships with the manufacturers of those products and competing ones.[21] Also, the data on CPG authors' associations with the pharmaceutical industry are atemporal because existing disclosure policies do not consistently require that authors specify the timing of their financial relationships with industry. Nonetheless, there are ethical considerations that are relevant whether the work group member's involvement in the drug industry occurred prior to or after publication of the CPG. Finally, it should be noted that before their dissemination, the

CPG are reviewed by both APA members and the Assembly of the APA. Although it was beyond the scope of the present study to address the entire process of guideline creation, it is recommended that future research address the usefulness of this multi-step review process with regard to how potential COI are identified and managed.

In light of the increasing evidence that financial associations between biomedical researchers and the pharmaceutical industry may result in the publication of imbalanced—and sometimes inaccurate—results and recommendations,[22–26] and based on the results of the present study, we recommend that the APA institute a more rigorous COI policy for the practice guidelines that the Association publishes and endorses. This could be accomplished by inviting only those individuals who are free of substantial COIs to author CPG. An ad hoc committee, also comprised of a majority of individuals who do not have substantial COI, could review financial disclosure statements made by prospective work group members. The ad hoc committee could provide oversight with regard to the inclusion of industry-funded research in the CPG. In keeping with Fava, we recommend that the criteria for what constitutes a substantial COI include: an employee of a private firm, a regular consultant or a member of the board of directors of a firm, a stockholder in a firm related to the field of research, and a holder of a patent or patent application directly related to the published work. In our study, we found that over 70 percent of CPG authors were consultants (although current disclosure guidelines make it impossible to discern the amount of money received from or the full extent of the consultancies), over 40 percent were on corporate boards, and over 30 percent served on speakers' bureaus of pharmaceutical companies.

Moreover, in the CPG for bipolar disorder, all of the studies cited as supporting evidence for the efficacy of olanzapine and the olanzapine/fluoxetine combination were authored by an employee and stockholder of the firm that manufactures that medication. Using the "substantial COI" criteria, individuals with these COI would not be eligible to serve on the ad hoc committee or as an author of a CPG.

Greater specificity in the language of disclosure policies is needed, such as: (1) an explicit statement listing all financial relationships (individual, family, or current research collaborators) regardless of whether or not an individual believes those relationships are relevant; this includes direct as well as indirect funding sources (e.g., industry funds given to one's department chair); (2) a disclosure of the total amount of money individuals received from each company; (3) an identification of the timing of the association; (4) no lower limit (e.g., $10,000) of disclosure—our recommendation not to include arbitrary thresholds on disclosing financial information is made because there is substantial evidence that even a small gift can influence behavior.[27–29]

Random audits will better ensure compliance with existing COI policies,[30] and recruiting some individuals with no industry ties to serve on these work groups may lead to a more balanced set of guidelines.[31] Additionally, future disclosure policies should require that authors report marketing ties as these may also represent potential COI. For example, in addition to their financial ties to companies that manufacture the medications recommended in the practice guidelines, CPG authors also had financial ties to consulting firms dedicated to marketing biopharmaceuticals.

ACKNOWLEDGEMENT

The authors thank Abilash Gopal, MD, Department of Psychiatry, University of California at San Francisco, for his helpful feedback regarding our summary of the medications identified and described in the CPG.

Chapter 14

SCIENCE IN THE SUNSHINE: TRANSPARENCY OF FINANCIAL CONFLICTS OF INTEREST[21]

SHELDON KRIMSKY
Tufts University, Department of Urban and Environmental Policy and Planning

INTRODUCTION

In May 2010, the U.S. Department of Health & Human Services (DHHS) proposed revised regulations, applicable to all grantee institutions and investigators, which set requirements for the disclosure and management of financial conflicts of interest (FCOIs). The new rules would be the first major revisions promulgated since 1995, when investigator-FCOIs were first regulated. In this paper I review the historical events leading up to current policies adopted by journals and federal scientific funding agencies on FCOIs. I discuss the trends among science and medical journals toward full disclosure of FCOIs by contributing authors and examine the changes in the newly proposed federal policy. Finally, I explore some shortcomings in the new proposed policy for achieving the government's goal of ensuring unbiased publicly funded scientific research.

21Originally published in *Ethics in Biology, Engineering and Medicine* 1, no. 4 (2010): 273–284.

HISTORICAL TRENDS

Concerns of FCOIs in the public sphere have their origins in the U.S. Constitution. The Founding Fathers, who had justifiable concerns that elected officials of the new Congress could be influenced by gifts or special favors, wrote into the Constitution some explicit prohibitions against egregious FCOIs. Article I forbids any person holding an office from accepting gifts, holding employment, or accepting titles from foreign governments without the consent of Congress. Also, no former member of Congress can assume a federal post that was created during his or her term of office.

Nearly 200 years after the ratification of the Constitution, Congress passed its most comprehensive regulations on COIs of government employees. The Ethics in Government Act of 1978 established the Office of Government Ethics and created rules for financial disclosure for federal employees. Members of the upper levels of all three branches of government (including the president, vice president, members of Congress, federal judges, and certain staff members in each branch) must file annual public financial disclosure reports that list the sources and amount of all earned income; all income from stocks, bonds, and property; any investments or large debts; and the same information for spouse and dependent children. They must also report any positions or offices held in any business or nonprofit organization whether or not they are compensated.

Scientists serving on federal advisory committees were largely outside of federal oversight until 1972. In that year, the Federal Advisory Committee Act (FACA) was passed. Scientists serving on what currently amounts to about 1,000 federal advisory committees are considered special government employees. According to FACA, no individual appointed to serve on an advisory committee can have a COI that is relevant to the functions to be performed, unless the conflict is promptly and publicly disclosed and the National Academies of Science determines that the conflict is unavoidable. It is now standard practice for scientists participating on federal advisory committees to disclose their competing interests at the start of their service. By the early 1980s, there was a significant cultural shift in academic science that brought COI concerns of the public and the scientific community to a new level.

At the start of the decade, a series of laws, executive orders, and tax policies designed to improve U.S. competitiveness in high technology were enacted and adopted. These policies were premised on the idea that if closer ties were developed between universities and industry, the rate of discovery would increase, technology transfer would expand rapidly, and the resulting innovations would create new industrial sectors and new wealth. Included in this new policy initiative were the enactments of the Bayh–Dole Act, the Stevenson–Wydler Technology Innovation Act, new tax policies, and Executive Order 12591 that stimulated university-industry partnerships. These policies gave universities and industry intellectual property ownership to discoveries funded by the government, tax credits to companies that contributed equipment to universities, tax incentives for limited partnerships between companies and universities, and funding for the formation of university-industry research centers at the National Science Foundation.

In 1980, *Nature* magazine asked a series of questions about the unintended consequences of those policies: "As industrial corporations become more involved in developing new biological techniques, where does this leave the scientist? How will university biology departments maintain their integrity and autonomy? How will individual scientists react to corporate demands?"[1] Journal editors, the so-called gatekeepers of certified knowledge, were among the first to respond.

MEDICAL JOURNALS: FIRST RESPONDERS TO AUTHOR COIS

By the mid-1980s, two leading medical journals introduced FCOI disclosures for authors. The *New England Journal of Medicine* editor-in-chief, Arnold Relman, wrote an editorial in the journals titled "Dealing with Conflict-of-Interest," which was a path breaker for the medical journal community. Relman explained the reasons behind the new policy:

> . . . in recent years, as the commercial possibilities of the new biomedical discoveries have become increasingly attractive, these connections [between industry and academic medicine] have become more pervasive, complex and problematic. Now, it is not only possible for medical investigators to have their research subsidized by businesses whose products they are studying, or act as paid consultants for them, but they are sometimes also principals in these businesses or hold equity interest in them.[2]

The very first journal policy was nothing more than a suggestion to authors that they list any funding or direct business interests that they considered to be related to the subject matter of their submitted article. Other types of FCOIs, such as patents and business consultancies, were a lower priority for the journal, which made a commitment to handling them on a case-by-case basis.

As the print media and Congress brought more attention to the links between academic scholars and industry, especially in drug research, the leading medical journals incrementally deepened and broadened their disclosure policies. Initially applied to original research, disclosure of FCOIs was extended in many journals to editorials, commentaries, meta-analyses, review articles, and book reviews. Some journals banned authors with FCOIs from publishing certain types of articles for which author bias was more difficult to detect, such as reviews of a field and commentaries.

For 6 years (1996–2002), the *NEJM* adopted a policy that prohibited editorialists and authors of review articles from having an FCOI with a company that could benefit a drug or medical device discussed in the article. In 2002, Editor-in-Chief Jeffrey Drazen withdrew the zero-tolerance policy and replaced it with a de minimis FCOI requirement applying a "significant conflict-of-interest standard" that was used to exclude certain authors from publishing in the journal.

Medical journals, more visible to the public through the eye of the media than journals in other fields, were the first to take FCOIs seriously. The general science journals followed their lead. After an initial resistance to requiring authors to make their FCOIs known to readers, the *Nature* journals were the last holdouts of the high-impact science journals to adopt a disclosure policy. In 1997, the editors of *Nature* wrote defiantly:

> This journal has never required that authors declare such affiliations, because the reasons proposed by others are less than compelling. It would be reasonable to assume, nowadays, that virtually every good paper with a conceivable biotechnological relevance emerging from west and east coasts of the United States, as well as many European laboratories, has at least one author with a financial interest—but what of it? The work published (*Science and Engineering Ethics* 2, 395; 1996) makes no claim that the undeclared interests led to any fraud, deception or bias in presentation, and until there is evidence that there are serious risks of such malpractice, this journal will persist in its stubborn belief that research as we publish it is indeed research, not business.[3]

Four years later with the same editor-in-chief, the journal reversed itself and reached the decision to adopt a disclosure policy for author competing interests. Editor-in-Chief Philip Campbell wrote: "There is suggestive evidence in the literature that publication practices in biomedical research have been influenced by the commercial interests of authors. . . . There are circumstances in which selection of evidence, interpretation of results, or emphasis of presentation might be inadvertently or even deliberately biased by a researcher's other interests."[4] Campbell was referring to the growing evidence that private funding of science had a biasing influence on its outcome.[5] A study of 47 refereed toxicology and 180 medical journals found that 87 percent and 84 percent, respectively, had written COI policies for authors in 2009.[6]

FEDERAL FUNDING AGENCIES ISSUE COI RULES

After a ten–year period during which scientific and medical journals were developing FCOI disclosure policies, two major federal science agencies, the Public Health Service (PHS), which includes the National Institutes of Health, in conjunction with the Office of the Secretary of DHHS and the National Science Foundation issued regulations after a year of public debate and input. The final rules were promulgated in 1995. In essence, this was a decentralized, locally managed system for addressing scientific COIs for investigators who received federal grants with federal walk-in rights to obtain information.[7] Under the 1995 rules, faculty were required to report external income to a designated agent at their university. Much of that information was not available to the general public, researchers, or the media, but could be accessed by federal funding agencies at their request.

For the purpose of the FCOI rules the DHHS defined "significant financial interest" (SFI) as anything of monetary value including consulting fees, honoraria, equity interests, and intellectual property that exceeded $10,000 over a twelve–month period. The reporting requirement excluded any salary or royalties from the applicant's institution; income from seminars, lectures, or teaching engagements sponsored by public or nonprofit entities; and income from an applicant's service on advisory committees or review panels for public or nonprofit entities. The definition of SFI was designed to capture ancillary income from profit-making organizations that included the investigator's spouse and dependent children.

The institution's responsibility under the 1995 rules was to maintain and enforce a written policy and establish guidelines on COL, to ensure that investigators who receive PHS grants follow the policy and guidelines, to designate an official to solicit and review financial disclosures from those awarded these grants, and to take the appropriate action for managing, reducing, or eliminating significant FCOIs.

The institution must also report to the DHHS the "existence of a conflicting interest (but not the nature of the interest or the details) found by the institution. . . ." They must assure the DHHS that a significant COI is managed properly. While the institution must make information available to DHHS upon request, it is not obligated to disclose the COIs to the press or the public.

The federal compliance mechanism is triggered when two conditions occur: first, when an investigator fails to comply with the institution's COI policy, and second, when an investigator in non-compliance "has biased the design, conduct, or reporting of the PHS-funded research . . ." The burden is on the institution to show that both conditions apply before DHHS will undertake action on compliance. The system was based largely on the trust of institutions and investigators. NIH took little oversight responsibility.

THE INSPECTOR GENERAL: DEFICIENCIES IN OVERSIGHT

In January 2008 the inspector general (IG) of the DHHS completed his report on the number and nature of FCOIs reported by grantee institutions to the NIH, and on the extent to which NIH oversees its grantee institutions' FCOIs.The investigation had two goals: first, to determine if NIH kept an accurate accounting of the reported FCOIs, and second, to ascertain the extent to which NIH oversees grantee institutions' FCOIs. The IG found two deficiencies in NIH's oversight of the rules on COIs at grantee institutions. First, the NIH could not give the IG an accurate account of the FCOI reports for FY 2004–2006. Nearly half of the institutes could not provide the IG with any financial disclosure reports for FY 2004–2006. Second, the IG felt that there was insufficient information about the COIs that were reported: "NIH is not aware of the types of financial conflicts of interest that exist within grantee institutions because details were not required to be reported and most conflict-of-interest reports do not state the nature of the conflict."[8]

As previously noted, the 1995 rules did not require the grantee institutions to report on the specific nature of the FCOIs, so it was not surprising to learn that "at least 89 percent of financial conflict-of-interest reports did not state the nature of the conflicts or how they would be managed." Only 30 of the 438 FCOI reports provided by NIH and reviewed by the IG included detailed information.

Another finding of the IG was that the individual institutes did not have a proactive method for ensuring that institutions had FCOI policies or for checking the accuracy and quality of the reporting by grantee institutions. It was based mostly on good faith. In response to the IG report, NIH did not agree that it should require grantee institutions to provide details on the FCOIs they report. The NIH director argued that such information should remain with the institution.

CONGRESSIONAL OVERSIGHT

In April 2007, Iowa Senator Charles (Chuck) Grassley hired Paul D. Thacker after Thacker had resigned as news reporter for *Environmental Science Technology* (EST), an American Chemical Society magazine. While at EST, Thacker had written a number of articles about COIs in the biotechnology sector. It was after he honed his investigative journalistic techniques on the corporate influence on environmentally related science, fields such as energy and chemicals, that he began running into opposition from the EST board and editors. Thacker wrote a story in EST about the Weinberg Groups, a scientific consulting firm operating in Washington, DC, hired by chemical companies to create uncertainty about scientific claims regarding the health and environmental effects of chemicals such as Teflon. He raised issues about whether environmental health science was for sale to the highest bidder. This brought Thacker criticism from EST for allegedly lacking proper training in investigative journalism and for failing to be balanced in his coverage of a story.[9] After his magazine editor limited his coverage of certain topics, Thacker submitted his resignation and accepted a position with Senator Grassley, a ranking member of the Senate finance and budget committees, to work on the senator's investigative oversight projects.

Thacker began investigating the reporting mechanism of universities in response to FCOIs. He discovered a number of cases where there was a significant discrepancy between what a university professor claimed to report and what drug companies disclosed that they had paid the individual for consulting services. Based on Thacker's investigations for Senator Grassley, in June 2008 the *New York Times* reported "Researchers Fail to Reveal Full Drug Pay."[10] The *Times* wrote that Senator Grassley found egregious violations in federal COI reporting requirements. They cited a Harvard child psychologist who promoted the use of antipsychotic medicines in children while earning at least $1.6 million over a period of seven years of consulting.

Grassley wrote on his website: "We all rely on the advice of doctors, and leading researchers influence the practice of medicine. . . . Taxpayers spend billions each year on prescription drugs and devices through Medicare and Medicaid. The National Institutes of Health distributes $24 billion annually in federal research grants. So the public has a right to know about financial relationships between doctors and drug companies."[11]

With the support of Paul Thacker's findings that prominent NIH awardees failed to disclose consulting or equity income and thus flagrantly violated federal regulations, Senator Grassley began a two-year campaign to tighten up the rules and improve their oversight both at the NIH and at the awardee institutions. He wrote to NIH Director Elias Zerhouni on June 4, 2008 expressing his concerns about the management of COIs in NIH-supported institutions. Director Zerhouni responded on June 20 in agreement that "we need to increase transparency and enhance NIH's system of oversight" and that he was hopeful "that we can significantly enhance identification and management of FCOIs to insure that undisclosed, and therefore unmanaged, conflicts do not bias the design, conduct, or reporting of NIH-supported research."[12]

On June 25, 2008, Grassley wrote to the chair (Robert C. Byrd) and ranking member (Thad Cochran) of the powerful Senate Committee on Appropriations alerting them to the problems at NIH and the need for accountability and greater transparency. Grassley wrote, "As you know, institutions are required to manage a NIH's grantee's conflicts of interest. However, I am discovering that these regulations may be nothing more than words with little if any teeth." Citing the 2008 DHHS Inspector General report (see next section), Grassley teamed up with Senator Herb Kohl and on July 7, 2009 issued a press release urging NIH to take steps to increase transparency of federally funded biomedical research. They filed amendments into new legislation that would have placed new requirements on institutions receiving NIH grants.

As a ranking member of the Senate Committee on Finance, Grassley issued press releases supporting changes in the NIH's COI policies. In one release he was quoted: "Letting the sun shine in and making information public is basic to building people's confidence in medicine. And with the taxpayer funding that's involved, people have a right to know. Public trust and the public dollars are at stake."[13]

During the confirmation hearings of Governor Kathleen Sebelius, seeking to be secretary of Health and Human Services, Grassley posed a series of queries to the candidate outlining the problems that he had identified and requesting her response. Sebelius' answers, while expectedly somewhat vague, did agree with the principles that Grassley had raised. "I support NIH's efforts and agree that it is time to reevaluate the existing FCOI regulation to assure that PHS supported research is conducted without bias."[14]

While building support from members of the Senate and keeping the pressure on DHHS and NIH by requesting data and urging policy change, Grassley continued to request information from universities in those cases where awardees had neglected to make proper disclosures. His efforts to raise visibility on COIs were greatly reinforced by two investigations of the IG.

IG REVIEWS DHHS RULES ON COI

In January 2008, the IG issued the final report of its investigation on the number and nature of financial COIs reported by grantee institutions to the NIH, which sought to ascertain the extent to which NIH plays an oversight role in ensuring that its grantee institutions abide by the 1995 rules of reporting, managing, and mitigating FCOIs.[15] The IG reviewed 438 reports and found a number of deficiencies in FCOIs in NIH grantee institutions. Prominent among them is that the IG could not obtain an accurate count of the FCOI reports for FY 2004–2006; 93 percent did not state the nature of the conflicts; 89 percent did not state how the conflicts would be managed, reduced, or eliminated.

This latter point about whether the NIH should require detailed records of the types of COIs occurring at its grantee institutions became contentious between the agency and the IG. The 1995 rules did not require any level of detail in the reporting by grantee institutions to the NIH. The IG wrote: "At least 89 percent of financial conflict-of-interest reports did not state the nature of the conflicts or how they would be managed."[16]

The IG also found that individual institutes did not have a proactive method of ensuring that grantee institutions had FCOI policies, or if they did, the accuracy and quality of the FCOI information. Nearly half of the institutes could not provide the IG with any financial disclosure reports for FY 2004–2006. According to the IG report, the quality of the information was based mostly on "good faith." NIH responded to the IG report by disagreeing that it should require grantee institutions to provide details on the FCOIs they report. In essence, by demurring, NIH refused to take on a policing role of its grantee institutions, but rather preferred that the information be decentralized and based on trust.

The IG issued a follow-up report in 2009 titled "How Grantees Manage Financial Conflicts of Interest in Research Funded by the National Institutes of Health." The IG cited inquiries of the Senate Finance Committee into payments from drug and device manufacturers to academic researchers and physicians. In nine cases, they found five awardees who allegedly failed to report that they received payments of $1 million or more. In this investigation the IG reviewed 41 grantee institutions that submitted 225 FCOI reports to the NIH in FY 2006. These institutions were surveyed on how FCOIs were managed, reduced, or eliminated and how grantee institutions ensured that their researchers complied with federal regulations. After excluding 41 grantee institutions from its analysis, the study was left with 184 reports involving 165 researchers. The most common type of conflict that the IG found among researchers (67 percent) was "holding an equity ownership" in companies whose financial interests were related to the investigator's research, while 40 percent of the grantees consulted for an outside company. Sixty-five grantees had some type of position with outside companies including executive office and membership on the board of directors, advisory board, or medical review board.

Of the 184 reports, 136 indicated that the researchers' conflicts were managed, 6 indicated that the conflicts were reduced, and a mere 6 were eliminated. The most cited method for managing COIs was disclosure, and for one-third of the reports (n=60), there was no evidence in the submitted documentation to show that management methods were fulfilled.

Once again, the IG report confirmed that "researcher discretion" in deciding what to report guided the management plans. "Ninety percent of the grantee institutions rely solely on the researchers' discretion to determine which of their significant financial interests are related to their research and are therefore required to be reported."[17] The IG also reported that none of the grantee institutions have a policy of full disclosure of SFIs, but rather allow the researcher to make the determination of whether disclosure is appropriate. Most of the institutions do not make an effort to verify information submitted by researchers.

The IG recommendations in the second report once again focused on the lack of detail in the information sent to NIH from grantee institutions. The IG recommended that "after a grantee submits a report identifying the existence of a conflict, NIH use [its authority] to request details about the conflict and how it was managed, reduced or eliminated." The IG also criticized the investigator "trust standard" in the reporting of FCOIs and recommended that "NIH require grantee institutions to collect information on all significant financial interests held by researchers and not just those deemed by researchers to be reasonably affected by the research." Under these criteria, all external income that exceeds the threshold would be reported.

REVISED PHS RULE ON REPORTING FCOIS.

The DHHS issued a new set of proposed rules pertaining to COIs in academic research on May 21, 2010. The goal set forth by DHHS in this proposed rule was to ensure objectivity in funded research. To fulfill this goal, institutions and investigators had to completely disclose COIs, develop appropriate review of faculty with FCOIs, and aggressively manage the conflicts that are disclosed. DHHS stated that the 2010 proposed rules represent "substantial revisions to the current regulations."[18] Many of the proposed revisions were direct responses to the IG recommendations and to Grassley's campaign for change.

The proposal changed the definition of "financial interest" from "anything of monetary value" to "anything of monetary value or potential monetary value." The new rules would require awardees to report intellectual property and equity in a start-up company, which currently may have no monetary value. The new rules broaden the meaning of *investigator* to include anyone who is responsible for the design, conduct, or reporting of research, including sub-grantees and contractors, and *research* as any activity for which research funding is available, including PHS awards, grants, cooperative agreements, contracts, or career development funds.

The definition of conflict of interest in the document is "a financial interest consisting of one or more of the [financial] interests of the Investigators (and those of the Investigator's spouse and dependent children) that reasonably appears to be related to the Investigator's institutional responsibilities." The interests include stock holdings, remunerations from companies that aggregate within twelve months to more than $5,000 and intellectual property such as patents and royalties. Thus, the new rules cut the reporting threshold for external funding in half, from $10,000 to $5,000. If an investigator holds an equity interest in a non-publicly traded company (i.e., a start-up) then an SFI would exist regardless of the value. Also, there was a change in the time period under which the SFIs aggregate. Under the 1995 rule, aggregated payments were supposed to be calculated "over the next 12 months," whereas under the new rules, they are calculated over the past twelve months. If investigators receive external monies that meet the SFI threshold after their disclosure statement has been made, they are required to update their statement within thirty days.

The new definition of SFI also introduces new criteria for disclosing some activities as SFI. In the past rules, the external funding had to be related to the current PHS grant. The new rules state that disclosure is required for any external funding classified as SFI that would "reasonably appear to be related to the Investigator's 'institutional responsibilities.'" This means that institutions must disclose external funding classified as SFIs that relate to any aspect of an investigator's scientific life, including teaching at the institution.

In what is likely to be the most controversial changes to the current federal COI policy are the institutional requirements for record keeping and reporting of FCOIs. Institutions will be required to post their COI policies on a publicly accessible website, to develop and implement training programs on their policies, and to require all PHS investigators to complete the training. Up to now, the investigator bore the responsibility for determining the relatedness of the SFI to his or her PHS-funded research. Under the proposed rules, it is the institution's responsibility, through a predesignated office, to make the relatedness determination. When a determination of SFI and its relatedness to an investigator's work has been made, the past rules required the institution to manage the conflict, while the new rules would require the institution to create a management plan to be submitted to the awarding agency.

Standing out among the institutional changes is the new public disclosure provision. When an institution has determined that an SFI of a senior investigator or key personnel of a PHS-funded project is related to the research, and that it involves an FCOI, then the institution is required to post information describing the FCOI on a publicly accessible website. This means that some investigators will have personal information about earnings and their (and their family's) financial relationships with private companies made public. The publicly accessible website would be updated annually or within 60 days of a change in the SFI status of an investigator The DHHS discussion of the proposed guidelines noted the significance of the change to personal privacy. "We recognize that the proposed public disclosure requirement would place an additional administrative burden on institutions, and would also impact the privacy of Investigators who have information related to their personal financial interests posted publicly to the extent such interests are determined to the FCOI." In balance, DHHS noted, the publicly accessible website has the advantage of offering the public more complete information; it is also consistent with public disclosures in journals and at professional meetings.

The problem associated with poor compliance of the regulations was a major factor in bringing public attention to the 1995 COI policy. Under the old rules, if an investigator failed to comply, the institution was required to inform PHS of the action it planned to take or had taken. The new rules expand the power of investigation of the awarding agency. The agency would be able to undertake a site review before, during, or after the award period, gain access to all relevant records of the awardee institution, and exercise enforcement action that includes suspension of funding or imposing special award conditions. The new rules stipulate that greater enforcement attention is given to research that evaluates the safety or effectiveness of drugs, medical devices, or treatment.

While remaining within the same general framework as the existing rules, the proposed rules on COI provide greater detail, close up loopholes in reporting, provide greater transparency to the public, shift responsibility from the investigator to the institution, and establish higher accountability standards for the awarding institution.

CONCLUSION

The 2010 proposed DHHS rules on COIs have responded to most of the criticisms and recommendations issued by public critics, Senate oversight committees, and the IG. Some highly visible cases have illustrated the extent to which some academic scientists and their institutions treated the 1995 rules as burdensome and intrusive. Some members of this group of scientists felt emboldened to flaunt the rules, which they viewed as needless regulatory impositions on science that erodes their established tradition of "academic freedom and autonomy."

Under the newly proposed rules, Senator Grassley's idea of "science in the sunshine" is extended from journal disclosure to all grantees of PHS awards. Transparency is now required, at least for publicly funded research, at the outset of receiving an award and not just at the time of publication. This gives the university an opportunity to raise the question of whether the investigator's significant FCOI could bias the results of the research. However, even with the revised rules, there remain significant omissions in addressing FCOIs in science.

First, we have to acknowledge that a significant funding of science, particularly as applied medical research, comes from the private sector. Investigators who do not receive public funds are not bound by the DHHS rules. Private funding of academic research has introduced systemic bias, perhaps more directly than public funding. The bias is often introduced at the outset in the contract of the study. For example, recently it was disclosed that the international energy corporation BP proposed to contract out research to scientists at the University of Alabama, which would have given the company rights over publication. According to a report in the *Press Register* of Mobile, Alabama, BP attempted to hire the entire marine sciences department at one Alabama university under a contract that "prohibits the scientists from publishing their research, sharing it with other scientists or speaking about the data that they collect for at least the next three years."[19]

Second, improving the management and disclosure of COIs docs not solve the problem of prevention.[20] The introduction of bias in research can be very subtle. It is not easy to determine whether an SFI biases the outcome of research unless there are telltale clues. The policy sets no boundaries on preventing an FCOI, such as by prohibiting a clinical investigator supervising a clinical trial from holding an FCOI.

Third, the DHHS's proposed rules do not address the problem of institutional COIs, which the agency has thus far found intractable. It is the universities who may negotiate contracts with secret covenants that trade off scientific autonomy in exchange for large grants or other largesse to the institution (including equity interests in the funder) that makes this issue so visible yet beyond the management of investigator COIs.

Finally, the newly proposed DHHS guidelines would replace the original guidelines, which have been in effect for fifteen years. The new guidelines herald a new age of science, one in which "disinterestedness" as formulated by Robert Merton in his norms of science, is replaced with "managed bias," namely, an acceptance that science is largely influenced by entrepreneurship and that all there is left is to maximize the use of "organized skepticism" and transparency to ensure the publication of reliable knowledge. It broadens the responsibility from journals and the community of scientists to officers at the university who must decide whether "there is any reasonable expectation that the design, conduct, or reporting of PHS-funded research, by any investigator FCOI will be biased by the financial interest of that investigator."

Chapter 15

COMBATING THE FUNDING EFFECT IN SCIENCE: WHAT'S BEYOND TRANSPARENCY?[22]

SHELDON KRIMSKY
Tufts University, Department of Urban and Environmental Policy

INTRODUCTION

Professional ethics in government and in fields such as law, engineering, and accounting have evolved to protect the public from employee abuses and misconduct. Among those protections are rules that define, manage, or proscribe conflicts of interest. The term "conflict of interest" has been defined by Thompson as "a set of conditions in which professional judgment concerning a primary interest (such as a patient's welfare or the validity of research) tends to be unduly influenced by a secondary interest (such as financial gain)."[1] Before 1980, little if any attention was given to conflicts of interest in science and medicine. Beginning around that time, a major shift was taking place in sector boundaries affecting the media, finance, banking, medicine, and academia.

22Originally published in *Stanford Law and Policy Review* 21 (2010): 101–123.

The missions of distinctive sectors of our society were blended or superimposed onto one another. This has led to a fusion of sector goals and the creation of hybridized institutions. As examples, the entertainment and news sectors have, at times, become indistinguishable; banks and investment houses have begun adopting each other's roles. And, more to the point of this article, universities have been investing in for-profit enterprises started by their faculty. The new partnership between academia and business was reinforced by the passage of the Bayh–Dole Act of 1980.[2] Under the new law, universities were accorded intellectual property rights from any discoveries that were made under government grants. Business and academia became intertwined through a mutually reinforcing body of legislative acts fostering technology transfer. These changes were the cultural counterpart to what was happening in the biological sciences when species barriers were broken with the discovery of recombinant DNA molecule technology.[3]

The well-established biological boundaries that distinguished different life forms and the special features that distinguished socioeconomic institutions were disappearing. The new blended institutions of academia raised questions about changes in the normative framework that guided research practice and the commercial ventures within academic-clinical medicine. Mark Cooper argues that the commercialization of the university affects the faculty's choice of research problems "by shifting the focus of academic life scientists to a greater interest in research that generates patents or commercializable findings and away from research based on scientific curiosity and potential contributions to scientific theory."[4] The Bayh–Dole Act, along with a series of new federal laws, state economic development initiatives, and Presidential executive orders supporting university-industry partnerships, provided incentives for the development of a new class of entrepreneurial faculty who held onto their academic positions while setting up independent companies.

This article examines the evolution of the public's concerns over conflicts of interest (COIs) in science (including medical science and the practice of medicine). I discuss the ethical foundations of COIs and the remedies that have been proposed by government, academic institutions, journals, and professional societies to address these concerns. The "funding effect" in science, an outcome in which commercial sponsorship of research influences its findings, will be explored. The role of transparency as an antidote to conflicts of interest will be examined. Also, the article will identify initiatives designed to prevent and proscribe conflicts of interest rather than accepting them as inevitable and adopting transparency as the primary response. The thesis of this article is that disclosure of a conflict of interest is a necessary but not sufficient response to address the most serious problems arising from blending academic science with commerce. I shall argue that when the autonomy of the scientist, the independence of the university, or the public's trust in academic research is compromised, conflicts of interest should be prohibited.

PUBLIC LAWS ON CONFLICTS OF INTEREST

Policies on conflicts of interest have slowly evolved in federal law and regulation from an initial emphasis on government employees, later extended to government contractors, and more recently covering academic scientists in institutions that receive government funding. The Founding Fathers had some clear ideas about conflicts of interest in public life. They wrote three provisions into the Constitution that restricted the conflicts of interest of those who held posts in the Executive and Legislative branches of government. First, federal officials were prohibited from accepting gifts, holding employment, or receiving titles from foreign governments.[5] Second, members of Congress were denied the opportunity of being appointed to a federal office that was created, or whose salary was increased, during the member's term in Congress.[6] Third, members of Congress were prohibited from receiving an increase in salary until they stood for re-election.[7] Not until the infamous Watergate affair on June 17, 1972, during the presidency of Richard M. Nixon, when five men were arrested for breaking and entering into the Democratic National Committee headquarters at the Watergate Office complex in Washington, DC, had Congress thought seriously about a comprehensive ethics law for government employees. In 1978 Congress passed the Ethics in Government Act, which required certain federal employees to disclose their finances, and established the Office of Government Ethics. Then, in 1989 the Ethics Reform Act was passed, which established post-employment restrictions for members of Congress and high-level congressional staff. It also banned honoraria for almost all government employees, and restricted federal employees from accepting gifts.

Scientists at academic institutions were largely outside the scope of federal conflict of interest regulations before 1972, when the Federal Advisory Committee Act was passed.[8] The main conflict of interest provisions applying to scientists serving on federal advisory committees (called Special Government Employees or SGEs) are found in 18 U.S.C. § 208(a). The statute prohibits SGEs from participating on federal advisory committees on a matter that could affect their financial interest or that of members of their family or an organization on which they serve. Waivers can be granted (and many have been) when an administrator finds that the need for the individual's service outweighs the potential for a conflict of interest.[9] In 1995 the National Institutes of Health (NIH) and the National Science Foundation issued guidelines to universities for managing and documenting faculty conflicts of interest.[10]

ETHICAL FOUNDATIONS OF CONFLICTS OF INTEREST

There are four ethical grounds for managing or proscribing conflicts of interest among university faculty. They can be characterized by the terms stewardship, transparency, consequentialism, and integrity of science. Stewardship pertains to the responsibility for the proper management of public funds and resources used in carrying out research. Transparency requires that the methods, sources of materials, background literature, contributions of authors to the research project, and limitations to the study are made available to the reviewers, journal editors, and readers. Consequentialism refers to the link between a behavior (such as a COI) and the quality of the research outcome (such as bias). Finally, integrity of science speaks to the public confidence in the scientific enterprise, which could be compromised despite complete transparency and an outcome of objective science.

The ethical grounds for government conflict of interest policies can most readily be traced to the concept of "stewardship." Because elected officials and government employees are the temporary stewards of the laws, lands, and properties that have been placed under the authority of government, and because they also bear the responsibility for promoting the health and welfare of the citizenry, it is in the public's interest that these officials are devoid of conflicts of interest. Without laws prohibiting conflicts of interest for public officials, the citizenry could never be confident that "private gain" rather than "public interest" was the motivating force behind a decision.

The stewardship concept has only limited application to academic scientists, who, after all, are not public employees. But they do receive federal research funds, and therefore they have a responsibility for the proper use (stewardship) of the funds. In 1994, the NIH issued a proposed rule on "objectivity in research," a term which became a euphemism for a conflict of interest policy. Under the rule, NIH stated: "prudent stewardship of public funds includes protecting federally funded research from being compromised by the conflicting financial interests of any investigator responsible for the design, conduct, or reporting of PHS-funded research."[11] The NIH rule, which became part of the *Code of Federal Regulations*, had as its explicit goal "to ensure there is no reasonable expectation that the design, conduct, or reporting of research funded under [Public Health Service] grants or cooperative agreements will be biased by any conflicting financial interest of an Investigator."[12] Thus, the government had connected the concept of "stewardship" with research, and in doing so linked conflicts of interest with biased science.

Scientists are stewards of the funds they receive from the federal government, and it is their responsibility to use these funds to generate "objective" knowledge untainted by bias and personal interest. The consequences of the NIH rule, which applied to all institutions that received NIH funding, was that the institutions were responsible for managing or preventing research conflicts of interest that could compromise the objectivity of science. But unlike government employees, scientists are afforded considerable autonomy within their institutions to undertake multi-vested activities that include teaching, research, service on public and private advisory committees, and consultancies. While good stewardship of research funds includes engaging in proper management of those funds (scientists are forbidden to use research dollars for unauthorized purposes), as well as conducting "objective" research, there is more consensus over the criteria for the former than the latter. Misuse of federal research funds has resulted in strong punitive actions against institutions.[13] There are no comparable penalties for biased research arising from conflicts of interest.

Thus, the NIH left it to the institutions to manage conflicts of interest as they saw fit. As an ethical foundation for regulating conflicts of interest in scientific research, the concept of "stewardship" falls short. There are no guidelines for stewardship of the "knowledge commons," namely, that the production of knowledge is protected from biasing commercial interests. If anything, the government has created incentives for scientists to partner with industry.

Science progresses through norms that are more relevant to a process that sociologist Robert Merton called "organized skepticism." Those appropriately trained in the discipline must have as much information about a study as possible in order to fully exercise their skepticism over whether the data are sound, the conclusions are reliable, and an experiment is properly executed. Organized skepticism, according to Merton, "is both a methodological and an institutional mandate" involving "[t]he suspension of judgment until "the facts are at hand."[14] Whereas "most institutions demand unqualified faith . . . the institutions of science make skepticism a virtue."[15] The exercise of skepticism over the reliability of scientific results is an essential quality control feature of science. The burden of proof in science is to demonstrate that a hypothesis or conjecture is true. Reviewers of a new study typically begin their review with skepticism. We assume the claims are false until the results of the inquiry melt our skepticism away.

Because the conflicts of interest held by scientists in the subject matter of their research are potential biasing factors, conflicts of interest should be as transparent as any other aspect of research. Scientists may have potentially biasing intellectual interests, such as a predilection for a certain theory or an association with certain advocacy groups. These interests are usually expressed by the authors' own writings or public activities. Financial COIs, however, have been traditionally more secretive, and therefore their biasing effects are less transparent.[16] A journal reviewer with knowledge of a conflicting interest of an author can ratchet up his or her skepticism and as a result pay closer attention to the reliability of the findings. Anything that can potentially bias the methods or outcome of a scientific study must, on ethical grounds, be available to anyone who is part of the scientific peer community.

There are several distinct ways that transparency is built into the scientific enterprise. First, authors must cite the evidence for their claims in a paper. The evidence must be accessible to others in the field. In some fields, readers and reviewers can get access to original data. Second, the methods of the experiment or investigation must be stated in sufficient detail to enable another investigator trained in the discipline the opportunity, where possible, to replicate the results and reviewers to evaluate the plausibility of the results. Third, in fields like biology, created cell lines are made available to other researchers.

Once it became clear to journal editors that author conflicts of interest were a potential biasing factor in scientific studies, reviews, and commentaries, beginning in 1984 journals began adopting COI disclosure requirements.[17] Transparency also became an important requirement for publishers of clinical practice guidelines in medicine[18] and professional manuals that provide diagnostic criteria for assessing illness (such as the *Diagnostic and Statistical Manual for Mental Disorders* or DSM).[19]

Surveys of science and medical journals taken between 1997 and 2009 have shown a rapid growth over that decade in the adoption by journals of author COI requirements, from 16 percent[20] to about 85 percent.[21] Currently, for English-language journals in science and medicine, transparency of author COIs has become the norm. The disclosure policies among journals, however, vary significantly. Some are highly specific in what they request and cover a broad scope; others are vague and much narrower in scope.[22] For example, one journal requires that authors are responsible for submitting "[a] statement of financial or other relationship that might lead to a conflict of interest."[23]

Another journal requires that authors report all financial relationships, including employment, consultancies, stock ownership or options, paid expert testimony, grants or patents received or pending, and royalties.[24] With the reporting of COIs in journals, transparency in science has been extended from describing the methodology, materials, and science that supports the hypothesis to the commercial interests of scientists.

Consequentialism in ethics is the view that moral conduct can be evaluated by the consequences of one's actions. Applied to scientists and physicians holding conflicts of interest in their research and practice, the moral significance of their actions should be gauged by the impact the COIs may have on the quality and integrity of the research. The relevant question is: are authors with COIs, whether revealed or not, compared to those without COIs, more or less likely to exhibit a deficiency in moral integrity such as bias, exaggeration, false claims, misconduct, and scientific fraud?

Before journals began requiring authors to disclose their COIs, questions about their impact on science were not being asked. By the early 1990s, social scientists began investigating the relationships between the source of funding in science (private versus public) and the outcome of studies. The purpose of this line of investigation was to determine whether there was a "funding effect" in science.[25] Can the difference in outcome in a group of similar studies be accounted for by the source of the funding? Within two decades, a body of research has confirmed the existence of the funding effect in certain fields that have been investigated.

The litigation by state attorneys general against the tobacco industry produced a wealth of documents through discovery that shed light on how tobacco companies influenced research findings pertaining to smoking and health.[26] Tobacco companies hired public relations specialists who played the role of "sponsors of science." Under the names of contracted academic scientists, they placed articles in the medical literature without revealing the source of support for the research.[27] Tobacco companies sponsored a large number of studies, literature reviews, and scientific conferences, which were conducted by pseudo-independent organizations sponsored by the tobacco industry.[28] One study found that "[s]cientists acknowledging tobacco industry support reported typically that nicotine or smoking improved cognitive performance while researchers not reporting the financial support of the tobacco industry were more nearly split on their conclusions."[29] In another report by the World Health Organization (WHO), the authors revealed the extensive campaign by the tobacco industry against WHO's scientific findings on tobacco health concerns.[30]

After learning about big tobacco's influence on science, public health advocates issued a "call for policymakers to demand complete transparency about affiliations and linkages between allegedly independent scientists and tobacco companies."[31] The call for transparency by itself does not correct the scientific record for bias and distortion. At most it allows readers to label the record: industry sponsored versus non-industry sponsored research. Moreover, transparency just shifts the problem from one of "secrecy of bias" to "openness of bias." Good public policy demands peer reviewed science. Once a study is published in a reputable refereed journal, it is not sorted out by its source of funding, nor perhaps should it be. We should expect the scientific community to do the sorting and quality control before the article gets into print. The question remains: will transparency improve the quality of review and thus result in less bias?

Some may argue that the tobacco industry is unique as a rogue industry that has stopped at nothing to promote its products. The pattern of bias in industry-funded research, however, can also be found in biomedical studies. In 2003, Bekelman et al. undertook a meta-analysis of thirty-seven original articles that investigated the extent, impact, and management of financial conflicts of interest in biomedical research. The authors concluded that "financial relationships among industry, scientific investigators, and academic institutions are widespread. Conflicts of interest arising from these ties can influence biomedical research in important ways,"[32] and that "evidence suggests that financial ties that intertwine industry, investigators, and academic institutions can influence the research process."[33]

They summarized their results by stating that "strong and consistent evidence shows that industry-sponsored research tends to draw pro-industry conclusions."[34] Among the original research supporting the funding effect, Kjaergard and Als-Nielsen found an association between competing interests and authors' conclusions in epidemiological studies of randomized clinical trials published in the *British Medical Journal*.[35] Stelfox et al. studied the relationship of funding and authors' views about the safety of calcium channel blockers. They found a strong association between the source of funding and the reporting of drug risks.[36] Djulbegovic et al. found a near balance in the effectiveness between new therapies and traditional ones in studies funded by nonprofit organizations, whereas the balance was tipped in the significant favor of new therapies for studies funded by profit-making institutions.[37] Rothman et al. assert that "both quantitative and qualitative research demonstrates [sic] the power of gifts to bias physicians' choices."[38]

Other journal articles cited an association "between funding and conclusions in randomized drug trials,"[39] "between competing interests and authors' conclusions in randomized clinical trials,"[40] between "researchers acknowledging tobacco industry support" and conclusions favorable to the tobacco industry,[41] between favorable results in drug studies and pharmaceutical company support,[42] and between for-profit financial support and positive outcomes for drugs in random clinical trials.[43]

Once the funding effect in science is established (at least for tobacco research and drug experiments), the ethical concerns about conflicts of interest reach beyond transparency as the sole norm of commercially funded science. Without conclusive evidence of a funding effect, transparency is little more than political correctness, and there is no reason to believe that the quality of science is affected by conflicting interests, particularly financial interests. Once the quality of science is at stake, transparency takes on a different meaning. Awareness of the multi-vested interests of authors and of the source of funding can guide reviewers, editors, regulators, and readers on how to weigh the significance of a study. Consequentialism and the "funding effect" warn us that not all studies are equal. But even without evidence of the "funding effect," there is one other factor that should be considered in examining the ethical foundations of conflicts of interest.

The appearance of objectivity is an important value in the scientific enterprise. Conflicts of interest in science distort that appearance even when the results of science are beyond reproach in their validity. Disinterestedness is the antipode of conflict of interest. Robert Merton cited "disinterestedness" as one of the pillars of the normative structure of science.[44] Others have extended the concept to a contemporary scientific milieu. Disinterestedness in science "requires that scientists apply the methods, perform the analysis, and execute the interpretation of results without considerations of personal gain, ideology, or fidelity to any cause other than the pursuit of truth."[45]

Of course, scientists have intellectual interests. They may show partiality to a hypothesis or theory. Or they may be partial to obtaining a positive outcome in demonstrating an effect, such as the efficacy of a drug, because it is easier to publish positive results. But you cannot remove passion, predilection toward a hypothesis, or the impulse to believe in an outcome from the practice of science. In contrast, there is nothing essential to doing science that impels one to have a commercial investment in a process or product. Scientists can just as easily cross the frontiers of stem cell research without having a patent on a cell line or equity in a company poised to commercialize stem cells.

The widely recognized Mertonian norms of science arose from observations the acclaimed sociologist of science made in the 1930s and 1940s when U.S. science was situated primarily in academic centers that were self-consciously independent of the industrial economy. Those norms are: universalism (certified scientific knowledge transcends the particularity of cultures), communalism (common ownership of the fruits of scientific investigation), disinterestedness (institutional requirements that keep personal interests from influencing one's work), and organized skepticism (suspension of judgment until the facts are at hand). Fifty years later, the model of academic business partnerships in medicine and science changed the social norms of practice. In this evolution, some have argued that "disinterest" has been supplanted by "multi-vested interest."[46] A single academic scientist may also be a consultant to a private company, a patent holder of an invention, or a principal in a start-up company.

The late John Ziman, a physicist, fellow of the Royal Society, and erstwhile sociologist of science, characterized the changes in academia by coining the term "post-academic university." According to Ziman, "disinterestedness" as an internal norm of scientific practice was no longer viable in the new milieu of the entrepreneurial (post-academic) university. Moreover, he wrote, it was not needed to protect scientific objectivity, which was sufficiently protected by other norms. "The production of objective knowledge then depends less on genuine personal "disinterestedness" than on the effective operation of other norms, especially the norms of communalism, universalism and skepticism. So long as post-academic science abides by these norms, its long-term cognitive objectivity is not in serious doubt."[47] Ziman was persuaded that the self-correcting function of science would overcome any bias, distortion, or misconduct arising from post-academic science. But in addition to "cognitive objectivity," Ziman also recognized the importance of "social objectivity," or the public's trust in science. The loss of disinterestedness could, he believed, have an irreversible effect on "social objectivity," and therefore public trust in science.[48] An ethical argument can be made that the reestablishment of such trust is morally obligatory and cannot be accomplished merely by the transparency of interests.[49] Without the public's trust in science, people will be inclined to support policies that disregard rational scientific conclusions in favor of less reliable sources of belief. The first step, however, is exposing COIs in medicine and science at the time of publication, since post-publication media revelations of commercial interests are likely to create suspicion and mistrust.

PHYSICIAN DISCLOSURE OF GIFTS AND HONORARIA

The pharmaceutical industry (Big Pharma) has a symbiotic relationship with research scientists and practicing physicians. Pharmaceutical companies hire academic clinicians to recruit and oversee patients for clinical trials in order to test the safety and efficacy of their new drugs. They also engage with practicing physicians to showcase and market their approved drugs through direct contact with "detail men" and through industry-funded programs in Continuing Medical Education.[50] Much has been reported of gift vacations and lucrative honoraria for service on speakers' bureaus of companies. U.S. Senator Chuck Grassley (R-IA) has taken on the challenge of creating greater transparency in the financial relationships between physicians and the drug, device, and biologic industries. He and Senator Herb Kohl (D-WI) have introduced the Physicians Payment Sunshine Act.[51] Grassley described his goal in backing the legislation:

> I'm working to shed light on financial relationships between drug companies and doctors. I've conducted oversight, and I'm working for passage of legislation that would require public reporting by drug companies of the money they give to doctors for consulting, travel, speeches, meals and other activities. The public interest is clear. We all rely on the advice of doctors and leading researchers influence the practice of medicine. Taxpayers spend billions of dollars each year on prescription drugs and

> devices through Medicare and Medicaid. The National Institutes of Health distributes $24 billion annually on federal research grants. So the public has a right to know about financial relationships between doctors and drug companies.[52]

Several states, including Minnesota, Vermont, Massachusetts, Maine, the District of Columbia, and West Virginia have already passed legislation with similar objectives.[53] One of the strongest of these laws was passed by Vermont, which not only requires public transparency for the payments that the pharmaceutical industry makes to physicians, but also bans drug companies and manufacturers of medical devices and biological products from paying for gifts, such as meals and travel, to physicians, hospitals, nursing homes, pharmacists, and health plan administrators. There are some allowable payments drug companies would be able to make to doctors that pertain to education. Starting in 2011 those payments have to be posted in a database on a public website hosted by the Vermont Attorney General. The goal of both the transparency provisions and the prohibitions established by the law is to limit the influence of drug companies on prescription behavior and treatments by physicians.

The federal Physician Payments Sunshine Act, if passed, would override physician sunshine laws at the state level by establishing preemptive national rules and regulations.[54] The federal law would require drug, biological, and medical device manufacturers with $100 million or more in annual gross revenue to participate in a national registry listing drug company payments to physicians. The registry would disclose the names and office addresses of every physician who receives a gift valued at more than $25 from one of these participating companies.

But what will these registries do besides create a public record of the mutually reinforcing quid pro quo relationships between physicians and drug companies? The media and the general public will have an opportunity to learn about physician honorarium, capitation payments for recruiting patients into clinical trials, vacation junkets, free drug samples, etcetera, but to what end? Will disclosures contribute to better, more consumer-oriented, or more accessible health care?

For example, the Vermont Attorney General issued a report for the 2007 fiscal year on the state's Physician Sunshine Act, which noted that seventy-eight pharmaceutical manufacturers spent $2.9 million on fees, travel expenses, and other direct payments to Vermont's physicians, hospitals, and universities as part of their marketing plan, and that from fiscal years 2004 through 2008, "there has been a decrease of nearly 30 percent in the amount of expenditures and an increase of over 40 percent in the number of manufacturers who have reported marketing expenditures."[55]

Will the physician sunshine laws change behavior? One consequence of the current bill pending in Congress is that it would allow officials at NIH to compare what companies say they are paying academic physicians with what doctors actually report they are receiving from consulting income. Investigations carried out by Senator Chuck Grassley's staff have revealed a number of cases where physicians were underreporting their consulting income. In the case of prominent Harvard psychiatrist Joseph Biederman, Grassley's investigation found that he received at least $1.6 million in consulting fees by drug makers from 2000 to 2007, but alleged that most of this income was not reported to Harvard officials for several years.[56]

Would patients ask doctors about their relationship with drug companies? If this were to become a norm for patients, and if it were to affect their choice of physicians, it could change physician behavior. Negative media attention directed at physicians who accept drug company gifts could feed patient mistrust and eventually turn physicians away from drug company influence. Thus far, studies indicate that few patients will reject a physician because of his or her conflict of interest. In a survey taken by the Community Catalyst's Prescription Project, a group funded by the Pew Foundation, about one thousand people were interviewed in 2008 and asked what the likelihood was that they would query their doctor to determine if he or she accepted gifts, free samples, speaking fees, or other financial support from pharmaceutical companies. About 55 percent of the respondents said that they were unlikely to directly ask their doctors about their relationship with drug companies. About 68 percent responded that they were likely to support legislation requiring drug companies to disclose gifts to doctors.[57] While the public seems to support disclosure, few people would be inclined to act on the information. In a study of 470 cardiac patients, a mere 5 percent of those who were informed about a hypothetical investigator's equity interest in the clinical trials said they would not participate in the trial for that reason alone.[58]

Among the states that have thus far passed physician conflict-of-interest laws, Massachusetts contains requirements that go well beyond transparency. Passed in 2008, the law states that "a pharmaceutical or medical device manufacturing company . . . shall disclose . . . the value, nature, purpose and particular recipient of any fee, payment, subsidy or other economic benefit with a value of at least $50 . . . to any covered recipient in connection with the company's sales and marketing activities,"[59] and prohibits "financial support for the costs of travel, lodging, or other personal expenses of non-faculty health care practitioners attending any [Continuing Medical Education] event."[60]

An ethical standard of preventing conflicts of interest, which is reflected in the provisions of the Act, does not depend on responses to COIs by the public or physicians for ending certain practices. It is conceivable that federal and state laws mandating the transparency of gifts to physicians could eventually lead to the prohibition of those gifts once public awareness grows. In small measure this has begun to happen in some medical schools as a result of ethics rules adopted by professional organizations.[61]

LIMITS OF TRANSPARENCY

Transparency of COIs responds to one of the core ethical issues in science and medicine. But unless transparency results in behavior change, it does not address the issues of bias and public trust discussed previously. Consider the case of COIs in the judicial system. According to the American

Bar Association's Code of Judicial Conduct, the appearance of a conflict of interest must be avoided.[62] In his essay Law's Blindfold, David Lisbon asks: why prohibit mere appearances of a conflict of interest? According to Lisbon, the theory is that the appearance of impropriety is almost as bad as impropriety itself, because—as the old saw puts it—justice must not only be done, but be seen to be done. Unless judges avoid the appearance of impropriety, public confidence in the fair administration of justice will be undermined.[63]

Consider the case of a judge who makes the following declaration to his courtroom prior to announcing the prison term a convicted felon will receive:

> I will be sentencing the defendant, who has now been tried by his peers, to be incarcerated in a for-profit prison in which I have an equity interest. The extra money I earn from this partnership between my court and a reputable penal institution helps to compensate my low salary and allows me to serve the public interest and render more thoughtful and objective decisions.

JUDGE I. D. CLARE

The reason most people would feel uncomfortable with Judge I. D. Clare's disclosure is that we expect judges to be fair and disinterested in applying the law. If a judge has an equity interest in a for-profit penal institution, he can no longer exhibit an appearance of disinterestedness. The fact that he chooses to disclose his financial interest does not ameliorate the conflict or assure the public that he can render a fair decision. There is no way we can understand whether his financial interest in the prison will affect his sentencing decision. The judge himself may not understand the effect his equity interest in the prison has on his choice of the venue of a prison or duration of the sentence for the convicted felon. As Andrew Stark noted, "Because we cannot prevent officials from mentally taking notice of their own interests, we prohibit the act of holding certain kinds of interests in the first place."[64]

Are there circumstances in science and medicine where disclosure does not resolve the ethical dilemma associated with conflicts of interest? In medicine we distinguish between medical research and clinical medicine, which does not involve research. The ethical issues pertaining to conflicts of interest differ between these roles in medicine, although both share the norm of the Hippocratic Oath ("Do no harm").

Among the ethical concerns involving physician and physician-scientists are the following: Should a physician who has a financial interest in a drug therapy he is studying be permitted to serve as clinical investigator on a clinical trial for that drug? Should conflicted scientists be permitted to write editorial and book reviews for journals? Should physicians be allowed to get capitation fees for finding candidates for clinical trials? Should conflicted scientists be permitted to serve on federal advisory committees?

THE JESSE GELSINGER CASE: PHYSICIAN-ENTREPRENEUR IN A CLINICAL TRIAL

This case exemplifies the conflicts of interest held by a clinical researcher and his host institution that were not adequately disclosed to the patient in a clinical trial. It raises the question of whether the COI should have even existed.

Human Gene Therapy research (HGT) experienced rapid growth in the 1990s. The number of English-language journal articles published in HGT grew steadily from 175 in 1990 to 1,550 in 2000.[65] Likewise, there was a spectacular rise in U.S. HGT grants from 159 in 1990 to 1932 in 2000. As the medical research subspecialty in HGT exploded, there was a parallel growth in its commercial interests. The number of patents awarded to HGT techniques grew from zero in 1990 to 111 in 2000. Twelve years after 1990, 156 biotechnology companies listed HGT as one of their primary research and development missions. Peak firm formations consisted of eighteen, twenty-four, and seventeen in 1992, 1997 and 1999 respectively.[66]

Heightened expectations for HGT can be found in the media and in scientific journals during the 1990s. Clinical trials involving somatic gene transfer were widely reported to have improved the conditions of children afflicted with X-linked Severe Combined Immunodeficiency Disease (SCID), a disease that strips away the immune system. To prevent deadly infection, SCID children must live in an artificial infection-free bubble.[67]

In 1999 Jesse Gelsinger reached his eighteenth birthday after surviving for sixteen years with a rare metabolic liver condition called "ornithine transcarbamylase (OTC) deficiency." When functioning correctly, the OTC gene provides instructions for making the enzyme ornithine transcarbamylase. If the gene is mutated (as in Jesse's case), excess nitrogen from protein sources is not converted to urea for excretion, which results in ammonia accumulating in the body. A high level of ammonia is toxic, especially to the nervous system. This accumulation can cause neurological problems such as seizures, poorly controlled breathing, and mental retardation. Jesse's disease was somewhat under control by diet and extensive medication.

At his physician's suggestion, Jesse entered a clinical trial conducted at the University of Pennsylvania Medical School. The trial involved a new gene therapy protocol that was not designed to help Jesse's situation, which, in relative terms, was mild compared to the fatal form of the disease.[68] Tragically, within a few days after his HGT treatment in September 1999, Jesse fell mortally ill; his organs stopped functioning and he died. At first his father Paul Gelsinger considered his son's death one of the tragic and unanticipated consequences in the heroic path toward advancing medical science. But after he investigated his son's death, Paul Gelsinger learned some things that neither he nor his son had understood about the clinical trial. The director of the Human Gene Therapy Institute at the University of Pennsylvania was a founder and equity holder in a biotechnology company poised to benefit from a successful outcome of Jesse's human experiment. The University of Pennsylvania also had an equity stake in the company. These relationships were not revealed to Jesse Gelsinger in the informed consent documents he signed prior to the initiation of the trial. Paul

Gelsinger filed a wrongful death lawsuit against the University of Pennsylvania, its private sector biotechnology collaborator, Genovo, and two local hospitals.[69] Based on claims of negligence and conflicts of interest, the complaint argued that the conflicts of interests of the clinical investigator and the university were not disclosed to Jesse prior to his involvement in the HGT trial.

The Gelsinger case raised questions about whether a university with an institutional conflict of interest in a therapy or drug should be permitted to host a clinical trial involving that therapy or drug. It also brought into debate whether clinical investigators with an equity interest in a drug or medical procedure (such as HGT) should be permitted to participate in any aspect of the human trial.

These issues were brought into the policy sphere in 2001 when an interim guidance document of the Department of Health and Human Services (DHHS) stated:

> The financial interest of the institution in the successful outcome of the trial could directly influence the conduct of the trial, including enrollment of subjects, adverse event reporting or evaluation of efficacy data. In such cases, the integrity of the research, as well as the integrity of the institution and its corporate partner, and the well-being of the research participants, may be best protected by having the clinical trial performed and evaluated by independent investigators at sites that do not have a financial stake in the outcome of the trial, or carried out at the institution but with special safeguards to maximally protect scientific integrity of the study and the research participants.[70]

Following the Draft Interim Guidance Policy, in 2004 DHHS issued a final guidance document, which recommended that clinical investigators make their financial interests in a human experiment transparent, but left the responsibility of how to protect subjects to the individual institution.[71]

In the spring of 2009, the Institute of Medicine (IOM) issued a report that went beyond the disclosure of COIs in clinical trials. The IOM, a division of the National Academies of Science, emphasized the need for prevention of COIs by limiting financial interests in clinical trials rather than simply disclosing the interests to research participants. "The disclosure of individual and institutional financial relationships is a critical but limited first step in the process of identifying and responding to conflicts of interest."[72] The IOM was perhaps responding to the rising tide of commentaries that spoke critically of conflicts of interest in academic medicine and proposed that medical schools sever their commercial ties. However, another cohort of medical science researchers spoke of the unavoidability of physician-industry relationships and advocated proper management of COIs.[73]

Even as the counterattack proceeded against advocates of stronger conflict-of-interest rules, a group of eighteen prominent physicians published a statement in the *Journal of the American Medical Association* proposing that professional medical associations (PMAs) wean themselves from industry funding. PMAs are private associations of medical specialties and subspecialties that offer continuing medical education courses, set diagnostic and treatment guidelines, and promote ethical norms for their members. In their recommendations, Rothman et al. wrote that "PMAs should work toward a complete ban on pharmaceutical and medical device industry funding ($0), except for income from journal advertising and exhibit hall fees."[74] Another recommendation of the physicians is: "Industry should not be allowed to provide a grant [to PMAs] for a project of its choosing or be associated with a specific project. Research funds from industry, like educational support from industry, should go to a PMA's central repository or committee . . ."[75] This group of physicians also would prohibit PMAs from accepting funding from industry for journal supplements or for developing practice guidelines or outcome measures. "Disclosure of industry relationships by committee members is not sufficient protection."[76] Rothman et al. ended their recommendations to physician-centered PMAs by drawing on the public trust argument as the rationale for zero-dollar tolerance. "Professional Medical Associations have such an important role to play in speaking for medicine, defining best practices, and promoting evidence-based decision making that they cannot allow relationships with industry to diminish the public's trust."[77]

Those advocating a strict financial firewall between academic science and medicine and commerce were once thought to be a fringe group with naïve views about the progress of medicine and the role of industry-university partnerships. But there may have been a sea change when IOM issued its report in April 2009 recommending an end to industry support for medical refresher courses.[78] The prestige of the National Academies of Sciences was now behind the idea that some COIs in medicine and science should be diminished or avoided. Writing in the *New England Journal of Medicine (NEJM)*, Steinbrook describes two of IOM's new recommendations. First, "academic medical centers, research institutions and medical researchers should *restrict participation of researchers with conflicts of interest in research with human participants*, except where an individual's participation is essential for the conduct of research."[79] Second, "groups that deliver clinical practice guidelines should restrict industry funding and conflicts of panel members."[80]

The terms "restricted participation" and "restrict industry funding" leave room for interpretation and balancing. While it is not a categorically zero tolerance prohibition, it nevertheless rejects transparency as the sole ethical response to conflicts of interest. This is a first step in creating a firewall between certain medical activities and drug company gifts and funding. It is premised on two ideas: first, even small gifts can bias scientist-physicians; second, the appearance of objectivity is every bit as important as objectivity itself in protecting public trust in medical science.

PREVENTING COI AUTHORSHIP

In medical publishing, the *New England Journal* of *Medicine* is among a few high profile medical journals to have taken a leadership role in first establishing a COI policy and subsequently in elevating the standards of that policy. Between 1996 and 2002, *NEJM* had a policy that prohibited editorialists and authors of review articles from having any financial interest with a company that could benefit from a drug or device discussed in the article.[81] The distinction made between original research articles versus reviews and commentaries was based on the fact that, in the latter submissions, authors have broader discretion to make editorial choices that open up opportunities for bias. Jeffrey Drazen, Editor-in-Chief of *NEJM*, revised the policy in 2002, eliminating a zero-tolerance prevention in favor of a *de minimis* COI requirement. Drazen informed readers and contributors that with regard to original articles and special articles, the policy was the same as it was in 1996. But he indicated that his editors were having difficulty finding expert reviews from the "small and shrinking pool of authors eligible to evaluate drugs for the journal."[82] His revised policy gives the *NEJM* editors the authority to use a "significant conflict of interest" standard. He wrote: "Because the essence of reviews and editorial is selection and interpretation of the literature, the journal expects that authors for such articles will not have any *significant financial interests* in a company (or its competitor) that makes a product discussed in the article."[83] Drazen argued that the change from zero tolerance of COIs to "significant COIs" will enable the editors to recruit the best authors, that is, people who have experience with new treatments, to write editorial and review articles. Physicians writing reviews for the journal could accept up to $10,000 a year from each drug company in speaking and consulting fees. In contrast, another leading journal, *The Lancet Oncology*, prohibits authors with financial interests in or contracts with a relevant company within the past three years from publishing any review, personal view, or health care paper in the journal.[84]

INDUSTRY-SPONSORED ACADEMIC RESEARCH WITH STRINGS ATTACHED

It has been a standard practice in the biomedical sciences for authors to disclose in publications the sources of funding for their research. However, grants and contracts that commercial entities negotiate with universities contain provisions that are rarely transparent. Some of these provisions give sponsors control over the data, veto power over publication, or some degree of editorial control over the interpretation of results.

After a few highly publicized cases in which the research sponsor took editorial control away from the investigator, a number of universities adopted a zero-tolerance standard for secret contract covenants that gave the commercial sponsor control over the research methods, data, or interpretation of results. In one notable case, Betty Dong of the University of California at San Francisco (UCSF) was the principal investigator of a drug company-sponsored contract to evaluate the bioequivalency of the generic and trade versions of a drug. When Professor Dong completed her study, her data showed that the two drugs were bioequivalent for the medical conditions they were approved to treat. In the small print of the contract, the sponsoring company was given the right to exercise control over publication. Under threat of personal litigation and left by her university to her own devices, Professor Dong had little option but to withdraw the paper from the galleys of *JAMA* after the paper had been refereed and was awaiting publication.[85]

There are no uniformly adopted policies among medical schools protecting principal investigators and universities from sponsor control of the published research findings. A commentary by an editor of *JAMA* alerted schools to the dangers of accepting contracts that restricted the autonomy of researchers to publish their findings whether or not a study favors the financial interests of the sponsoring organization. In cases where medical schools fail to take leadership in preventing sponsor control over data and research findings, journal editors have stepped in. For example, a group of thirty-seven editors of heart journals signed consensus documents on the responsibility of scientific authors, which state: "Authors must give final approval of the version to be submitted and any revised version to be published."[86]

As a greater percentage of clinical trials is being carried out by medical investigators operating through private organizations known as Contract Research Organizations (CROs), rather than by teams of scientists contracted through medical schools accountable to academic deans, ethics committees, and university administrators, the contracts between drug company sponsors and CROs are more likely to be responsive to sponsor-oriented covenants.

Withholding data from publication has been one of the outcomes of sponsor-controlled contracts. As an example, the German pharmaceutical company Bayer A. G. hired a CRO to test its drug Trasylol, which was given to patients before surgery to reduce the risks of blood loss. Bayer did not release the results of a trial that was not in their financial interests. The Food and Drug Administration learned of the trial and issued a public health advisory stating that "[the] use of Trasylol may increase the chance for death, serious kidney damage, congestive heart failure and strokes."[87] Medical schools that accept contracts that permit sponsor control of data, interpretation of results, or publication status compromise the scientific autonomy and independence of their scientists. Moreover, such covenants are almost always secret.

CONCLUSION

Public concerns over conflicts of interest in biomedical science and medical practice have spawned support for increasing transparency by scientists and physicians who hold competing interests. Universal disclosure of COIs will not address several core ethical issues including the systemic bias in commercially-funded research and the public's loss of confidence in biomedical scientists and physicians who balance their Hippocratic Oath and commitment to scientific standards with commercial interests. There are many types of bias resulting from COIs that are too subtle for referees to pick up in their reviews. When a field has been dominated by a few funding sources, the scientists funded by these sources may not even be aware that their framing of issues and interpretation of results has been influenced by the financial interests of their commercial patrons. The autonomy afforded to academic scientists and independent physicians makes this a challenging issue for government and university oversight. The 1995 guidelines of COI management issued by the National Institutes of Health and the National Science Foundation were designed for local management of conflicts of interest by and within institutions. Recently, investigations by Senate staff members show that transparency rules have not fully succeeded in keeping scientists honest about their conflicts of interest. The current debates among journals, medical schools, government agencies, and professional organizations are about the extent to which certain COIs should be proscribed in order to protect the integrity of scientific and medical institutions and the knowledge they produce. While increased transparency may not reduce the bias inherent in certain conflicts, it has allowed social scientists to document the bias associated with the "funding effect." As a consequence, some scientific groups and institutions are beginning to see the moral rationale for banning conflicted activities for which there is little public tolerance.

Chapter 16

BEWARE OF GIFTS THAT COME AT TOO GREAT A COST: DANGER LURKS FOR STATE UNIVERSITIES WHEN PHILANTHROPY ENCROACHES ON ACADEMIC INDEPENDENCE[23]

SHELDON KRIMSKY
Tufts University, Department of Urban and Environmental Policy

America's public universities risk compromising their autonomy and better judgment when faced with major budget deficits from declining taxpayer revenue, they grasp at opportunities to land external funding from private donors. The financial landscape makes institutions vulnerable to ideological predators who, under the cloak of philanthropy, wish to take control of what is taught and by whom.

The issue has been highlighted by the recent controversy over the 2008 decision by Florida State University (FSU) in Tallahassee to accept U.S. $1.5 million from the Charles G. Koch Charitable Foundation in Arlington, Virginia. Like many public universities, FSU has found it harder to attract high-level faculty members in a financial landscape dominated by state

23Originally published in the journal *Nature* 474 (June 9, 2011): 129.

budget cuts, an economic downturn that has hit endowments, and limits placed on tuition fees.

The Koch foundation is an example of private philanthropy with an ideology. Its billionaire founder, Charles Koch, is an advocate of minimalist government (a vestige of a nineteenth-century-style free-market economic system), personal responsibility in lieu of social safety nets, privately financed education, and an end to the government-run social-security system. Koch and his brother David have been among the leading funders of the libertarian Tea Party and support its organizations and political candidates.

Let's be clear. It is not unusual for private donors to support university faculty positions in certain fields. But the FSU case is remarkable for the strings that came attached to the money. I have examined many such agreements, but the one that FSU signed with the Koch foundation breaks troubling new ground.

First, and most publicly discussed, the university agreed to give the foundation the authority to decide the selection criteria used to fill the economics faculty positions that it paid for, and the right to veto candidates of whom it did not approve. This agreement is a marked departure from the well-established separation between private academic philanthropy and faculty hiring decisions.

In my view, the university was at the very least naive, and at most it turned a blind eye to a compromising agreement. FSU should tear the deal up and hand back the cash. This is no idle academic exercise, and there are more problems with the deal than who gets to decide who is hired.

The stated objective of the FSU-Koch agreement, of which I have a copy, is "to advance the understanding and practice of those free voluntary processes and principles that promote social progress, human well-being, individual freedom, opportunity and prosperity based on the rule of law, constitutional government, private property, and the laws, regulations, organizations, institutions and social norms upon which they rely." The phrase of most concern is the "practice of those free voluntary processes and principles." Students of political economics will recognize similar phrasing in the nineteenth-century anarchist writings of Peter Kropotkin and Pierre-Joseph Proudhon. Neither classical anarchists nor radical libertarians have any use for strong central authorities that oversee social-welfare programs. I see no problem with funding professorships in the study of classical anarchism or twenty-first-century libertarianism, any more than I would with funding a Marxist scholar. But the autonomy of the university is transgressed when the criteria for funding seek to advance the *practice* of a political ideology.

According to the agreement, performance objectives for the program will be reviewed by a three-member advisory board, chosen by the Koch Foundations, which will monitor the performance of faculty members and check whether they remain true to the program's mission. The agreement also states that "Individuals holding the sponsored professorship positions will be treated similarly to all other FSU faculty of similar rank." Really? It is inconceivable that the faculty handbook of FSU or any other state university uses "advancement of the practice" of political ideology to measure academic success. The agreement also stipulates that an "Undergraduate Political Economy Committee" should be set up in the FSU economics department, with one outside member chosen by the foundation. The purpose of this committee is to shape the undergraduate curriculum to ensure that it meets the goals of the agreement. These conditions are unacceptable at any respectable university.

Let there be no mistake: the controversy over the FSU-Koch agreement is not about the diversity of views on economics at America's universities. It is not even, as the university likes to portray, about whether it hired the staff it wanted to. It is about the wider threat to the independence and autonomy of academic appointments, and the proper boundaries between philanthropy and a university's choices about faculty and curriculum. Compromising these values, even under conditions of financial exigency, will turn a university against itself and corrupt its integral value to society.

Chapter 17

DO FINANCIAL CONFLICTS OF INTEREST BIAS RESEARCH? AN INQUIRY INTO THE "FUNDING EFFECT" HYPOTHESIS[24]

SHELDON KRIMSKY
Tufts University, Department of Urban and Environmental Policy and Planning

INTRODUCTION

The philosopher Charles Sanders Peirce claimed that of all ways of fixing our beliefs, science is the most dependable. He wrote in 1877, "Scientific investigation has had the most wonderful triumphs in the way of settling opinion."[1] Not only have we come to believe in the "dependability" of scientific claims, we have come to depend upon them for making important life decisions. It is generally understood that the production of scientific knowledge is accompanied by quality controls that are designed to filter out errors and bias. By errors I shall mean those assertions or calculations in a study that are factually incorrect and which would be recognized as such by anyone trained in the discipline. These can include errors in statistical analysis, citations, recording of data, or the application of measuring devices. Bias, on the other, is a more complex term.

24Originally published in *Science, Technology and Human Values* vol. 38, no. 4 (2012): 566–587.

As distinguished from error, bias is not as simple as an oversight or a mistake. Bias can be conscious or unconscious. It can be structural (by the choice of method) or nonstructural (by the interpretation of data). By "structural bias," I mean the adoption of certain norms or methods that would distort (over- or underreport) the effects being studied. This term has been used in media studies where a structural bias is said to be the result of a preference of journalists for some type of story or frame that leads them to pay more attention to some events over others.[2]

Bias could involve proper or improper (scientific misconduct) behavior. In his book *The Bias of Science*, Brian Martin considers "biased" research as synonymous with "value-laden" research "conditioned by social and political forces and dependent on judgments and human choices."[3] Under this definition, science, according to Martin, might never be unbiased or value-free. Resnik argues that a bias is an invalid assumption: "The person who conducts biased research is more like the person who defends a hypothesis that is later proven wrong than a person who makes a mistake or attempts to deceive his audience."[4]

I am using "bias" in a different sense. By research bias, I shall mean the use of a method, data collection, data analysis, or interpretation of results that, in the consensus view of scientists of a discipline, tends to yield results that distort the truth of a hypothesis under consideration, diminishing or negating the reliability of the knowledge claim. Bias must be viewed in terms of the current operating norms of science. Since "bias" distorts the truth, scientists must be aware of its presence and where possible prevent or diminish it. I leave open the question of whether research considered unbiased in one time period could be viewed as biased by scientists during another time period.

The function of our system of peer review is to identify error or bias before scientific studies are accepted for publication. After a study is published, it may still be criticized or corrected. Moreover, if an empirical finding cannot be replicated, the article may be withdrawn by the journal editors. Unlike other sources of establishing belief, science is considered to be a self-correcting enterprise where truth claims are kept open to new evidence. No one doubts, however, that bias can enter into published scientific work. While bias can be built into scientific methodology (structural), sometimes its subtlety can elude even the most careful reviewer and journal editor.

Only recently have government and journals turned their attention to Conflict of Interest (COI) as a source of bias. The first federal guidelines on scientific COI, issued simultaneously by the Department of Health and Human Services' (DHHS) Public Health Service (PHS) and the National Science Foundation were titled "Objectivity in Research."[5] The stated purpose of the regulation was "to ensure that the design, conduct, or reporting of research funded under PHS grants, cooperative agreements or contracts will not be biased by any conflicting financial interest of those investigators responsible for the research." And while the DHHS focused on financial COIs (FCOIs), it is generally recognized that interests other than direct financial interests can also play a potentially biasing role in science.[6] Writing in the journal *Cell Stem Cell* about the ethics of stem cells, Jeremy Sugarman noted: "Both nonfinancial and financial conflicts of interest may adversely affect good judgment regarding stem cell research."[7] But Sugarman also wrote that "financial conflicts of interest in research may be easier to identify, simply because financial interests can be measured and more easily described than those associated with nonfinancial interests, such as the advancement of scientific and professional concerns."[8]

Following the maxim "study what you can measure," social scientists began investigating the relationship between FCOIs and bias in the mid-1980s, when author disclosures of author FCOIs were still in their infancy. Most of the studies investigating a link between author FCOIs and private funding of science were carried out in the field of medicine, specifically medical pharmacology. The concept of a "funding effect" was coined after a body of research revealed that study outcomes were significantly different in privately funded versus publicly funded drug studies.[9,10] The funding effect was also identified in tobacco, pharmacoeconomic, and chemical toxicity research. This article examines the strongest evidence for the "funding effect," and explores the question of whether the "funding effect" is an indicator of scientific research bias, based on a previously stated criterion of "bias." To begin, I shall discuss sources of evidence behind the "funding effect." I shall argue that the "funding effect" is a symptom of the factors that are responsible for outcome disparities in product assessments and that social scientists should not, without further investigation and the elimination of other explanations, chose bias as the default hypothesis.

EVIDENCE OF THE "FUNDING EFFECT" IN SCIENCE

Beginning in the mid-1980s, scientists began testing the hypothesis that the source of funding from for-profit companies compared to nonprofit institutions and government can be correlated with the outcome of research, such as safety and efficacy in drug studies. This has been called the "funding effect" in science.[11] The assumption has been that where there is a "funding effect" there must be bias. I shall begin with the evidence for the "funding effect," largely from a group of studies in drug trials, and then discuss the possible causes of the effect.

Badil Als-Nielsen et al. tested the hypothesis that industry-sponsored drug trials tend to draw pro-industry conclusions.[12] The authors selected a random sample of 167 Cochrane reviews and found 25 with meta-analyses that met their criteria. From the meta-analyses, they studied 370 drug trials. After coding and numerically scoring the trials' conclusions and applying a logistic regression analysis, the authors found that "conclusions were significantly more likely to recommend the experimental drug as treatment of choice in trials funded by for-profit organizations alone compared with trials funded by nonprofit organizations."[13] The authors ruled out as an explanation of industry favored outcomes both the magnitude of the treatment effect and the occurrence of adverse events reported. They also noted that the clinical trial methods between for-profit and nonprofit organizations were not of the same quality. "Trials funded by for-profit organizations had better methodological quality than trials funded by nonprofit organizations regarding allocation concealment and double blinding."[14] The authors do not report on the sponsor involvement and influence on the conduct and reporting of a trial. Such information could help us understand whether the external funder influences the scientist running the trial. The effects they observed between funding and outcome occurred whether the sponsor's contribution was minimal (provided the drug) or maximal (funded the study).

The authors distinguish between potential biases in the empirical trial results (collection of data) and in the interpretation of those results, particularly in the recommendations they make about the experimental drug. As previously noted, bias can enter into any or all the stages of a study: the methodology, execution of the study, interpretation of results and recommendations (whether the experimental drug is better than the existing drug).

It is also possible that industry-funded studies, having been identified as being of higher quality, have gone through more internal (company-sponsored) study and analyses, than one would expect of a nonprofit organization. This study found statistically significant outcome differences in a class of studies, but not necessarily bias—although systemic bias is one hypothesis.

John Yaphe et al. selected for their study randomized controlled trials (RCTs) published between 1992 and 1994 of drugs or food products with therapeutic properties appearing in five journals: *Annals of Internal Medicine*, *BMJ*, *JAMA*, *Lancet*, and *NEJM*. A total of 314 articles met their inclusion criteria.[15] Of the 209 industry-funded studies, 181 (87 percent) and 28 (13 percent) had positive and negative findings, respectively, while of 96 non-industry-funded studies, 62 (65 percent) and 34 (35 percent) had positive and negative findings, respectively. What can account for this disparity in the outcomes of industry and non-industry trials? Clearly, the bias of an investigator internalizing the financial interests of the sponsor is one potential hypothesis.

Paula Rochon et al. investigated the relationship between reported drug performance and manufacturer association. They adopted a broad definition of "manufacturer association," which included supplying the drug or sponsoring a journal supplement where the publication of the study appeared. The authors selected as their study sample randomized drug trials (identified in Medline between 1997 and 1990) of nonsteroidal anti-inflammatory drugs used in the treatment of arthritis.[16] The authors found 1,008 articles published within that period but only 61 articles representing 69 individuals met their inclusion criteria. All the trials in their study had a "manufacturer association," because they reported there was a scarcity of non-manufacturer-associated trials. Therefore, they could not compare trials funded/supported by private companies with those funded/supported by nonprofit organizations. The authors also used several rating systems to estimate drug efficacy. The critical outcome measure was whether the drug being tested was superior, the same, or inferior to a comparison drug.

The results of the study showed the "the manufacturer-associated drug is always reported as being either superior to or comparable with the comparison drug" and that "these claims of superiority, especially with regard to side-effect profiles, are often not supported by trial data."[17] It is logically possible that head-to-head testing of new versus old drugs always shows the new drug superior. After all, that is the impetus for developing new drugs. But in this case, the framing of the tests can bias the outcome. Marcia Angell explains the process with an illustration from statins—drugs that lower blood cholesterol levels. "There is little reason to think one is any better than another at comparable doses. But to get a toehold in the market, me-too statins were sometimes tested for slightly different outcomes in slightly different kinds of patients, and then promoted as especially effective for those uses."[18]

In a study by Benjamin Djulbegovic et al., the investigators explored whether the reports of pharmaceutical-industry sponsored randomized trials result in biased findings.[19] They selected 113 articles published from 1996 to 1998 that described 136 randomized trials on multiple myeloma.[20] The authors compared the new therapy versus the standard therapy in the trials and then analyzed the outcome according to whether the sponsors were nonprofit or for-profit organizations. Nonprofit organizations showed a 53 percent versus 47 percent support for new therapies, but when the trials were sponsored by for-profit organizations the ratio was 74 percent to 26 percent, a statistically significant difference.

Friedman and Richter investigated whether sources of funding could be correlated to reported findings.[21] The authors analyzed original contributions in *NEJM* and *JAMA* published in 2001. They classified the presentation of results as positive (statistically significant clinical benefit from a treatment or absence of suspected side effects), mixed (clinical benefits but adverse side effects), negative (absence of clinical benefits), or other (unclear significance). They located 193 original articles in *NEJM*, 76 (39.4 percent) with a COI and 205 articles in *JAMA*, 76 (37.1 percent) with COI. The authors found 119 studies that investigated drug treatments and 174 studies for all treatments. They observed a "strong association between positive results and COI among all treatment studies" with an odds ratio of 2.35 and for drug studies alone an odds ratio of 2.64. The odds ratio is the ratio of probability of an event occurring in one group to the probability of it occurring in another group. An odds ratio of 2.35 for the drug studies is the probability of a positive result in a drug treatment study conducted by individuals with a FCOI divided by the probability of a positive result from a similar drug treatment conducted by individuals without a financial conflict of interest. In other words, an odds ratio of 2.35 means that investigators with an FCOI are more than twice as likely to produce positive results in a drug treatment study.

Another interesting finding is that the probability of reporting negative results in cases where an author had a FCOI was very low. One negative study of the sixty drug studies with FCOIs versus twenty-one negative studies of the fifty-nine drug studies without FCOIs were reported. The authors conclude that "the odds are extremely small that negative results would be published by authors with COI."[22]

The authors cannot provide an explanation for their observed association between FCOI and reported findings in medical treatments. They can only theorize about the cause. "One could surmise that drug companies are selective and only want to invest in treatments proven to produce positive results and that early clinical trials filter out the most promising treatments, which could explain the small number of studies funded by private corporations presenting negative findings."[23] But they also consider the possibility of bias and "spin." The question arises as to whether an investigator with a conflict of interest may be more inclined to present findings in order to gain favor with the sponsor or achieve any other extraneous objective—for example, to "spin." Notwithstanding the fact that the cause of the association is not apparent in their data, they state that:

> The observation that negative findings are less commonly reported among studies funded by private corporations raises troublesome ethical questions. Researchers appear to be failing to promote both the benefits and negative side effects of commercial products they review or simply failing to submit negative studies for publication because they are viewed as uninteresting.[24]

For social scientists studying the funding effect, the issue in this case is less a question of bias in the reported studies than it is an issue of bias in a failure of reporting negative studies, that is, in subverting the complete scientific record. Not all studies testing a hypothesis that there is an association between trial outcome or study quality and funding source reached positive findings. Tammy Clifford, Barrowman, and Moher selected a convenience sample of RCTs published between 1999 and 2000 by hand-searching five high impact general medical journals—*Annals of Internal Medicine*, *BMJ*, *JAMA*, *The Lancet*, and *NEJM*.[25] The quality of the trial report was evaluated according to the Jadad scale, which included randomization, allocation concealment, and withdrawals. The authors classified the trials according to funding source in four categories: entirely industry, entirely non-for-profit, mixed, and not reported. Sixty-six of the 100 trials reviewed were funded in whole or in part by industry; 6 did not disclose their source of funding. Of the 100 trials, 67 favored the new therapy, 6 favored conventional treatments, 19 reported neutral findings, and for 8 the outcome was unclear. Of the 67 trials that favored the new treatment, 30 came from "industry only," 15 came from "not-for-profit only," and 16 came from mixed sources; of the 6 trials that favored the conventional treatment, 4 came from "industry only," 1 came from "not-for-profit only," and 1 came from mixed sources.

The numbers for "favored conventional" were so low that statistical findings were not relevant. Also, this study only focused on funding and not on the financial ties of individual faculty associated with the trials. The authors noted limitations of their results. "Our failure to detect any significant association may result from a type 2 error that indicates inadequate statistical power. Although our results do not even hint at a trend . . . the potential for type 2 error is real."[26] Perhaps one conclusion can be drawn: of the 100 trials, 66 percent were funded in whole or in part by industry and 67 percent favored the new therapy. Thus, it appears that industry trials are dominant and driving the advocacy of new drugs over old treatments even without adding author FCOI.

Finally, I shall summarize the first meta-analysis that explored the "funding effect." Bekelman et al. culled 1,664 original research articles and ended up with 37 studies that met their criteria. They concluded: "Although only 37 articles met [our] inclusion criteria, evidence suggests that the financial ties that intertwine industry, investigators, and academic institutions can influence the research process. Strong and consistent evidence shows that industry sponsored-research tends to draw pro-industry conclusions."[27] Bekelman et al. were convinced that the "funding effect" is real.

I shall now turn to the relationship between FCOI and pharmacoeconomics, defined as the discipline that evaluates the clinical, economic, and humanistic aspects of pharmaceutical products, services, and programs.

PHARMACOECONOMIC STUDIES

A few studies have examined whether the results of economic analyses of drugs are correlated with the funding source. Because there is greater discretion in developing the methodology for economic studies of drugs, any inferences of bias must be addressed through the modeling, the stakeholder interests, and the specific parameters used in cost–benefit analysis rather than the omission or manipulation of clinical data. Johnson and Coons note that "Many different guidelines have been proposed for conducting pharmacoeconomic studies. The differences among the various versions reflect the diverse and sometimes conflicting views of those who specialize in economic evaluations."[28]

Mark Friedberg et al. (2010) searched the Medline and Health Star databases for articles published between 1985 and 1998 on cost or cost-effectiveness analyses of six oncology drugs.[29] They found forty-four eligible articles whose texts were analyzed for qualitative and quantitative conclusions and the funding source, based on predetermined criteria. Of the forty-four articles, twenty-four were funded by nonprofit organizations and twenty were funded by drug manufacturers. The authors found a statistically significant relationship between funding source and qualitative conclusions. Unfavorable conclusions were found in 38 percent (9/24) of the nonprofit-sponsored studies and 5 percent (1/20) of company-sponsored studies. Studies funded by pharmaceutical companies were almost 8 times less likely to reach unfavorable qualitative conclusions than nonprofit-funded studies and 1.4 times more likely to reach favorable qualitative conclusions.

C. M. Bell et al. undertook a systematic review of published papers on cost-utility analyses.[30] The authors found that industry-funded studies were more than twice as likely to report a cost-utility ratio below $20,000 per quality adjusted life year (QALY) as compared to studies sponsored by non-industry sources. A similar study reported in the *International Journal of Technology Assessment in Health Care* assessed the relation between industry funding and findings of pharmacoeconomic analyses.[31] The authors searched Pub Med for articles on cost-effectiveness and cost utility, performed during 2004–2009 on single drug treatments. They found 200 articles that met their criteria. They divided the articles into two groups based on whether or not the authors had financial support from the pharmaceutical industry. "Studies co-signed by at least one author affiliated to a pharmaceutical company and/or studies that declared any type of company funding were considered sponsored."[32] The authors also classified the main conclusions as favorable, doubtful, or unfavorable toward the drug. Of the 200 articles, 138 (69 percent) were sponsored by a pharmaceutical company. Sponsored articles reported a favorable conclusion 95 percent of the time as against 50 percent of the time for non-sponsored articles. They claimed that "the presence of a pharmaceutical sponsorship is highly predictive of a positive conclusion."[33] According to Krimsky:

> The differences observed between [pharmacoeconomic] studies funded by industry and nonprofit organizations may be the result of methods chosen, prescreening, or bias due to the source of funding. By following the traditions of professional societies, such as those of engineering and psychiatry in setting guidelines of practice, pharmacoeconomists can attain a special role in the health care policy community in developing independent studies that are based on accepted canons that meet the highest standards of the profession. Canada and the United Kingdom have developed national guidelines for cost effectiveness studies.[34]

K. S. Knox et al. reported on data collected in Friedberg et al. in comparing practices of pharmaceutical-sponsored and nonprofit-sponsored pharmacoeconomic studies.[35] They found that nonprofit studies more likely make an explicit statement of the significance of the findings (38 percent vs. 20 percent), provide a source of cost data (67 percent vs. 45 percent), and make a clear statement about the reproducibility of the findings in other settings (58 percent vs. 35 percent). As in Friedberg et al., Knox et al. considered only one type of economic relationship between industry and researchers, namely, direct funding of a study and omitted many other types of financial relationships. Had they broadened their criteria, some of the 42 pharmacoeconomic analyses they studied might be reclassified as "pharmaceutical associated" thus changing the statistical results.

Some of the authors who found a "funding effect" were cautious about inferring a bias from the data, although it was included in the list of hypotheses they considered. The next section explores alternative explanations.

EXPLANATIONS OF THE "FUNDING EFFECT" OTHER THAN BIAS

In Yaphe et al., the authors note that "the higher frequency of good outcomes in industry supported trials may stem from a decision to fund the testing of drugs at a more advanced stage of development."[36] In other words, industry has already done a lot of internal studies weeding out ineffective drugs. Thus, by the time a private company funds a trial, it would likely do better than a drug has not gone through its internal review. To fully understand this process, we need to know the extent to which companies test and reject drugs internally before funding a study by an academic group and whether the outcome results of "new drugs are always better" would be found in trials of the same drugs but funded by nonprofit organizations.

The methodologies of industry-funded as compared to nonprofit-funded trials may differ. For example, comparison of new drugs with a placebo may be more prevalent among industry-financed studies compared to non-industry-financed studies. "Comparison with placebo may produce more positive results than comparison with alternative active treatment."[37] Unless we have a profit organization and nonprofit organization using the same or very similar methods to test the same drugs, drawing an inference about bias can yield false conclusions. The appearance of low negative outcomes from private sponsors could be the result of company screening for low probability drugs before they sponsor the trial or the "reticence of investigators to submit negative findings for publication, fearing discontinuation of future funding."[38] These caveats speak against a conclusion that bias can be inferred from the data that show outcome differences.

Some tests use different doses of the new drugs and compare them to lower doses of the old drugs. This is corroborated by Rochon et al. in their study. "When we evaluated the relative range of dosing of the manufacturer-associated drug and the comparison agents in the trials on the basis of the recommended dosage suggested in standard tests, there was a considerable mismatch. In the majority of cases where the doses were not equivalent, the drug given at the higher dose was that of the supporting manufacturer."[39]

The authors surmise that higher doses "bias the study results on efficacy in favor of the manufacturer-associated drug."[40] This illustrates that bias may enter into the "funding effect" in subtle and complex ways that deal with how the trial is organized. Some authors try to explain the "funding effect" by maintaining that most industry studies use a placebo and as a result are more likely to show a positive outcome. Also, the method of drug delivery used by companies may have been different than that used in non-profit sponsor trials.

Others have questioned whether industry trials are of lower quality and thus are likely to produce more favorable results. Djulbegovic et al. rated the trial quality and concluded that "trials funded solely or in part by commercial organizations had a trend toward higher quality . . . than those supported by the governmental or other nonprofit organizations."[39] Thus, the outcome effect found in the industry-funded work of this group was not related to poor quality trials.

In Frieberg's pharmacoeconomic study, the authors offer several possible explanations for the "funding effect." First, for-profit companies are more likely than nonprofit companies to get "early looks" at the drugs, preliminary trial results, and economic data, weeding out those that would fail a cost-effectiveness standard. Companies might censor unfavorable studies by not funding them. Second, they surmise that funded studies with unfavorable results are less likely to be submitted for peer review and published.

A third explanation for the disproportionate favorable results could arise from "unconscious bias that could influence study conclusions" from scientists who have a financial conflict of interest—such as being paid by the company or holding an equity interest in the drug. As previously noted, the economists engaged in the study may internalize the values of the study sponsor, which could translate into a methodology that is more likely to yield a positive economic analysis.

And the final explanation suggested by the authors is that "the pharmaceutical companies can collaborate directly with investigators in devising protocols for economic analyses and indirectly shape the economic evaluation criteria."[42] The assessment of bias requires a standard or norm for pharmacoeconomic analysis against which one can compare different outcomes.[43] Several studies have addressed the quality of pharmacoeconomic analysis of drugs.[44,45] Currently, no standardization or best practice for pharmacoeconomic analyses exists. Because the choice of method can have a significant effect on outcome, a method that systematically yields outcomes consistent with the private sponsor's financial interest may be biased.

SINGLE PRODUCT ASSESSMENT: TOBACCO

The studies of funding effects in pharmaceutical products include many types of drugs in order to develop aggregate statistics. Companies may do in-house studies before sponsoring extramural studies. The type of drug studied is generally considered not relevant to the findings of a funding effect. However, investigators may have different histories with the products they are testing. Nonprofit investigators may have seen the product for the first time. By eliminating product variability, investigators of the funding effect can more precisely judge the possible linkage between the source of funding and outcome findings such as product quality, safety, or economic efficiency. Two product studies for a funding effect meet these criteria: tobacco and the chemical bisphenol A (BPA). I shall begin with a discussion of tobacco research.

Turner and Spilich investigated whether there was a relationship between tobacco industry support of basic research and the conclusions reached by authors of the study.[46] They utilized a comprehensive review of the literature on tobacco and cognitive development and used that to obtain their reference studies. Beginning with 171 citations, the authors selected 91 studies fulfilling their selection criteria that investigated the effects of tobacco and nicotine upon cognitive performance. They coded the conclusions of the papers as positive, negative, or neutral on the question of whether tobacco enhances performance and segmented the papers into those that acknowledged corporate sponsorship and those that did not. When one or more of the authors was an employee of a tobacco company, the article was coded as industry-supported. All other articles were coded as "noncorporate sponsorship," even in cases where one or more of the authors had previously received industry support.

For those papers reporting a negative relationship between tobacco and cognitive performance, sixteen were coded "non-industry supported," and one was coded "industry-supported." For those reporting a positive relationship, twenty-nine came from non-industry supported papers and twenty-seven from tobacco industry-supported papers. Among those papers reporting a neutral effect, eleven were from non-industry studies and seven from industry-supported studies. In this study, the industry/non-industry demarcation in the papers shows a disparity in negative results compared to positive results. Why did so few studies funded by the tobacco industry report negative effects on performance from tobacco use? Because the study methodologies were different, we cannot say that investigator bias played a role. It may just be that the industry-funded studies used a method that yielded fewer negative outcomes compared with an alternative method(s) used by the non-industry-funded studies. There is a phenomenon known as "bias in the study design," but that was not examined in the study. As previously mentioned, systematic bias in a study design seeking to test the toxicity of a chemical would be introduced by animal models that are inherently insensitive to the chemical in question.[47]

Deborah Barnes and Lisa Bero investigated whether review articles on the health effects of passive smoking reached conclusions that are correlated with the authors' affiliations with the tobacco companies.[48] Since tobacco is a relatively homogenous product, differences in outcome cannot be attributed to product variability or company pre-testing. Just as in pharmacoeconomic studies, there is no canonical method in undertaking a review article. Authors make a selection of articles that become part of the review. Some reviewers make their selection algorithm transparent. Others may not. Any two studies may use a different selection algorithm and they may weigh studies differently. "Ultimately, the conclusion of any review article must be based on the judgment and interpretation of the author."[49]

For this study, the authors adopted a search strategy use by the Cochrane Collaboration to select review articles from 1980 to 1995 on the health effects of passive smoking from the databases Medline and EMBASE. They located additional review articles from a database of symposium articles on passive smoking. Articles were evaluated on quality and were classified as concluding that passive smoking was either harmful or not harmful. The authors found that 94 percent (29/31) of reviews by tobacco-industry affiliated authors concluded that passive smoking is not harmful compared with 13 percent (10/75) of reviews without tobacco-industry affiliations. The influence of tobacco-industry affiliation on the finding of "safety of passive smoking" was very strong. "The odds that a review article with tobacco with tobacco industry-affiliated authors would conclude that passive smoking is not harmful were 88.4 times higher than the odds for a review article with nontobacco affiliated authors, when controlling for article quality, peer review status, article topic, and year of publication."[50] The authors reported that the "only factor that predicted a review article's conclusion was whether its author was affiliated with the tobacco industry." In this study, the authors had no alternative hypotheses other than the inherent bias of authors with industry affiliation.

Because there is a great deal of discrepancy among authors in how a review is carried out, including the selection and weighting of articles that form the basis of the review, there are a number of ways that the conclusion can be made to favor the funder's interests, not the least of which is to set a high bar for establishing evidence of causality. The authors impute conscious intentionality of bias to the funders in their statement that "the tobacco industry may be attempting to influence scientific opinion by flooding the scientific literature with large numbers of review articles supporting its position [which they paid for] that passive smoking is not harmful to health."[51] From tobacco, I shall now turn to an industrial chemical used in many products—bisphenol A.

SINGLE PRODUCT ASSESSMENT: BPA

While there are different variants of tobacco that depend on where the tobacco plant is grown, and even greater variation in cigarettes because of chemicals added to the tobacco and the paper, there is still greater homogeneity in studying tobacco than in studying different types of drugs. BPA, on the other hand, is a synthetic chemical that has a precise chemical structure. It was first reported synthesized in 1905 by a German chemist. In 1953, scientists in Germany and the United States developed new manufacturing processes for a plastic material, polycarbonate, using BPA as the starting material. In the 1990s, scientists began studying the toxicological effects of BPA leaching from plastic food and water containers. Despite the fact that some scientists claimed there was extensive evidence that BPA can disrupt mouse, rat, and human cell function at low part per trillion doses and that disruption at the same low doses is also found in snails and has profound implications for human health, other scientists disagreed. Vom Saal and Welshons[52] divided the studies into those funded by industry and those funded by nonprofit organizations. Of the 119 studies funded by the federal government, 109 showed harmful toxicological outcomes while 10 had outcomes which showed no harm. Of the studies funded by the chemical companies, there were zero with outcomes showing harm and 11 with outcomes of no harm.

The authors write: "Evidence of bias in industry-funded research on BPA." Is it systematic bias and if so what form does it take? Is industry using a different methodology than most of the federally-supported studies? If so, is their methodology sound or is it designed to get a "no harm" outcome?

Vom Saal and Welshons argue that industry-funded studies have a built in bias (what I have referred to as structural bias) against finding positive effects of BPA. They maintain that "To interpret whether there is a positive or negative effect of a test chemical, such as BPA, appropriate negative and positive controls also have to be examined."[53] Vom Saal argues that the industry-supported tests omitted a positive control and without positive control findings, one cannot interpret a reason for purely negative results. The authors also noted that some industry-funded BPA studies used test animals that had very low sensitivity to exogenous estradiol and thus would not be expected to exhibit effects from BPA. Other industry-funded investigators used a type of animal feed, which because of its estrogenic activity, would give a false result. "Inclusion of an appropriate positive control . . . would have allowed a determination of whether the failure to find effects of BPA was due to the lack of activity of BPA or to a lack of sensitivity of the animal model and/or estrogenic contamination of the feed that was used."[54]

In his classic work, *The Logic of Scientific Discovery*, Karl Popper developed the philosophical foundations of scientific methodology.[55] Science, Popper argued, is not an inductivist enterprise, where truth is built up from data that are consistent with a hypothesis. Scientists must seek to falsify a hypothesis, and only when a hypothesis is recalcitrant against a rigorous attempt at falsification can it be accepted as truth. The critical point is that deduction and not induction is the logical grounding of empirical science. In the latter case, scientists would be given: A1 is B, A2 is B . . . An is B therefore All A is B. In the former case, scientists seek to falsify "All A is B" by trying to find a disconfirming instance (Ax is not B).

For example, one can reach the conclusion that "all crows are black" by observing crows in certain parts of Africa. Or you could imagine a geographical location that would most likely nurture a non-black crow, such as the North or South Pole. If after all the seeking for a falsifying instance none appears, then, under the Popperian program, you can claim that the hypothesis "all crows are black" is confirmed. Vom Saal and Welshons illustrate this point in the toxicology of BPA.

> . . . it is a common event in toxicological studies conducted by the chemical industry for purposes of reporting about chemical safety to regulatory agencies to provide only negative results from a study in which no positive control was included but from which positive conclusions of safety of the test chemical are drawn.[56]

As Peirce noted, "We are, doubtless, in the main logical animals, but are not perfectly so."[57] Both he and Popper understood that knowledge claims drawn inductively can be easily distorted by the social context of scientists. This is most notably the case in the field of toxicology, which is composed of academic scientists and contract toxicologists working on behalf of for-profit companies. These scientists are usually paid by chemical companies to fulfill the information needs of their regulatory requirements. The standards for doing toxicological research may vary, especially in new subfields like low-dose, endocrine toxicology. Thus, until the norms of good scientific practices are adopted across the subfield and by the government regulators, contract toxicologists may perform studies that have structural biases because they are more likely than not to produce false negatives. This is the take-home message from the criticism by vom Saal and Welshons of private-company-sponsored studies. They are looking to confirm the null (no effect) hypothesis rather than trying to falsify the null hypothesis, which would provide more confidence in the claim that the chemical is not harmful.

CONCLUSION

This analytical review of studies of studies that investigate an association between funding source and study conclusions has revealed several important results. First, there is sufficient evidence in drug efficacy and safety studies to conclude that the funding effect is real. Industry-sponsored trials are more likely than trials sponsored by nonprofit organizations, including government agencies, to yield results that are consistent with the sponsor's commercial interests. Second, there is some circumstantial evidence that this effect arises from two possible causes. Either the drugs sponsored by industry have gone through more internal testing and less-effective drugs are screened out, or the methods used in industry-sponsored drug testing have a structural bias that is more likely to yield positive outcomes.

Third, a small number of pharmacoeconomic studies also show evidence of a funding effect. Without standardization of economic studies or the use of third-party "economic auditors" who have no economic ties to a company, it is difficult to account for the factors that explain this effect.

A person who files his income tax is likely to use whatever discretionary decisions at his disposal to reduce his tax obligation. Similarly, a company that performs its own economic analysis of a new drug is likely to choose a model and use inputs that are advantageous to it. When a company hires an independent agent to undertake the economic analysis, little is known about what influence the company has in shaping the study. Also, little is known about drugs that are kept out of the testing pool by companies because they have already done the economic analysis.

When we turn to studies of the funding effect on individualized commodities, the results are less ambiguous. There is an extensive body of research on tobacco, both primary (smokers) and secondary (secondhand smoke) exposures. This research shows a clear demarcation between studies funded by the cigarette industry and studies funded by nonprofit and governmental organizations. From this body of research, it is reasonable to conclude that the tobacco industry hired scientists to play a similar role as their contracted lawyers, namely, to develop a brief, in this case a scientific argument, that provides the best case or their interest. If that interpretation of tobacco-funded research is correct, it could explain the funding effect in tobacco studies.

The second homogenous product discussed in this article is BPA. However, with only one study of this compound found that addresses the funding effect, a generalization cannot be drawn. But the scientists who published the study help the reader understand why a funding effect is a probable outcome. They show the systemic bias involved in the industry-funded studies that ordinarily do not appear in studies funded by nonprofit organizations.

What I have argued in this article is that the "funding effect," namely the correlation between research outcome and funding source, is not definitive evidence of bias, but is prima facie evidence that bias may exist. Additional analyses of the methodology of the studies, interpretation of the data, interviews with investigators, and comparison of the products studied can resolve whether the existence of a funding effect is driven by scientific bias. Social scientists should follow Robert Merton's norm of "organized skepticism" when they frame an initial hypothesis about the cause behind the "funding effect" phenomenon.[58] The notion of bias based on possessing a financial conflict of interest is certainly one viable hypothesis. But there are others. Social scientists must be equipped to compare the methods used across a cluster of studies funded by for-profit and not-for profit companies to determine whether a particular method biases the results toward "no detectable outcome" while other more sensitive methods yield positive results. Certain chemical effects may show up in animal fetuses and not on the adult animals.

In addition, social scientists must gain an understanding of the entities being tested across a series of studies to determine whether the differences in the entities can account for the "funding effect." Calcium channel blockers represent a class of drugs. It is important to understand whether the partition of studies between for-profit and not-for-profit funders coincides with a random distribution of the entities being studied. Drugs that have passed a prescreening test are more likely to show more favorable outcomes than similar drugs that have not. This potential confounder can be eliminated when the entities are relatively homogenous, like tobacco or a chemical like BPA.

In some cases, ethnographic studies can determine whether for-profit companies have made internal decisions about drugs before they send them out to academic laboratories for study and how that compares with drug studies funded by not-for-profit organizations. Ethnography can also help social scientists ascertain when investigators reach beyond the data when they interpret results and whether the frequency of such over-interpretation (claiming benefits not found in the data) is more likely in studies funded by for-profit funders. Interviews with academic investigators, who are funded by private for-profit companies, and company executives, can reveal whether and how the funding organization helps frame the study, contributes to the interpretation of the data, and plays a role in deciding whether the results get sent for publication. The "funding effect" is merely a symptom of the factors that could be driving outcome disparities. Social scientists should not suspend skepticism and choose as the default hypothesis that "bias" is always the cause.

ACKNOWLEDGMENT

This research was supported in part by funding from the International Center for Alcohol Policies.

DECLARATION OF CONFLICTING INTERESTS

The author declared no potential conflicts of interest with respect to the research, authorship, and/or publication of this article.

FUNDING

The author received funding for an earlier version of this paper from the International Center for Alcohol Policies.

Chapter 18

A COMPARISON OF DSM-IV AND DSM-5 PANEL MEMBERS' FINANCIAL ASSOCIATIONS WITH INDUSTRY: A PERNICIOUS PROBLEM PERSISTS[25]

LISA COSGROVE

Harvard University, Safra Center for Ethics and University of Massachusetts, Department of Counseling Psychology

SHELDON KRIMSKY

Tufts University, Department of Urban and Environmental Policy and Planning

INTRODUCTION

All medical subspecialties have been subject to increased scrutiny about the ways by which their financial associations with industry, such as pharmaceutical companies, may influence, or give the appearance of influencing, recommendations in review articles[1] and clinical practice guidelines.[2] Psychiatry has been at the epicenter of these concerns, in part because of

25Originally published in *PLOS Medicine* 9, no. 3 (March 2012): e1001190

high-profile cases involving ghostwriting[3,4] and failure to report industry-related income,[5] and studies highlighting conflicts of interest in promoting psychotropic drugs.[6,7] The revised *Diagnostic and Statistical Manual of Mental Disorders* (DSM), scheduled for publication in May 2013 by the American Psychiatric Association (APA), has created a firestorm of controversy because of questions about undue industry influence. Some have questioned whether the inclusion of new disorders (e.g., Attenuated Psychotic Risk Syndrome) and widening of the boundaries of current disorders (e.g., Adjustment Disorder Related to Bereavement) reflects corporate interests.[8,9] These concerns have been raised because the nomenclature, criteria, and standardization of psychiatric disorders codified in the DSM have a large public impact in a diverse set of areas ranging from insurance claims to jurisprudence. Moreover, through its relationship to the International Classification of Diseases,[10] the system used for classification by many countries around the world, the DSM has a global reach.

After receiving criticism that DSM-IV had no financial disclosure of panel members, to its credit the APA instituted a mandatory disclosure policy.[11] The DSM-5 panel members are required to file financial disclosure statements, which are expected to be listed in the publication, and the APA has made a commitment to improve its management of financial conflicts of interest (FCOIs).

This new APA requirement makes the DSM's disclosure policy more congruent with most leading medical journals and federal policies on FCOI. FCOIs are widely recognized as problematic because of the data showing a clear connection between funding source and study outcome whereby results are favorably biased toward the interests of the funder[12,13]—what has been referred to as the "funding effect."[14] Some have argued that greater transparency of financial interests may facilitate a decline in FCOIs and a decrease in the potential bias that accompanies them, and that it may encourage professionals and consumers to more critically evaluate medical information.[15] Others are not sure that disclosure will reduce FCOIs and the potential for bias, because transparency alone just "shifts the problem from one of 'secrecy of bias' to 'openness of bias.'"[16] Additionally, there is the concern that disclosure may open the door for subterfuge.[17] That is, when researchers or panel members list every affiliation that they have ever had, including funding from federal agencies, it can create a "signal-to-noise problem," thereby obscuring the truth about deeply problematic financial relationships with industry.

We have reported elsewhere on industry relationships with DSM-5 task force members.[18] Although the composition of the task force has changed slightly since its formation in 2007 (e.g., Pilecki et al.[19] found 72 percent of the members had ties in early 2011) industry relationships persist despite increased transparency. Currently, 69 percent of the DSM-5 task force members report having ties to the pharmaceutical industry. This represents a relative increase of 21 percent over the proportion of DSM-IV task force members with such ties (57 percent of DSM-IV task force members had ties). This finding is congruent with emerging data from fields outside of psychiatry suggesting that transparency of funding source alone is an insufficient solution for eliminating bias.[20–23]

In 2006, we analyzed all DSM-IV panel members' financial associations with industry.[24] We have undertaken a similar analysis for DSM-5 panels, which allowed us to compare the proportions of DSM-IV and DSM-5 panel members who have industry ties. There are 141 panel members on the 13 DSM-5 panels and 29 task force members. The members of these 13 panels are responsible for revisions to diagnostic categories and for inclusion of new disorders within a diagnostic category.

Three-fourths of the work groups (Figure 18.1) continue to have a majority of their members with financial ties to the pharmaceutical industry. It is also noteworthy that, as with the DSM-IV, the most conflicted panels are those for which pharmacological treatment is the first-line intervention. For example, 67 percent (n=12) of the panel for Mood Disorders, 83 percent (N=12) of the panel for Psychotic Disorders, and 100 percent (n=7) of the Sleep/Wake Disorders (which now includes "Restless Leg Syndrome") have ties to the pharmaceutical companies that manufacture the medications used to treat these disorders or to companies that service the pharmaceutical industry.

GAPS IN APA'S DISCLOSURE POLICY

Although the APA has made the disclosure of FCOIs of DSM panel members more transparent, there are important gaps in the current policy that need to be addressed. The current APA disclosure policy does not require panel members to specifically identify speakers' bureau membership but rather cloaks it under "honoraria." (A speakers' bureau usually refers to an arrangement between a commercial entity or its agent whereby an individual is hired to give a presentation about the company's product. The company typically has the contractual right to create and/or control the content of the presentation.) Therefore, despite increased transparency, it remains unclear how many individuals participate on speakers' bureaus, because panel members may simply list "honoraria." None of the DSM panel members identified participation on a speakers' bureau. When we did an internet search of the 141 panel members, we found that 15 percent had disclosed elsewhere that they were members of drug companies' speakers' bureaus or advisory boards. These internet searches were conducted for sources published in the years 2006 (one year before the task force was appointed) to 2011, a time period congruent with published research on financial conflicts of interest. Searches included peer-reviewed articles, conferences, participation in continuing medical education events (i.e., courses and/or seminars for health professionals) and self-reporting of any industry ties following interviews with the media. Speakers' bureau and advisory board participation were included in our analysis only when there was unambiguous information (e.g., "Dr. Smith discloses that he serves on the speakers' bureau for Eli Lilly and Pfizer") and both authors (L. C., S. K.) were in agreement. The nature of these relationships needs to be spelled out more precisely; speakers' bureau participation is usually prohibited elsewhere (e.g., for faculty in medical schools), as it is widely recognized to constitute a significant FCOI. Pharmaceutical companies refer to individuals who serve on speakers' bureaus as "key opinion leaders" (KOLs) because they are seen as essential to the marketing of diseases as well as drugs.

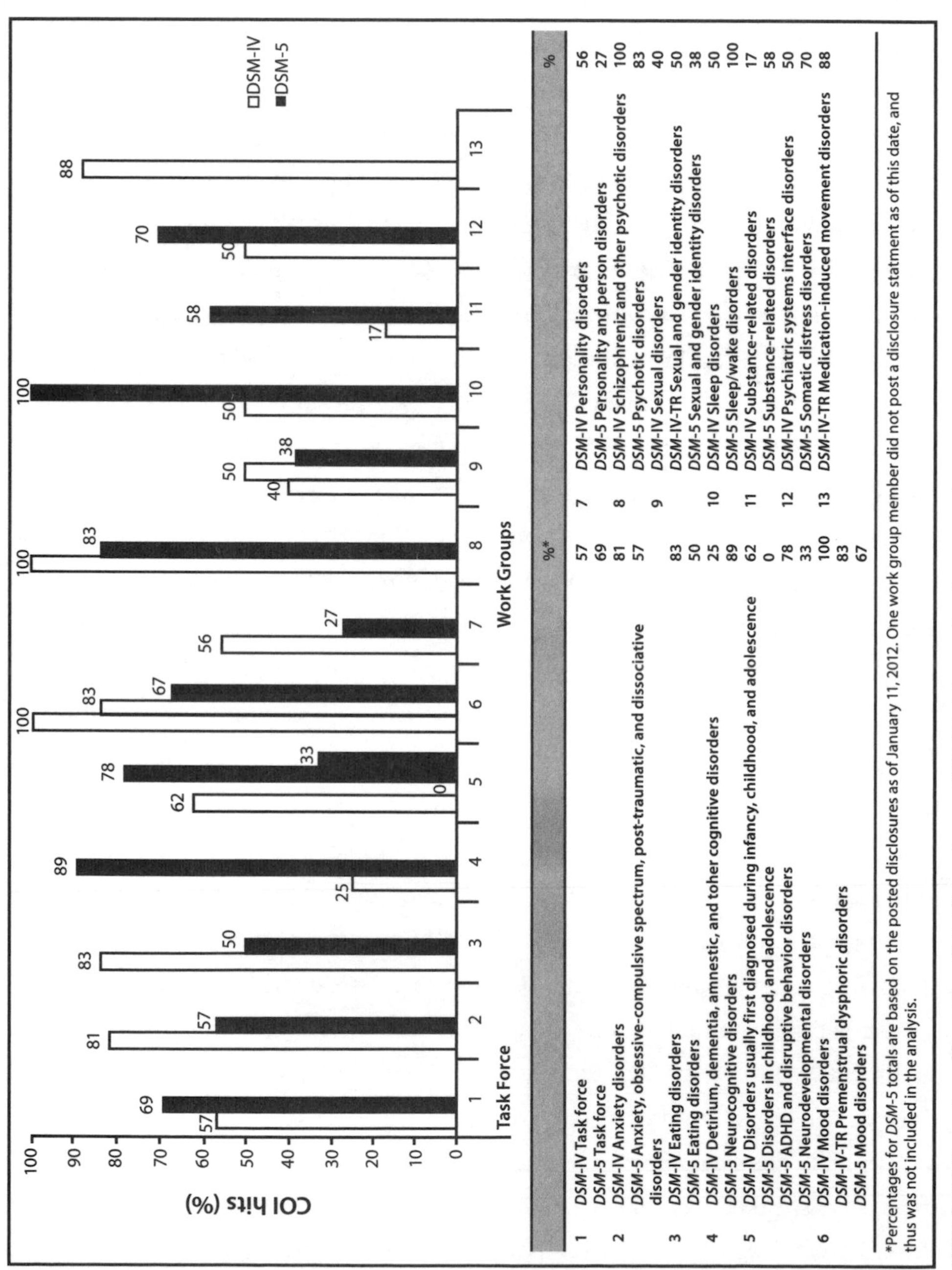

		%*			%
1	*DSM-IV* Task force	57	7	*DSM-IV* Personality disorders	56
	DSM-5 Task force	69		*DSM-5* Personality and person disorders	27
2	*DSM-IV* Anxiety disorders	81	8	*DSM-IV* Schizophreniz and other psychotic disorders	100
	DSM-5 Anxiety, obsessive-compulsive spectrum, post-traumatic, and dissociative disorders	57		*DSM-5* Psychotic disorders	83
3	*DSM-IV* Eating disorders	83	9	*DSM-IV* Sexual disorders	40
	DSM-5 Eating disorders	50		*DSM-IV-TR* Sexual and gender identity disorders	50
4	*DSM-IV* Detirium, dementia, amnestic, and toher cognitive disorders	25		*DSM-5* Sexual and gender identity disorders	38
	DSM-5 Neurocognitive disorders	89	10	*DSM-IV* Sleep disorders	50
5	*DSM-IV* Disorders usually first diagnosed during infancy, childhood, and adolescence	62		*DSM-5* Sleep/wake disorders	100
	DSM-5 Disorders in childhood, and adolescence	0	11	*DSM-IV* Substance-related disorders	17
	DSM-5 ADHD and disruptive behavior disorders	78		*DSM-5* Substance-related disorders	58
	DSM-5 Neurodevelopmental disorders	33	12	*DSM-IV* Psychiatric systems interface disorders	50
6	*DSM-IV* Mood disorders	100		*DSM-5* Somatic distress disorders	70
	DSM-IV-TR Premenstrual dysphoric disorders	83	13	*DSM-IV-TR* Medication-induced movement disorders	88
	DSM-5 Mood disorders	67			

*Percentages for *DSM-5* totals are based on the posted disclosures as of January 11, 2012. One work group member did not post a disclosure statment as of this date, and thus was not included in the analysis.

FIGURE 18.1: Comparison of financial conflicts of interest among DSM-IV task force and work group members.

1. Exclusions to the APA DSM-5 disclosure policy include unrestricted research grants;[11] that is, panel members are not required to disclose unrestricted research grants from industry. However, we would argue that this exclusion allows for commercial interests to be reflected in the revision process: there is no evidence to suggest that simply because money comes in the form of a large "unrestricted" research grant it does not create an obligation to reciprocate or invoke an implicit bias.

2. The current policy places high and arbitrary threshold limits on monies allowed from industry: DSM panel members are allowed to receive U.S. $10,000 per year from industry (e.g., for consultancies), and panel members are allowed to have up to U.S. $50,000 in stock holdings in pharmaceutical companies.

3. In contrast to other disclosure policies (e.g., the Physician Payments Sunshine Act of 2007 and the 2011 U.S. National Institutes of Health policy on conflicts of interest), APA's policy does not require disclosure of the amount of money received from industry.

However, transparency alone cannot mitigate bias. Because industry relationships can create a "pro-industry habit of thought,"[25] having financial ties to industry such as honoraria, consultation, or grant funding is as pernicious a problem as speakers' bureau participation. Over four decades of research from social psychology clearly demonstrates that gifts—even small ones—create obligations to reciprocate.[26–28] Also, because of the enormous influences of diagnostic and treatment guidelines, the standards for participation on a guideline development panel should be higher than those set for an average faculty member.[29,30]

CONCLUSION

The DSM-5 will be published in about fourteen months, enough time for the APA to institute important changes that would allow the organization to achieve its stated goal of a ". . . transparent process of development for the DSM, and . . . an unbiased, evidence-based DSM, free from any conflicts of interest."[31] Toward that goal we believe it is essential that:

1. As an eventual gold standard and because of their actual and perceived influence, all DSM task force members should be free of FCOIs.

2. Individuals who have participated on pharmaceutical companies' Speakers' Bureaus should be prohibited from DSM panel membership.

3. There should be a rebuttable presumption of prohibiting FCOIs among the DSM work groups. When no independent individuals with the requisite expertise are available, individuals with associations to industry could consult to the DSM panels, but they would not have decision-making authority on revisions or inclusion of new disorders.

These changes would accommodate the participation of needed experts as well as provide more stringent safeguards to protect the revision process from either the reality of or the perception of undue industry influence.

AUTHOR CONTRIBUTIONS

Analyzed the data: L. C., S. K. Wrote the first draft of the manuscript: L. C. Contributed to the writing of the manuscript: L. C., S. K. ICMJE criteria for authorship read and met: L. C., S. K. Agree with manuscript results and conclusions: L. C., S. K.

Chapter 19

TRIPARTITE CONFLICTS OF INTEREST AND HIGH STAKES PATENT EXTENSIONS IN THE DSM-5[26]

LISA COSGROVE
Harvard University, Safra Center for Ethics and University of Massachusetts, Department of Counseling Psychology

SHELDON KRIMSKY
Tufts University, Department of Urban and Environmental Policy and Planning

EMILY E. WHEELER
JENESSE KAITZA
SCOTT B. GREENSPANA
NICOLE L. DIPENTIMAA
University of Massachusetts, Department of Counseling and School Psychology

KEYWORDS

Conflict of interest, Mental disorders, Clinical trials, Drug therapy, Diagnostic and Statistical Manual of Mental Disorders

26Originally published in *Psychotherapy and Psychosomatics* 83 (2014): 106–113.

Previous research documented the financial ties between the panel members for the fourth edition of the *Diagnostic and Statistical Manual of Mental Disorders* (DSM-IV) and the drug companies that manufacture the medications used to treat the disorders identified in this manual.[1] To its credit, the American Psychiatric Association (APA) instituted a conflict of interest policy requiring all panel members on the DSM-5 to file financial disclosure statements. This policy resulted in some changes in work group composition; compared to DSM-IV some DSM-5 work groups had fewer individuals with industry ties. Elsewhere we reported[2] that this new APA requirement rendered the DSM's disclosure policy more congruent with most leading medical journals and federal policies on financial conflicts of interest (FCOI). DSM panel members were required to list any FCOI for three years prior to their appointment on the DSM, and they could not accept more than $10,000 from industry (e.g., for consultancies) per year or hold more than $50,000 in stock in a pharmaceutical company during their tenure on the DSM.[2] Although APA's increased transparency was an important step forward in restoring public trust, the revision process for (and recent publication of) the DSM-5 ignited debates about the taxonomy of mental illness and the widening of diagnostic boundaries. The fact that the pharmaceutical industry had a major financial stake in the outcome of these debates raised additional concerns. Thus, the issue of trustworthiness in the revision process is a critical one. In 2010, the APA issued an official policy document, approved by the Board of Trustees, in which the APA leadership stated that:

> We affirm our support of the Institute of Medicine report [*Conflict of Interest in Medical Research, Education, and Practice*]. Members involved in clinical practice, education, research, and administration must be diligent and aware in identifying, minimizing, and appropriately managing secondary (personal) interests (financial, contractual, career-centered) that may inhibit, distract, or unduly influence their judgment or behavior in a manner that detracts from or subordinates the primary interest of patients and may be perceived by some as undermining public trust.[3]

Clearly, the perception of trustworthiness in relation to FCOI is critical in the medical field, especially in terms of maintaining confidence in professional judgment. Harvard philosopher Dennis Thompson's work in this area has been highly influential (see e.g., the 1993 decision made by *The New England Journal of Medicine* to develop an FCOI policy), and he emphasizes the fact that the conflict is not an indictment of wrongdoing but rather points to a generic risk: "The point is to minimize or eliminate circumstances that would cause reasonable persons to suspect that professional judgment has been improperly influenced, whether or not it has."[4] Congruent with both the APA's and Thompson's concern that FCOI may undermine public trust, we investigated how FCOI function in these new diagnostic categories during this period of transparency.

The DSM-5, which was published in May 2013,[5] introduced new or revised diagnoses such as Binge Eating Disorder, Autism Spectrum Disorder, Disruptive Mood Dysregulation Disorder in children, Mild Neurocognitive Disorder, and Premenstrual Dysphoric Disorder. In addition to the newly included diagnoses, one of the most controversial revisions in the DSM-5 is the elimination of the bereavement exclusion from the diagnostic criteria for a Major Depressive Episode. With this change, individuals who are actively grieving a loss may be diagnosed with Major Depressive Disorder (if they present with symptoms of depression two weeks after the loss). Some clinicians maintain that this change is a positive one in that now individuals who are actively grieving a loss may receive the diagnosis and treatment that they need. Others have argued that people who are going through the normal process of grieving would now be diagnosed with depression. Indeed, pharmaceutical companies were already operating clinical trials of drugs that could be used to treat new DSM-5 disorders before the publication of the manual in May 2013. Certainly, these companies have a fiduciary responsibility to serve their shareholders' interests by working to increase their shareholder value. Although questions of potential bias may be raised with any treatment modality, if the heavy emphasis on the use of psychotropic medications to treat new DSM-5 disorders is linked to the financial interests of APA panel members and researchers who test the safety and efficacy of drugs, then the objectivity of scientific findings will be questioned. The purpose behind federal and professional conflict of interest rules is to reduce the probability of bias entering into the decision-making process.[6]

In fact, concerns about preventing bias and producing high-quality science led the Institute of Medicine to recommend that only independent experts (i.e., individuals without commercial ties) be involved in clinical guideline decision-making.[7] Questions about the potential for bias when making judgments about the validity of new DSM disorders, and about what interventions should be developed to treat these conditions, are rendered even more salient when drugs being investigated as treatments for them are under patents that have expired or will soon expire. Without patent protection, companies lose considerable profit to generics, providing a strong incentive to find new indications that will effectively grant extended patent protection to a drug. In light of this incentive, it is critical that researchers charged with the responsibility of making decisions about psychiatric diagnosis and treatment do not have FCOIs that could increase the probability or appearance of bias in clinical decision-making. Over diagnosis in the mental health field can have significant adverse public health consequences because it leads to unnecessary drug treatment.[8] This is the first study that investigates FCOIs with ongoing clinical trials, showing the three-part relationship among DSM panel members, principal investigators (PIs) of clinical trials for new DSM-5 diagnoses, and drug companies.

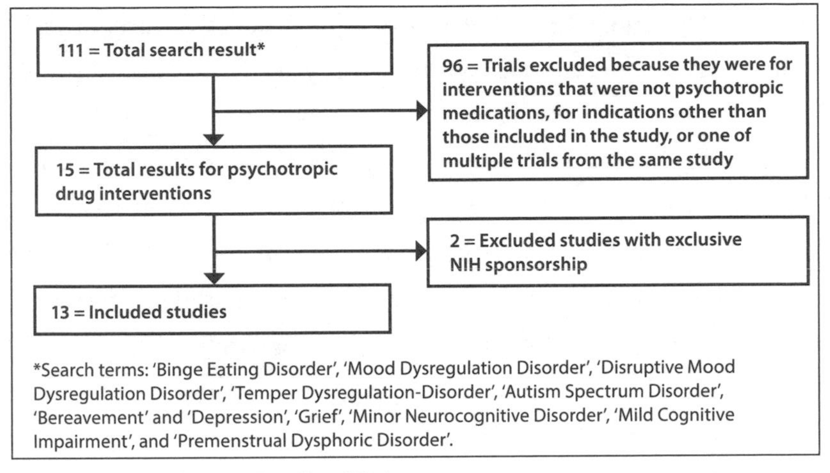

FIGURE 19.1: Results from searching Clinical-Trials.gov.

METHODS

We examined the FCOI of DSM panel members responsible for decisions about the inclusion of five new DSM disorders and one major revision (elimination of the bereavement exclusion for Major Depressive Disorder) and the pharmaceutical companies conducting clinical trials for drugs to treat these new disorders. We also examined the FCOI of PIs for the clinical trials of treatments for these newly included disorders, whereby FCOI is defined in this study as financial associations with the manufacturers of trial medications. Congruent with previous research,[9–11] financial associations are defined in our study as consultancies, honoraria, speakers' bureau membership, expert testimony, research funding, and stock holdings.

The disorders investigated were: Bereavement-Related Depression, Binge Eating Disorder, Disruptive Mood Dysregulation Disorder, Autism Spectrum Disorder, Mild Neurocognitive Disorder, and Premenstrual Dysphoric selected because of the questions raised regarding their validity,[12–15] concerns that these diagnoses lack specificity and will result in unnecessary diagnostic inflation,[16] and documented problems with reliability.[17,18]

We searched ClinicalTrials.gov for the six disorders of interest (Figure 19.1). Because previous research has found that industry-affiliated clinical trials are more vulnerable to bias than government-funded ones,[19] we excluded trials that were exclusively funded by one of the National Institutes of Health (NIH). It is possible that receiving NIH or National Institute of Mental Health funding also presents a conflict of interest (financial and/or intellectual), although probably a much subtler one. There are ties between NIH-funded investigators and grant reviewers and possibly DSM panel members. However, these ties are not the focus of our study. Industry sponsorship of the trials was identified by the sponsors and collaborators listed on the trial page. Manufacturers of the drugs and patent status information were identified using the U.S. Food and Drug Administration's (FDA) Orange Book. (www.accessdata-fda.gov/scripts/cder/ob/default.cfm).

There are two main groups who serve on the DSM and are charged with decision-making authority: Task Force members and Work Groups. Task Force members provide oversight for the entire manual, and Work Group members are teams of individuals who review a specific diagnostic category (e.g., Eating Disorders). Following previous research, we use "panel members" to refer collectively to both Work Group and Task Force members included in the study. Posted disclosure statements from the DSM-5 website for the included members of the DSM panel were reviewed to identify: (1) financial ties to pharmaceutical companies, and (2) any DSM panel member who was also a PI for one of the clinical trials. Members of the DSM-5 Work Groups that were responsible for the five new disorders and one major revision included in the search (e.g., Eating Disorders Work Group for Binge Eating Disorder) were screened for FCOI using their posted disclosure statements on the DSM-5 website (www.dsm5.org), accessed between March 15, 2013 and March 25, 2013. Because of their importance in clinical decision-making, all DSM-5 Task Force members were also screened for FCOI. Task Force members, who include Work Group chairs, played a critical role in the revision process by shaping the panel through nomination of other Work Group members, contribution to the draft criteria, and review of the final revisions to the draft before its final approval.

Additionally, we conducted Internet searches to determine if PIs of the clinical trials had financial associations to manufacturers of trial drugs. Internet searches were conducted for sources published 3 years prior to the start of the clinical trial, a time period congruent with published research on FCOIs and consistent with the APA's own FCOI policy. Searches included ProPublica, peer-reviewed articles, conferences, participation in continuing medical education events (i.e., courses and/or seminars for health professionals), and self-reporting of any industry ties following interviews with the media. Internet searchers were also conducted for speakers' bureau participation of DSM panel members because speakers' bureau membership was not an identified FCOI category in the DSM-5 disclosures. Speakers' bureau participation was included in our analysis only when there was unambiguous information.

RESULTS

Thirteen clinical trials met inclusion criteria. These clinical trials were designed to investigate ten patented drugs and one investigational new drug. Nine of these trials were testing blockbuster drugs with patents that had expired or would expire in the next two years. Table 19.1 provides a summary of trial drugs, their patent status, and their 2012 revenue (obtained from the drug manufacturers' 2012 annual reports). The trial drug manufacturer was one of the sponsors or collaborators for 8 of the 13 trials (62 percent).

Financial Ties between DSM Panels and Drug Manufacturers

Of the fifty-five Work Group members, fifteen (27 percent) reported at least one FCOI to a trial drug manufacturer, while nineteen of thirty-one (61 percent) of the Task Force members similarly reported at least one FCOI to a trial drug manufacturer. In three of the thirteen trials (23 percent), a DSM panel member reported speakers' bureau participation (i.e., company X sponsored a clinical trial for a new indication and a panel member responsible for decisions about inclusion of the new disorder served on the speakers' bureau of company X).

There were three instances in which DSM panel members were also PIs (i.e., an individual was both a DSM panel member responsible for making decisions about including a new disorder and a PI for a trial for a drug to treat the new disorder); each of these three panel members reported an FCOI to the trial drug manufacturer (see Table 19.1 for a summary of DSM panel member FCOI data by trial).

Financial Ties between PIs of Clinical Trials and Trial Drug Manufacturers

In five (38 percent) out of thirteen trials, at least one of the trial PIs reported an FCOI other than research grant funding to the trial drug manufacturer (i.e., in addition to the pharmaceutical company sponsoring the trial, the PI reported an additional FCOI to the company).

Because some of these thirteen clinical trials had more than one PI, and one individual was a PI on multiple trials, there were a total of forty-one PIs. Twelve out of the forty-one (29 percent) PIs reported research funding from the trial drug manufacturer and eight (20 percent) had ties other than grant funding to the trial drug manufacturer, including three PIs that reported participating on the speakers' bureau for the company (see Table 19.2 for a summary of the PI FCOI data by trial).

TABLE 19.1: Summary of FCOI among the included DSM panel members by trial

Trial	Trial New DSM-5 diagnosis	Trial drug manufacturer	Work group	Work group FCOI to any pharmaceutical company	Work group FCOI to trial drug manufacturer	Task force FCOI to any company	Task force FCOI to pharmaceutical trial drug manufacturer
1	Major depressive episode, bereavement exclusion eliminated	Eli Lilly	Mood Disorders	8/12	5/12	20/31	15/31
2	Major depressive episode, bereavement exclusion eliminated	Forest	Mood Disorders	8/12	1/12	20/31	5/31
3	Binge eating disorder	Cephalon	Eating Disorders	6/12	0/12	20/31	2/31
4	Binge eating disorder	Eli Lilly	Eating Disorders	6/12	3/12	20/31	15/31
5	Binge eating disorder	GlaxoSmithKline	Eating Disorders	6/12	3/12	20/31	5/31
6	Binge eating disorder	Shire	Eating Disorders	6/12	1/12	20/31	1/31
7	Autism spectrum disorder	Seaside	Neurodevelopmental Disorders	5/13	2/13	20/31	1/31
8	Autism spectrum disorder	Noven	Neurodevelopmental Disorders	5/13	0/13	20/31	0/31
9	Disruptive mood dysregulation disorder	Shire	Child and Adolescent Disorders	0/10	0/10	20/31	1/31
10	Disruptive mood dysregulation disorder	Janssen	Child and Adolescent Disorders	0/10	0/10	20/31	5/31
11	Premenstrual dysphoric disorder	GlaxoSmithKline	Mood Disorders	8/12	2/12	20/31	5/31
12	Mild neurocognitive disorder	Eisai	Neurocognitive Disorders	7/8	2/8	20/31	1/31
13	Mild neurocognitive disorder	Janssen	Neurocognitive Disorders	7/8	3/8	20/31	5/31

TABLE 19.2: Summary of FCOI data among trial PIs by trial

Trial	New DSM-5 diagnosis	Trial drug manufacturer	PI FCOI to any pharmaceutical company	PI research funding to trial drug manufacturer	PI all other FCOI to trial drug manufacturer
1	Major Depressive Episode, bereavement exclusion eliminated	Eli Lilly	0/1	0/1	0/1
2	Major Depressive Episode, bereavement exclusion eliminated	Forest	5/5	3/5	2/5
3	Binge Eating Disorder	Cephalon	1/2	0/2	0/2
4	Binge Eating Disorder	Eli Lilly	1/1	1/1	0/1
5	Binge Eating Disorder	GlaxoSmithKline	1/1	0/1	0/1
6	Binge Eating Disorder	Shire	17/21	4/21	3/21
7	Autism Spectrum Disorder	Seaside	5/8	3/8	1/8
8	Autism Spectrum Disorder	Noven	0/1	0/1	0/1
9	Disruptive Mood Dysregulation Disorder	Shire	1/1	1/1	1/1
10	Disruptive Mood Dysregulation Disorder	Janssen	1/1	0/1	1/1
11	Premenstrual Dysphoric Disorder	GlaxoSmithKline	0/0	0/0	0/0
12	Mild Neurocognitive Disorder	Eisai	0/1	0/1	0/1
13	Mild Neurocognitive Disorder	Janssen	0/0	0/0	0/0

1) Revenue for individual drug not found. $3,694 million reflects total revenue for all Forest Pharmaceuticals central nervous system drugs. 2) Revenue for individual drug not found. $71 million reflects 2009 revenue data from Shire Pharmaceuticals, which held licensing rights between 2003 August 2010. Total 2012 sales of all products for Noven Pharmaceutical's parent company, Hisamitsu Pharmaceutical Co., were $1,707 million. 3) Revenue for individual drug not found. $2,874 million reflects total revenue for all Johnson & Johnson neuroscience drugs except for Concerta, Invega, and Invega Sustena.

DISCUSSION

In all but one trial, FCOIs were found between DSM-5 panel members and the pharmaceutical companies that manufactured the drugs that were being tested for the new DSM disorders. The financial associations of panel members included research grants, consultation, honoraria, speakers' bureau participation, and/or stock. Seven out of the ten patented drugs included in the trials either are currently or have been blockbusters for their manufacturers. (A blockbuster drug is defined as a drug that earns over $1 billion in revenue in 1 year).[20] Our data show that there are financial ties between some DSM panel members and pharmaceutical companies that have a vested interest in finding a new indication for their drugs. A new indication allows the drug manufacturer to obtain an additional three years of exclusivity for that drug. Pharmaceutical companies have used "exclusivity" as an informal mechanism to effectively extend patent protection for that time period.[21] However, it should be emphasized that trials examining off-label indications conducted after a patent has expired are not necessarily meant to obtain a secondary indication.

The fact that in three out of thirteen (23 percent) of the trials the PIs were also DSM panel members raises questions about the potential of such multi-vested interests for implicit bias when making decisions about inclusion of new DSM disorders and their respective treatments. These questions are pressing in light of the fact that there are no biological markers for the majority of psychiatric disorders; the use of subjective discretion to widen diagnostic boundaries becomes more likely when there are no biological tests to ground clinical decision-making.

For example, Binge Eating Disorder may be diagnosed in individuals who do not have Anorexia or Bulimia Nervosa and who have the following three "symptoms" one time per week for three months: (1) eating more rapidly than normal, (2) eating until uncomfortably full, (3) and eating large amounts of food when not physically hungry. Mild Neurocognitive Disorder may be diagnosed based on "concerns of the individual, a knowledgeable informant, or the clinician that there has been a modest decline in cognitive function." These cognitive deficits "did not interfere with capacity for independence in everyday activities" and the decline may be based on a "clinical evaluation" (i.e., formal testing is suggested but not required for the diagnosis). Certainly some individuals consistently overeat and some individuals struggle with age-related cognitive decline. However, both researchers and clinicians have expressed concerns about "diagnostic inflation" when non-specific diagnoses such as Binge Eating Disorder and Mild Neurocognitive Disorder are identified as specific mental disorders. In fact, a former president of the APA writing about the revisions to DSM-5 noted that:

> The flexible boundaries of many psychiatric diagnostic categories, in the absence of definitive diagnostic tests, may encourage expansive definitions of affected populations and create opportunities for industry to promote treatments for people who would not previously have been seen as having a disorder.[22]

Indeed, our study shows that increased transparency (e.g., registration on ClinicalTrials.gov) and mandatory disclosure policies (such as APA's disclosure policy for DSM-5 panel members) may not be robust enough to prevent the appearance, if not the reality, of bias in both the DSM revision process as well as clinical decisions about appropriate interventions for DSM disorders. In fact, a 2012 comparison between DSM-IV and DSM-5 panel members showed that despite increased transparency, commercial ties remained strong. Although some work groups had decreased the number of individuals with industry ties, overall, 69 percent of the DSM-5 task force members reported financial ties to industry, representing a 21 percent increase in the proportion of DSM-IV task force members with such ties. Also, three-fourths of the work groups continued to have a majority of members with ties to drug firms, and it is noteworthy that, as with the DSM-IV, the most conflicted panels are those for which pharmacological treatment is the first-line intervention.

In light of the decrease in government funding of clinical trials over the past two decades, it is not surprising that 29 percent of the PIs of trials in this study reported research funding from a trial drug manufacturer. However, 20 percent of all of the PIs in our sample had financial ties other than research funding with the trial drug manufacturer, and three were on speakers' bureaus for the manufacturers of the drugs they are investigating. Many policy makers, medical journal editors, and bioethicists have raised concerns that the line between marketing and research has become blurred[23,24] when researchers have ongoing, close, and lucrative ties with industry such as speakers' bureau participation.

Our findings suggest that there may be a risk of industry influence on the DSM revision process. Additionally, our findings of FCOI of PIs running the clinical trials suggest that there also may be a risk of industry influence on the clinical decision-making process for identifying interventions to treat these new "disorders." Of particular note is the fact that in 3 of the clinical trials, PIs reported that they participated on company speakers' bureaus. Such participation may have a biasing effect. Transparency of FCOI and of clinical trial data are important first steps in strengthening public and professional trust in evidence-based medicine. However, the improvements facilitated by transparency are insufficient. Disclosure alone is not a satisfactory response to prevent bias in the revision process for psychiatric diagnostic guidelines or for maintaining integrity of psychotropic drug research.

The present study has several limitations. Our study did not include all of the revised or new DSM-5 diagnoses and thus our findings for the six new or modified disorders should not be overgeneralized. The sample size is small and caution should be exercised when interpreting the data. Also, our metric for assessing independence in clinical decision-making (DSM panel members' and PIs' financial associations with industry) is an indirect measure and thus no conclusion can be drawn about actual bias in decision-making. Moreover, the complexity of the debate over FCOI and the potential for bias is compounded by the fact that trials that are commercially funded often report negative findings. For example, researchers found that half of the studies on the efficacy of antidepressants failed to show an advantage over placebo (and over older tricyclic antidepressants) even though many of these were industry-funded studies.[25]

Despite these limitations, our examination of financial ties among DSM panel members, PIs of drug trials, and trial drug manufacturers suggest that the public, clinicians, and policy makers should be concerned about the way in which new diagnoses in the DSM-5 may provide an opportunity for pharmaceutical companies to effectively extend their patents on blockbuster drugs. For example, Eli Lilly is listed on ClinicalTrials.gov as a collaborator for a clinical trial to test the efficacy of one of Lilly's antidepressants (Cymbalta) for "bereavement-related depression," and Eli Lilly is listed as a sponsor for a clinical trial testing Cymbalta for "Binge Eating Disorder." The patent for Cymbalta expires in December 2013. Five of the twelve members of the Mood Disorders Work Group and three of the twelve members of the Binge Eating Disorder Work Group have ties to Eli Lilly. If the FDA approves Cymbalta for these new indications, Lilly will benefit by obtaining another three years of market exclusivity for this drug. It has been one of Lilly's recent blockbuster drugs: In just the fourth quarter of 2012, Lilly reported revenue of $1.42 billion from Cymbalta alone (24 percent of total revenue for that quarter).[26]

There are also three clinical trials for "Binge Eating Disorder," testing an antidepressant, a "mood stabilizer," and a psycho-stimulant as potential treatments for this new condition. (The three trial drugs, Cymbalta, Lamictal, and Nuvigil, made $5 billion, $937 million, and $347 million in revenue in 2012, respectively.) The FDA requires at least two trials to obtain authorization to market a drug for a new indication. Although more trials are needed before the FDA would grant authorization, it is important to note that the pharmaceutical companies that manufacture these three drugs would clearly benefit financially if they received such authorization.

A Call for Drug Trials That Are Not Sponsored by For-Profit Entities

Our FCOI findings show the tripartite interrelationship among DSM panel members, PIs of clinical trials for new DSM-5 diagnoses, and drug companies. These findings suggest that FCOI may function subtly, but powerfully, to shift the direction of the research, focusing on interventions that are the most commercially attractive but that do not necessarily represent the best science. Indeed, as was recently noted, when NIH decreased funding of clinical trials for new drugs, "turning new drug development over to industry, many clinically important clinical trials . . . were simply not done."[27] Hence, there must be systemic valuing and support of disinterested experts and their scientific contributions, and there is a clear need for drug trials that are not sponsored by and managed by industry. In our opinion, PIs should be prohibited from participating on a speakers' bureau for a company whose drug they are testing. Speakers' bureau participation is usually prohibited elsewhere (e.g., for faculty in medical schools), as it is widely recognized to constitute a significant FCOI. Pharmaceutical companies refer to individuals who serve on speakers' bureaus as "key opinion leaders" because they are seen as essential to the marketing of drugs.

Finally, as a policy objective, it is critical that the APA recognize that transparency alone is an insufficient response for mitigating implicit bias in diagnostic and treatment decision-making. Specifically, and in keeping with the Institute of Medicine's most recent standards, we recommend that DSM panel members be free of FCOI. In the future, DSM panel members should also be prohibited from serving as PIs of trials for any disorder being considered for inclusion in the DSM.

ACKNOWLEDGMENTS

We thank Professors Larry Lessig and Mark Somos and the lab fellows at the Edmond J. Safra Center for Ethics, and Allen Shaughnessy, Robert Whitaker, and Elia Abi-Jaoude for their insights, support, and helpful feedback. We also thank Rachel Tyrell for her assistance with preparing the manuscript.

DISCLOSURE STATEMENT

No funding was used to support this work. The authors have no competing interests to declare.

Chapter 20

CONFLICTS OF INTEREST AMONG COMMITTEE MEMBERS IN THE NATIONAL ACADEMIES' GENETICALLY ENGINEERED CROP STUDY[27]

SHELDON KRIMSKY

Tufts University, Department of Urban and Environmental Policy and Planning

TIM SCHWAB

Food & Water Watch

NTRODUCTION

During the mid-1980s, the first journals introduced requirements that authors disclose any financial interests they have on the subject matter of their submitted papers. By the early 1990s, the world's leading science journals and the majority of medical journals had adopted conflict of interest (COI) policies for authors. Soon thereafter, COI policies were established by the National Science Foundation and the Public Health Service for their grant recipients.

27Originally published in *PLOS ONE* 12, no. 2 (2017): e0172317.

Financial COI disclosures by individuals serving on scientific panels responsible for developing guidelines, such as clinical practice guidelines,[1] or writing reports on behalf of professional organizations became another arena for transparency of financial COIs. After a study was published highlighting the undisclosed conflicts of interest of panel members who established guidelines in the Diagnostic and Statistical Manual of Mental Illness (DSM-IV), considerable media attention was directed at the American Psychiatric Association, publisher of the DSM volumes, calling for greater transparency.[2] In the next edition of the DSM (DSM-5), conflict of interest disclosures by DSM-5 panel members were cited in the volume.

Disclosure of conflicts of interest by individuals and institutions has become a widely accepted ethical norm and a legal requirement in science and medicine. Failure to disclose medical financial conflicts of interest has even been grounds for a lawsuit against a prominent university.[3] The growth of COI disclosure policies for university researchers was influenced by a body of social science research that discovered a new effect, termed the "funding effect."[4,5] This finding showed that studies funded by private companies, compared to independent nonprofit and government sources as controls, tended to produce outcomes consistent with the financial interest of those companies.[6,7] The effect can be found in data collection and in the interpretation of results.

The "funding effect" has been reported across a number of fields, including pharmaceuticals,[8] chemical toxicology,[9] tobacco,[10] surgery,[11] mobile phones,[12] nutrition,[13] and biotechnology. Diels et al. (2011) examined research on genetically modified crops and found that "COIs through professional affiliations or direct research funding are likely to influence the final outcome of such studies in the commercial interest of the involved industry."[14] Another meta-analysis documented widespread conflicts of interest in published research about efficacy and durability of genetically engineered crops, and also found that the presence of conflicts was associated with favorable outcomes to agricultural biotechnology companies.[15]

FEDERAL ADVISORY BODIES

The federal government had long established conflict of interest requirements for its employees and for members of Congress, but scientists and experts serving on federal advisory committees were not covered under those regulations prior to 1972 when the Federal Advisory Committee Act (FACA) was passed. Under FACA, scientists invited to serve on federal advisory committees are prohibited from holding substantial conflicts of interest on matters before the committee on which they serve. This applied to all federal agencies including the Food and Drug Administration, the National Institutes of Health and the Centers for Disease Control.

The Federal Advisory Committee Act set certain principles for government agencies when setting up advisory committees. The act stipulated that balance, independence and transparency were requirements for establishing a federal advisory committee. The General Services Administration (GSA) was given the responsibility for developing regulations and guidance for federal agencies that establish advisory committees under FACA and the Office of Government Ethics (OGE) was responsible for issuing regulations on conflicts of interest for advisory committee members (see 18 U.S.C §208). In addition to the requirements of GSA and OGE, federal agency heads have responsibility for issuing guidelines and management procedures for addressing conflicts of interest on advisory committees they appoint. Members who serve on federal advisory committees are defined as special government employees, appointed by a government agency to perform certain temporary duties, with or without compensation.[16]

THE NATIONAL ACADEMIES

The NASEM, a private, nonprofit chartered by Congress, has always functioned as an advisory body to federal agencies and Congress since it was established in 1863. It recruits scientists from academia, industry and nonprofit organizations, who serve on an unpaid basis to author studies and provide technical consultation to the government. Nevertheless, the organization did not fall under the Federal Advisory Committee Act in 1972. The Academies receive hundreds of millions of dollars in annual contributions from private and public sources,[17] and it has long maintained close ties to industry that have raised conflicts-of-interest allegations at times directed at the committees of invited scientists that author its reports. A 1970 committee examining the health effects of lead came under scrutiny because several committee members were employed by companies producing lead additives.[18] Shortly after, a committee examining food chemicals and cancer drew controversy because some members were consultants to the food industry, which was also providing funding to the committee.[19]

In amendments made to FACA in 1997, specific COI requirements were laid out for committees of the NASEM, at that time called the National Academy of Sciences.[20] Congress noted that federal agencies could not utilize the scientific advice of the Academy unless the following instructions were heeded:

1. No individual appointed to serve on the committee has a conflict of interest that is relevant to the functions to be performed, unless such conflict is promptly and publicly disclosed and the Academy determines that the conflict is unavoidable.

2. The committee membership is fairly balanced as determined by the Academy to be appropriate for the functions to be performed.[21]

The NASEM, noting its requirements under FACA, has developed its own conflicts-of-interest guidelines that define a conflict as "any financial or other interest which conflicts with the service of the individual because it (1) could significantly impair the individual's objectivity or (2) could create an unfair competitive advantage for any person or organization." The interest in question must be closely related to the work of the committee.

According to the NASEM, removing financial conflicts eliminates potentially compromising situations in which others "could reasonably question, and perhaps discount or dismiss, the work of the committee simply because of the existence of such conflicting interests."[22] This language appears in line with a National Academies' guidebook of best practices for scientists, "On Being a Scientist" (now in its third edition), which notes that "even the appearance of a financial conflict of interest can seriously harm a researcher's reputation as well as public perceptions of science."[23]

Despite its attention to conflicts of interest, The Academies' work continues to raise questions, stemming, in part, from the continued recruitment of scientists who have financial interests in the field studied by committees on which they serve. By 1998, the year after FACA was amended to include The Academies, 24 percent of their committee members worked for industry.[24] A 2006 investigation by a public-interest group examined twenty-one NASEM committees, finding that nearly one in five committee members had financial ties to industry.[25,26]

Under FACA, the NASEM may appoint scientists with financial COIs, but must make every effort to avoid doing so—and may only do so when it is an "unavoidable" conflict of interest. If the NASEM appoints a committee member with a financial COI, it is obligated to "promptly and publicly" disclose it. Critics have argued that the NASEM does not follow FACA guidelines by virtue of the number of panel members with financial COIs, often undisclosed.

One area of the NASEM's work that has repeatedly drawn criticism around conflicts of interests over the last two decades concerns agricultural biotechnology, as numerous media outlets have reported on a variety of alleged conflicts related to several reports.[27–29] In 2016, the NASEM issued one of its most comprehensive studies on GE agriculture, a topic that has drawn considerable public debate. Because of the importance of this study and the fact that the NASEM has long had a conflicts-of-interest policy in place, we chose to examine this report.

GOALS AND METHODS OF THE STUDY

In our study, we examine financial COIs among committee members of the NASEM publication issued in May 2016 titled *Genetically Engineered Crops: Experiences and Prospects* (hereafter the NASEM2016 report). One of the principal objectives of the NASEM2016 report was to assess the "basis of evidence for purported negative effects of GE crops and their accompanying technologies, such as poor yields, deleterious effects on human and animal health, increased use of pesticides and herbicides, the creation of 'super-weeds,' reduced genetic diversity, fewer seed choices for producers, and negative impacts on farmers in developing countries and on producers of non-GE crops, and others, as appropriate."[30]

Our study sought to determine whether there were undisclosed financial COIs among committee members in the NASEM2016 report. We compared our results on financial COIs with the NASEM's guidelines and its recommendations on how conflicts of interest among members of its committees are handled. We restricted our analysis to financial COIs—and disregarded other conflicts such as conflicts of commitment or ideological and political positions held by committee members—because an analysis of financial COIs allowed us to draw on well-established criteria from the NASEM and a variety of other guidelines.

We examined the conflicts of interest of twenty committee members who developed, wrote, and approved the analysis and made recommendations in the NASEM2016 report. We applied a multi-modal method of ascertaining whether panel members held a financial conflict of interest that involved examining funding sources, patents, company advisory boards and consulting.[31]

The following criteria were used to determine the existence of a financial COI. If within three years prior to the start of the study in 2014, a panel member was found to have one or more of the following financial interests related to the subject matter of the report, such findings were recorded as a financial COI for that panel member: (1) holds a patent or patent application on a genetically modified crop or a process involved in producing genetically modified crops. (2) holds equity in a company with a financial interest in the success or failure of genetically engineered (GE) crops; (3) serves or served on an advisory committee of such a company; (4) received research funding, in-kind contributions, or research supplies (other than donated seeds) for work related to GE crops from such a company; (5) employed by such a company, or by a nonprofit that is primarily funded by such a company; (6) consults for such a company.

These six categories are found in the NASEM's own definitions of financial COIs, however our criteria differ from the NASEM in two ways. First, the NASEM only considers financial investments, that is, equity holdings, greater than $10,000 to represents a disclosable financial COI, whereas we placed no threshold on financial investments. The U.S. Department of Health and Human Services, which oversees the National Institutes of Health, recently reduced its reporting threshold from $10,000 to $5,000.[32] The National Academy of Medicine, in recommendations it issued for biomedical institutions in 2009, determined that no thresholds should be used in financial COI forms, and that all financial relationships should be disclosed for review.[33] It also noted that "Because bias is unintentional and not a matter of corruption, however, small gifts may still produce results and therefore should not be assumed to be benign." Given the scientific literature showing that the size of a financial relationship may be irrelevant to whether it creates bias, we chose to put no threshold on the value of a committee member's financial investments.[34]

Second, our criteria may also differ from the NASEM's in the scope of our financial COI review. The NASEM only examines committee members' "current interests," noting that it "does not apply to past interests that have expired, no longer exist, and cannot reasonably affect current behavior." This language is included in the instructions given to committee members on financial COI disclosure forms, who apparently must interpret for themselves whether any of their previous financial relationships could "reasonably affect current behavior."[35]

A number of contemporary guidelines recognize that previous financial relationships can reasonably affect judgment and sometimes require five years of financial COI reporting.[36] The International Committee of Medical Journal Editors (ICMJE) uses three years as the disclosure interval for authors submitting publications to journals.[37] The American Psychiatric Association requires disclosure covering a period of the last three calendar years and the current year.[38] The National Academy of Medicine in 2009 noted that scientific institutions typically review any financial COIs from the previous year, and sometimes review as many as five years of information.[39] We chose a three-year review period.

Holding a patent or patent application that is active during one's tenure on the NASEM2016 committee was sufficient to meet our criteria and also appears to meet the NASEM's "current interests" standard. However, we cannot be confident that our three-year reporting period for other categories of financial interests—such as previously having consulted for or received research funding from a company—is what the NASEM and committee members interpreted as a previous financial COI that could "reasonably affect" judgment.

The NASEM2016 study began in late 2014, so we examined financial COIs of committee members starting in 2012. Each of the twenty listed committee members was evaluated for financial conflicts of interest from 2012–2016 through reviewing published studies in journals with COI disclosures, committee members' online biographies or curriculum vitae, databases that list scientist-industry collaborations, news articles that highlight scientists' company activities, and the United States Patent and Trademark Office's patent database.

INSTITUTIONAL CONFLICT OF INTEREST AT THE ACADEMIES

Though much of the literature and discussion surrounding the funding effect concerns individual researchers, there is now a growing appreciation of the potential bias introduced through institutional conflicts of interest (institutional COIs). The National Academy of Medicine notes that an institutional COI exists when an "institution's own financial interests or those of its senior officials pose risks of undue influence on decisions involving the institution's primary interests."[40]

The Association of American Medical Colleges and American Association of Universities have issued guidelines for human-subject research that enumerate sources of potential institutional COIs, which include "substantial gifts (including gifts in kind) from a potential commercial sponsor."[41] These guidelines also recommend that institutional leaders establish an organizational culture around conflicts of interest, "demonstrating" to the academic community and to the public that compliance with these policies, including full disclosure of financial conflicts of interest, is an imperative reflecting core institutional values." The American Association of Universities makes the very broad, unqualified recommendation around institutional COIs to "disclose always."[42]

The NASEM's conflicts-of-interest policy does not address institutional COIs, and it is not known if or how The Academies addresses this issue. Nevertheless, because the NASEM is a private institution, which receives funding from private entities that may have a financial interest in the NASEM's work, we examined the role that private entities play in the NASEM's operations to determine whether institutional COIs exist and if or how they were disclosed.

FINDINGS

The NASEM2016 report discusses its finding that no committee members had financial conflicts of interest.

> Every Academies committee is provisional until the appointed members have had an opportunity to discuss as a group their points of view and any potential conflicts of interest related to the statement of task. They also determine whether the committee is missing expertise that may be

> necessary to answer questions in the statement of task. As part of their discussion, committee members consider comments submitted by the public about the committee's composition. The discussion takes place in the first in-person meeting of the committee. The committee is no longer provisional when it has determined that no one with an avoidable conflict of interest is serving on the committee and that its membership has the necessary expertise to address the statement of task.

The report stated, without qualification, that the NASEM "did not identify" any conflicts of interest among the twenty panel members. "The Committee on Genetically Engineered Crops did not identify any conflicts of interest among its members. However, in light of comments received from the public before its first meeting and because of two resignations around the time of the first meeting, one new member with experience in molecular biology and two new members with international experience and expertise in sociology were added to the committee. Those appointments brought the committee's membership to 20. That is a large committee for the Academies, but it ensured that diverse perspectives were represented in committee discussions and in the final report."

By contrast, our inquiry found that six out of twenty committee members had financial COIs (Table 20.1). Five individuals received research funding from for-profit companies related to the subject matter of the report and five had patents or patent applications on the subject matter of GE crops. Four panel members had two financial COIs. In total, there were ten financial COIs among the six committee members, and it would appear that most of these conflicts meet the NASEM's own standards for financial COIs. Five of the six committee members with financial COIs had patents or industry research funding during the time that they served on the NASEM2016 study, which would appear to meet the NASEM's standard for "current" financial COIs. The only committee member who may not fall under the NASEM's disclosure criteria was Carol Mallory-Smith, who reported receiving industry research funding in a 2012 journal article. This funding falls within our three-year window, but did not occur while Mallory-Smith served on the NASEM2016 committee and may or may not have met the NASEM's criteria for a financial COIs. In all cases, the financial COIs concerned private interests that have a financial interest in the promotion of genetic engineering, not in opposition to it. Table 20.1 shows the range of financial COIs found for the panel members and their sources and Table 20.2 (omitted here) provides the documentation for the financial COI determination.

INSTITUTIONAL COI FINDINGS

Just as the NASEM did not disclose any financial COIs among its committee members, it also did not disclose institutional COIs. At the time the NASEM was developing its 2016 GE crop report, it was receiving money from agricultural biotechnology companies that have a financial interest in the study. The organization's annual financial reports do not give exact figures but note that three leading agricultural biotechnology companies (Monsanto, Dupont, and Dow) have given up to $5 million each to the NASEM.[43] Some of the companies' donations have been directed to NASEM projects focused on agricultural biotechnology. This includes a 2015 NASEM workshop on how to communicate the science about GE agriculture to members of the public, which was co-organized by the committee chair of the NASEM2016 report. The NASEM did not disclose this funding on the printed agenda handed out at this workshop, but months later issued a report that did disclose the funding.[44–45]

The NASEM also disclosed the funders of the NASEM2016 report—which came from foundations and the USDA—but the NASEM failed to disclose the fact that it has received millions of dollars in institutional support from agricultural biotechnology companies.

Another potential institutional COI relates to the institutions boards at the NASEM that oversee research projects on genetic engineering, which have long included representatives from companies with a financial interest in the outcome of these studies.[46–48] The NASEM Board on Agriculture and Natural Resources, which helped oversee the NASEM2016 report, included a representative from Monsanto and representatives from several other food and feed companies including Cargill, Nestle, Purina, and Novus International.

The front matter of the NASEM2016 report disclosed this, but it is unclear how The Academies managed these potential institutional COIs. The Association of Governing Boards of Colleges and Universities has addressed such institutional COIs by asking board members to recuse themselves from matters in which they have a financial interest.[49]

TABLE 20.1: Financial Conflicts of Interest of NASEM Panel Members:

Committee on Genetically Engineered Crops: Past Experience and Future Prospects.

Committee Member	EH	AC	RF	PPA	EMP	CON
Fred Gould, North Carolina State University, Raleigh, NC						
Richard M. Amasino, University of Wisconsin–Madison, Madison, WI				X2–D		
Dominique Brossard, University of Wisconsin–Madison, Madison, WI						
C. Robin Buell, Michigan State University, East Lansing, MI			X4–C	X 4–D		
Richard A. Dixon, University of North Texas, Denton,TX			X5–C	X5–D		
Josė B. Falck-Zepeda, International Food Policy Research Institute (IFPRI), Washington, DC						
Michael A. Gallo, Rutgers-Robert Wood Johnson Medical School (retired), Piscataway, NJ						
Ken Giller, Wageningen University, Wageningen, The Netherlands						
Leland Glenna, Pennsylvania State University, University Park, PA						
Timothy S. Griffin, Tufts University, Medford, MA						
Bruce R. Hamaker, Purdue University, West Lafayette, IN						
Peter M. Kareiva, The Nature Conservancy Washington, DC						
Daniel Magraw, Johns Hopkins University School of Advanced International Studies, Washington, DC						
Carol Mallory-Smith, Oregon State University, Corvallis, OR			X14–C			
Kevin Pixley, International Maize and Wheat Improvement Center (CIMMYT), Texcoco, Mexico						
Elizabeth P. Ransom, University of Richmond, Richmond, VA						
Michael Rodemeyer, University of Virginia (formerly), Charlottesville, VA						
David M. Stelly, Texas A&M University and Texas A&M AgriLife Research, College Station, TX			X18–C	X18–D		
C. Neal Stewart, University of Tennessee, Knoxville, TN			X19–C	X19–D		
Robert J. Whitaker, Produce Marketing Association, Newark, DE						

EH = equity holding; AC = advisory committee; RF = research funding; PPA = patent or patent application; EMP = employed by a company; CON = consulting for a company

DISCUSSION

Our finding that the NASEM failed to disclose financial COIs held by committee members raises questions about The Academies' implementation of the FACA requirements and the effectiveness of its financial COI policy. Whether this is due to weak enforcement of financial COI policies by the NASEM, a failure of committee members to disclose financial COIs to the NASEM, weaknesses within the language of FACA, or other explanations is a subject of further inquiry.

Congressional language in FACA gives The Academies substantial discretion to assess and manage financial COIs. As stated in the legislation, "The Academy shall make its best efforts to insure" that financial conflicts of interest are avoided, or disclosed if unavoidable, and that committees are balanced.

The NASEM maintains an official conflicts-of-interest policy, which speaks to its requirements under FACA, and all committee members are required to submit a form about financial COIs and other "background information."[50] The form asks a broad range of questions about committee members' professional work related to the subject matter of the study, including any public statements they have made, and financial COIs, including investments, research funding, patents, consulting and several other activities. These disclosures are confidential and not available for public review.

Potential weaknesses in the NASEM's financial COI review include, as noted previously, its high threshold for financial investments ($10,000 or greater). Additionally, the NASEM's disclosure forms instruct committee members to focus on disclosing "current" financial COIs, which may not capture recent financial relationships that could reasonably affect judgment.

Even if committee members did not disclose all of the financial COIs we documented, The Academies were apprised of conflicts from other sources. When the NASEM undertakes a study, it allows stakeholders to submit public comments. Under FACA requirements, the NASEM is supposed to make the public comments it receives available for public review, but the NASEM informed us that it does not allow public review of comments related to committee composition, including those that allege financial COIs. (William J. Skane, pers. comm., October 16, 2016).

Nevertheless, we were able to secure and review several letters submitted to the NASEM that discuss problems with the committee. One letter, co-signed by twenty-three individuals, mostly from academia, criticized the committee composition, noting that "the panel includes staff and representatives from institutions and agencies that have an established history and institutional orientation toward seeking technological approaches to intensifying agricultural production and prioritizing yield increases over addressing the complex amalgam of factors that contribute to achieving authentic food security" and that this makes it "more challenging for those committee members to bring an independent, critical eye to their assessment project."[51] Another letter to The Academies provided documentation alleging financial COIs of numerous committee members, including the patent interests and research funding of two committee members identified in our financial COI analysis.[52]

The final authority on financial COI decisions for NASEM reports rests with the chair of the National Research Council, the research arm of the NASEM. However, an additional and apparently integral part of the NASEM review process includes committee members openly discussing amongst themselves issues of balance and financial COIs. The NASEM notes that at the first committee meeting, committee members, who are still considered to be "provisional," "are asked to discuss the issues of committee composition and balance and conflict of interest, and the relevant circumstances of the individual members."

Professional colleagues may be reluctant to candidly and openly judge each other's fitness to serve on the committee, however, especially around a sensitive subject like financial COIs. Likewise, it is unclear why the NASEM waits until after the committee has been announced and held its first public meeting to finalize its financial COI reviews, as this puts The

Academies in a potentially awkward position of having to publicly announce one or more "provisional" committee members being removed from the committee because of financial COIs. It is not clear how often the NASEM removes provisional committee members for reasons of financial COIs.

The National Academy of Medicine has issued recommendations about conducting financial COI reviews that appear to differ from the NASEM's approach, recommending that biomedical institutions create special conflict-of-interest committees to carry out reviews and noting that "... accountability is generally enhanced if public representatives serve on institutional panels that review individual relationships that may present conflicts of interest."[53] Such a committee dedicated to analyzing financial COIs does not appear to exist at the NASEM.

The NASEM's prestigious journal, *Proceedings of the National Academy of Sciences (PNAS)*, also has different financial COI policies in place. For example, *PNAS* imposes sanctions on authors that "deliberately or recklessly failed to disclose conflicts of interest ... including being banned from publishing in *PNAS* for a period of time."[53]

A more conservative financial COI assessment. It is noteworthy that several members of the committee that did not meet the financial COI criteria adopted in this analysis had ties to companies that a broader analysis may have cited as potentially introducing bias or imbalance. For example, one committee member directs a research project that accepts money from the Syngenta Foundation (associated with the agricultural biotechnology company Syngenta) and has recently published research that received in-kind contributions from agricultural biotechnology companies.[54–57] Another committee member serves as the director of a university-industry research center that receives money from an agricultural biotechnology company.[58] While these financial relationships may present potential financial COIs, they did not meet our strict criteria.

Similarly, two committee members we identified as having financial COIs, Richard Dixon and C. Neal Stewart, both have consulted extensively with agricultural biotechnology companies, but we did not identify this as a financial COI in Table 20.1 because we could not definitively document any consulting that met our criteria of having taken place within three years of the NASEM2016 report.[59, 60]

In contrast, our search found no evidence of any committee members being financially associated with private entities that can be construed as having a financial interest in restricting or opposing genetic engineering. We found only one remote association.

One committee member serves as an unpaid advisor to a nonprofit organic farm that has an apprenticeship program funded by the dairy industry.[61,62] The funders include Stonyfield, a company that promotes its avoidance of GE ingredients and that has played a prominent role in political campaigns to require labeling of food containing GE ingredients.[63] Another funder is the corporate foundation of Stonyfield's parent company, Danone, which sells some products containing GE ingredients.[64,65]

The various ties to industry described above did not meet our conservative definition of a financial COI, though a less conservative analysis may have viewed them differently. It is noteworthy that one public-interest group examining the NASEM2016 committee took a broader definition of COIs than our analysis took.

Citing the Academies' requirements under FACA to form "fairly balanced" committees, that analysis identified financial COIs and also evidence of ideological positions that committee members had in favor of agricultural biotechnology in journal articles, statements to the press, or political advocacy work.[66] The NASEM's COI policy notes: "An individual may have become committed to a fixed position on a particular issue through public statements . . . through publications . . . through close identification or association with the position or perspectives of a particular group, or through other personal or professional "activities."

The NASEM indicates that this would "ordinarily constitute a potential source of bias, but not a conflict of interest," though it acknowledges it could, in some circumstances, present a COI. Given the lack of clear definitions and guidelines on non-financial COIs from the NASEM or other scientific bodies, our analysis explicitly narrowed the focus to financial COI disclosure and did not consider committee members' intellectual interests, professional associations, or public statements on biotechnology.

POTENTIAL EFFECTS OF COIs

The NASEM recruited a highly multi-disciplinary committee of scientists, including experts in wide-ranging fields like molecular biology, sociology, law and ecology for its NASEM2016 report. This array of expertise reflects the breadth of questions that the committee was asked to address in its report.

It is notable that the committee members we identified as having financial COIs comprised all of the committee's expertise on key topics, including plant biotechnology, molecular biology, plant breeding, weed science and food science.[67] Presumably, committee members were asked to author the sections of the report relevant to their expertise, meaning entire chapters may have been written by committee members with financial COIs.

INSTITUTIONAL COIs

Several recommendations around institutional COIs discuss establishing review boards to manage conflicts, but also note the difficulty forming review bodies that have both the independence and the power to effectively manage institutional conflicts.[68] The NASEM differs from other academic institutions in that Congress has laid out specific requirements around financial COIs under FACA. These requirements do not currently speak to institutional COIs, though Congress has the power and independence to expand its requirements. Even the appearance of institutional COIs can affect the public perception of science.

CONCLUSION

Our results show that The Academies failed to follow FACA requirements in making transparent the financial conflicts of interests of members of its Committee on Genetically Engineered Crops in its May 2016 publication. We also showed that the omitted disclosures may not have met the standards established by The Academies' own guidelines or by contemporary standards of financial COI disclosure.

Among the limitations of this study are that we were able to identify financial COIs that were available in the open literature. We had no privileged sources. We also were not able to review all committee members' research grants because most universities do not make these public. We were also not able to review committee member's personal financial investments. As a result, we may have missed some financial COIs.

Our analysis utilized slightly different criteria for assessing financial COIs than the NASEM, though five of the six committee members we identified as having financial COIs would appear to meet both the NASEM's criteria and our own. Because the NASEM does not make public its internal financial COI reviews, we were limited in our ability to assess the organization's rationale for not disclosing any of the financial COIs we identified. Likewise, because the NASEM shields from public review any public comments that stakeholders submit about financial COIs, we were limited in understanding the extent to which the NASEM was apprised to financial COIs from other sources. It is unclear why the NASEM does not make these public comments open for review, but, as noted, we located one public comment that provided the NASEM with documentation about several undisclosed financial COIs identified in this paper.

Between the financial COI disclosure forms submitted by committee members to the NASEM and the public comments the NASEM received, many or most of the financial COIs identified in our analysis should have been available for the NASEM to review. Transparency of conflicts of interest is one of the foundational principles for ethical science because it gives readers a basis for drawing their own conclusions of bias and the confidence level of the paper or study.[69,70] The NASEM's apparent failure to disclose financial COIs does not conform to best practices widely employed by the scientific community or to Congress's requirements under FACA.

Disclosure is a first, critical step toward addressing the potential bias stemming from financial COIs. Removing potential bias from research also requires management of conflicts of interest, a topic that Congress partially addresses by requiring the NASEM to form "fairly balanced" committees of experts. While this issue is beyond the scope of this paper, it is noteworthy that every financial COI and institutional COI we identified in this analysis concerns financial interests related to private interests that favor the use of agricultural biotechnology, not companies with a financial interest in restricting the use of agricultural biotechnology. Disclosure is not an elixir for preventing bias from financial COIs. The intent of FACA is to avoid financial COIs first and foremost, unless it is absolutely necessary to accept them. As Krimsky 2010 notes: ". . . unless transparency results in behavior change, it does not address the issues of bias and public trust."

1. The NASEM, as one of the world's most prestigious research institutions, is in a position to set a high standard for disclosure and management of financial COIs that other institutions can look to for guidance. By failing to disclose conflicts, The NASEM sends a message to the broader scientific community that its prestige is sufficient to forego rigorous standards of financial COI disclosure. A report as important and influential as "Genetically Engineered Crops: Past Experience and Future Prospects," which may help shape federal rules and regulations around agricultural biotechnology, should aspire to the highest level of transparency in order to achieve the greatest public confidence in its objectivity.

2. In the light of our findings, we offer the following recommendations for improving the NASEM's financial COI guidelines and their implementation.

3. Create a dedicated body within the NASEM to carry out financial COI reviews of committee members and board representatives, ideally with the assistance of public stakeholders to enhance accountability.

4. Revise the NASEM's COI policy to require committee members to submit three years of financial COI disclosures.

5. Eliminate the financial COI threshold of $10,000 for financial investments.

6. Disclose all financial COIs of committee members that the NASEM deems unavoidable.

7. Make all public comments submitted to the NASEM committee part of the public record, as the federal government does in its policy making.

8. Disclose in reports any institutional funding the NASEM received in the previous three years from companies that could benefit from the NASEM's findings. Require the NASEM board representatives who have a financial interest in NASEM matters to recuse themselves of any decisions regarding reports related to their financial interests.

AUTHOR CONTRIBUTIONS

- Conceptualization: S. K. T. S.
- Funding acquisition: S. K.
- Investigation: S. K. T. S.
- Methodology: S. K. T. S.
- Project administration: S. K. T. S.
- Supervision: S. K. T. S.
- Validation: S. K. T. S.
- Visualization: S. K. T. S..
- Writing—original draft: S. K. T. S.
- Writing—review and editing: S. K. T. S.

Chapter 21

CONFLICT OF INTEREST POLICIES AND INDUSTRY RELATIONSHIPS OF GUIDELINE DEVELOPMENT GROUP MEMBERS: A CROSS-SECTIONAL STUDY OF CLINICAL PRACTICE GUIDELINES FOR DEPRESSION[28]

SHELDON KRIMSKY
Tufts University, Department of Urban and Environmental Policy and Planning

LISA COSGROVE
EMILY E. WHEELER
SHANNON M. PETERS
MADELINE BRODT
University of Massachusetts, Department of Counseling and School Psychology

ALLEN F. SHAUGHNESSY
Tufts University, Department of Family Medicine and Tufts University Family Medicine Residency at Cambridge Health Alliance

28Originally published in *Accountability in Research* 24, no. 2 (2017): 99–115.

BACKGROUND

The medical community relies on clinical practice guidelines (CPGs) to advance the practice of medicine, conform to standards of care, and improve patient outcomes. Over the last ten years, not only have guidelines proliferated, but concerns have also been raised that some look more like industry "marketing and opinion statements."[1] Because CPGs have become so influential—they are the primary mechanism for "communicating . . . standard of care" expectations to practicing physicians"[2]—their validity and trustworthiness are critical for public confidence and improving medical outcomes.

In 2009 the Institute of Medicine (IOM) recommended that guideline development groups address the issue of trustworthiness, stating, "Financial ties between medicine and industry may create conflicts of interest. Such conflicts present the risk of undue influence on professional judgments and thereby may jeopardize the integrity of scientific investigations, the objectivity of medical education, the quality of patient care, and the public's trust in medicine."[3] Two years later, the IOM[4] issued a report that specifically addressed the issue of transparency and management of financial conflicts of interest (FCOI), in addition to researchers who have provided recommendations for addressing the prevalence of FCOI in research.[5] The IOM called for guideline development groups to develop policies regarding FCOI of panel members, specifically recommending that the chair or co-chair should not have conflicts of interest and that, ideally, all guideline panel members should be free of commercial ties.[6] Organizations with conflict of interest policies may help send the message that these conflicts can produce an unacceptable influence on CPG quality.[7] Other groups, including the National Institute for Clinical Excellence (NICE)[8] and Guidelines International Network (G-I-N)[9] have issued similar standards to control the influence of industry on the guideline development process. It is also noteworthy that the U.S. Agency for Healthcare Research and Quality (AHRQ)[10] recently required stricter standards (e.g., a systematic literature search of the evidence base must be conducted) for clinical practice guidelines to be included on the National Guideline Clearinghouse, a publicly available database produced by AHRQ. The stricter inclusion criteria provided another signal that more safeguards are needed to ensure the trustworthiness of guidelines.

The impetus behind the call for greater independence of individuals involved in guideline development arises in part from the documented "funding effect" (i.e., corporate funded research tends toward outcomes that support the funder's financial interests)[11–14] and the "third person effect" (i.e., the tendency for scientists to believe that their colleagues, but not themselves, are susceptible to unconscious bias when their research is industry funded;[15] see Tereskerz (2003) for review of conflict of interest research[16]). A recent report by the Cochrane organization, the leading independent source of systematic reviews, concluded that the funding effect in both drug and medical device studies is a pernicious problem and one that "cannot be explained by standard 'risk-of-bias' assessments" (i.e., blinding, randomization),[17] highlighting the fact that unconscious biases are common and can distort the medical literature.[18] Thus, it is important to attend to the ways organizations that produce guidelines attempt to prohibit and/or manage conflicts of interest, and assess whether characteristics of guideline developers are associated with differences in FCOI management.

CURRENT STUDY

The purpose of the present study is to expand on previous research in this area to include an international set of guidelines for another disease category, major depressive disorder.[19] Specifically, we examined the extent and type of industry associations of panel members, and the relationship between type of guideline producer (i.e., government vs. non-government and U.S. versus non-U.S.) and extent of FCOI. Additionally, guidelines were assessed for their requirements for transparency and management of FCOIs. Guidelines for depression were chosen based on our previous work in which we identified inconsistencies in both treatment recommendations and disclosure requirements in a subset of guidelines on major depressive disorder.[20]

METHODS

This is a cross-sectional study of clinical practice guidelines for major depressive disorder. We examined a) the disclosure requirements for panel members and b) the extent and type of their FCOI.

SAMPLE

CPGs were identified using several databases. Medline was searched using PubMed and OvidSP (see Appendix for search terms and strategy). We also searched for guidelines using the Trip search engine[21] and the International Guideline Library of the Guidelines International Network (2015). All databases were initially searched for guidelines published between January 1, 2009 and December 31, 2014. We later updated the search in March 2016. We adopted IOM's definition of a guideline: "Clinical practice guidelines are statements that include recommendations intended to optimize patient care that are informed by a systematic review of evidence and an assessment of the benefits and harms of alternative care options."[22]

Guidelines were included if they were issued by national organizations; medical specialty organizations; professional associations; or government, nonprofit, or commercial entities, and were intended to guide the treatment of unipolar depression (i.e., Major Depressive Disorder) in adults. Organization type was determined according to the categories of guideline producers used by the AHRQ's National Guideline Clearinghouse. For those guidelines not listed (and thus not categorized) in the National Guideline Clearinghouse, the second author (S. K.) assigned producer categories (government vs. non-government) after reviewing their statements. Congruent with previous research,[23] in order to limit heterogeneity in the sample and best ensure that guidelines could be compared fairly, the following exclusion criteria were applied to the guidelines retrieved from our search: an updated version of the same guideline was also retrieved, relevant for special populations or subgroups only (e.g., pregnant women), or limited to a particular form of treatment (e.g., psychotherapy). Non-English guidelines were included if a professional translation was available.

DISCLOSURE REQUIREMENTS OF PANEL MEMBERS

Fifteen disclosure categories were identified for COI policies. These categories were identified by one of the researchers (S. K.). They were based on previous research and current guidelines for FCOI management (e.g., IOM 2009;[24] International Committee of Medical Journal Editors' [ICJME] conflicts of interest disclosure guidelines;[25] and National Institutes of Health [NIH] FCOI disclosure criteria).[26] Each guideline was searched line-by-line for policies regarding disclosure and management of FCOI. For any guideline that did not include a disclosure policy, the developing organization's website was searched and if a policy could still not be located, the organization was contacted by a research team member and asked to provide its relevant COI policy if any existed. Policies were copied in their entirety and sponsorship and panel members' disclosures were de-identified. One researcher (S. K.) reviewed the de-identified policies and the disclosure statements and matched them against these categories.

MAIN OUTCOME MEASURE

We defined FCOI based on the International Committee of Medical Journal Editors definition.[27] This definition includes industry financial ties such as consultancy, honoraria, speakers' bureau membership, expert testimony, research funding, stock holdings, advisory board membership, unrestricted education grants, or patent holding. Industry sponsorship of the CPGs and the types of commercial associations of their panel members were identified and recorded in a template.

IDENTIFICATION OF FCOI

Two sources of data were used for identifying FCOI: a) any disclosure statements provided in the CPGs, and b) a multi-modal search of several databases to identify financial conflicts of interest of the panel members that were not disclosed in the guideline. We searched Medline, PsycINFO, CINAHL, and Academic Search Complete; ProPublica's "Dollars for Docs" database; and the U.S. Patent and Trademark Office patent database for the three years on either side of the guidelines' publication dates; these searches and time frames are congruent with previous research methodology.[28–30] Searches in databases for peer-reviewed journals (e.g., Medline) were terminated once a disclosure statement was found that met standards set by the International Committee for Medical Journal Editors[31] (e.g., "Dr. Smith reports receiving honoraria, research funding, and consulting fees from company X"). We also used Internet search engines (Google), combining each panel members' names with the name of each of the major drug manufacturers that developed antidepressants, to find other disclosures (e.g., author disclosures provided for peer-reviewed conferences).

ANALYSIS

Descriptive statistics were used for most outcomes. Two-sided chi-square tests with a .05 significance level (SPSS v.22) were used to examine the proportion of guideline panelists with FCOI according to the guideline characteristics of type of sponsoring organization (government vs. non-government) and national origin of the sponsoring organization (U.S. vs. non-U.S.).

RESULTS

A total of fourteen guidelines met our inclusion criteria. (See Figure 21.1 for the results of the systematic review and Table 21.1 for a list of the included CPGs.) Six were from the U.S., two from the U.K., one from Singapore, one from Germany, one from Brazil, one from Canada, one from Finland, and one from Spain. Ten of these guidelines were originally published in English; the others provided English-language translations. Statements about the funding of guidelines were found in five out of the fourteen CPGs. Three of the five guidelines were developed by non-U.S. organizations and two of the five were by government organizations. For these 5 guidelines, the statement clarified that the sponsoring organization funded the guideline and no external funding was sought or used. CPG panels ranged in size from 4 to 23 (Mean = 12.3, Median = 11). These panels were comprised of a total of 172 panel members and 171 different individuals (i.e., 1 individual participated in 2 different panels).

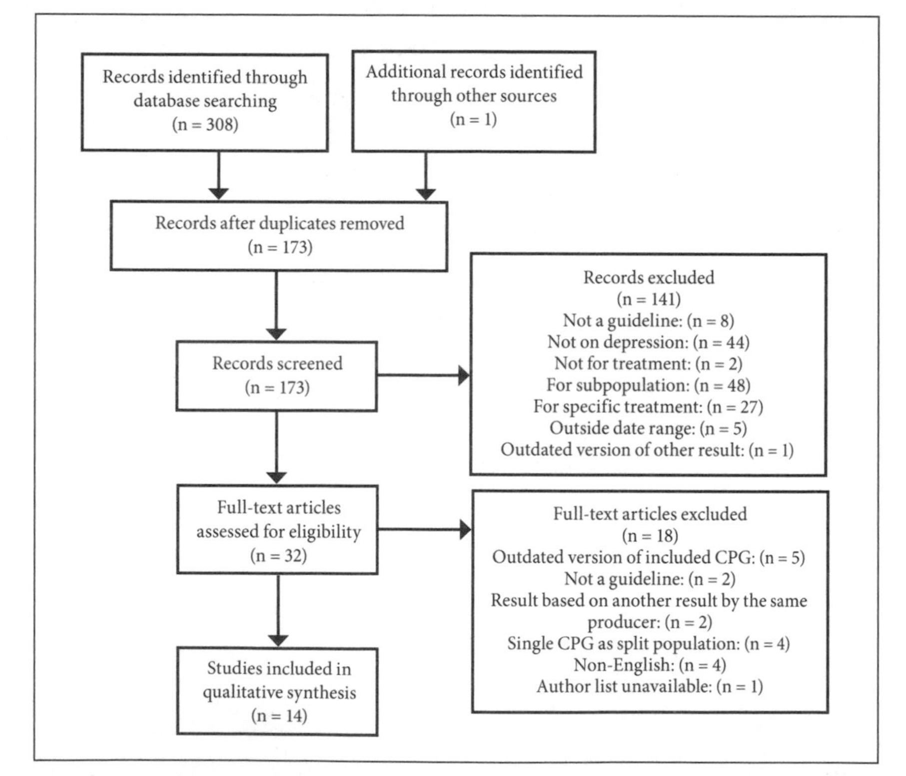

FIGURE 21.1: Systematic review of flow diagram

FCOI POLICIES OF THE GUIDELINES

Six of the fourteen guidelines included a FCOI policy in the guideline. The research team contacted the organizations that developed the remaining eight guidelines and policies for five additional CPGs were identified, for a total of eleven policies.

The policies varied in their disclosure requirements. For example, while eight of the eleven policies required disclosure of research funding received from industry, only two of them specified disclosure of authorship of an industry-funded study. Among the guidelines that required any of the fifteen disclosure categories, only three required eight or more categories. (See Table 21.2 for complete results of CPGs with policies.)

ANALYSIS OF FCOI

Eight of the fourteen guidelines (57 percent) had at least one panel member who had an industry financial tie to at least one company that manufactures antidepressant medication. In seven of these guidelines, there was at least one member who participated on a drug company's speakers' bureau. Of the eleven CPGs with identified chairs, six of these chairs (55 percent) had FCOI. (See Table 21.1 for a summary of results.)

In the total sample of 172 panelists, 31 (18 percent) were found to have FCOI. Of the 70 panel members who had the opportunity to declare FCOI, 23 reported FCOI and only 1 additional member was found to have an undeclared commercial tie. Seven of the 102 members (7 percent) who did not have the opportunity to declare conflicts were found to have FCOI. The most commonly found conflict, among the 31 individuals with industry affiliations, was speakers' bureau membership (n=23, 74 percent). Frequencies of other FCOI categories are: research funding (n=20, 64 percent); advisory board members (n=17, 55 percent); honoraria (n=10, 32 percent); consultants (n=10, 32 percent); and one (3 percent) had another type of FCOI (reported as "salary"). See Figure 21.2.

TABLE 21.1: Included guidelines with FCOI results.

CPG	Title	Year Published	Producer	Country	FCOI	Chair FCOI	SB/AB FCOI
1	Depression: The NICE Guideline on the Treatment and Management of Depression in Adults	2010	National Collaborating Centre for Mental Health	UK	3/17	Yes	3/17
2	VA/DoD Clinical Practice Guideline for Management of Major Depression Disorder (MDD)	2009	Department of Veteran Affairs; Department of Defense	USA	0/23	No	0/23
3	Practice Guideline for the Treatment of Patients with Major Depressive Disorder	2010	American Psychiatric Association	USA	6/6	Yes	5/6
4	Depression in Adults	2012	Map of Medicine	UK	¾	N/A*	¼
5	Depression	2011	University of Michigan Health System	USA	0/5	No	0/5
6	Major Depression in Adults in Primary Care	2013	Institute for Clinical Systems Improvement	USA	0/11	No	0/11
7	Primary Care Diagnosis and Management of Adults with Depression	2016	Michigan Quality Improvement Consortium	USA	0/17	N/A*	0/17
8	Depression	2012	Singapore Ministry of Health	Singapore	1/21	Yes	1/21
9	Adult Depression Clinical Practice Guideline	2012	Kaiser Permanente Medical Care Program	USA	0/22	No	0/22
10	Unipolar Depression: Short Version	2015	German Association for Psychiatry and Psychotherapy; German Medical Association; National Association of Statutory Health Insurance Physicians; Association of the Scientific Medical Societies	Germany	1/4	Yes	1/4
11	Review of the Guidelines of the Brazilian Medical Association for the Treatment of Depression (Full version)	2009	Brazilian Medical Association	Brazil	3/8	No	3/8
12	Canadian Network for Mood and Anxiety Treatments (CANMAT) Clinical Guidelines for the Management of Major Depression Disorder in Adults	2009	Canadian Network for Mood and Anxiety Treatments	Canada	12/14	Yes	12/14
13	Depression	2014	Finnish Medical Society Duodecim	Finland	2/11	Yes	0/11
14	Clinical Practice Guideline on the Management of Depression in Adults	2014	Galician Health Technology Assessment Agency; GuiaSalud; Ministry of Health	Spain	0/9	N/A*	0/9

Note. FCOI reported as number of panelists with reported conflicts out of total number of panelists for each guideline. AB = advisory board membership, SB = speakers bureau membership.

*No chair was identified for the CPG.

TABLE 21.2: Assessment of conflict of interest policies in eleven guidelines with policies

Disclosure categories	CPG #1	CPG #2	CPG #3	CPG #4	CPG #5	CPG #6	CPG #7	CPG #10	CPG #12	CPG #13	CPG #14
1. Speakers' bureau of advisory board membership	X	X	X	X	X	X	X	X	X		
2. Holds office in org or advocacy group with a direct interest in depression	X			X		X					
3. Non-financial interest, i.e., public statements made about depression.	X			X							
4. Grants or fellowships from companies or foundations connected to depression	X	X	X	X		X	X	X	X		
5. Family associated with health care industry paid or unpaid	X		X			X					
6. Stock holdings in companies that manufacture or distribute antidepressants	X	X	X	X			X	X			
7. Declare interests before being appointed	X	X	X				X	X			
8. Declare interests before each development meeting or periodically as needed	X	X			X	X		X			
9. Limit panel membership of those with industry financial ties/ balanced membership			X			X					
10. Reference to 10M policies on FCOI					X	X					
11. No industry employees accepted on panel			X								
12. Recusal as determined by GDG member, organization board or independent review panel		X				X					X
13. Author of industry-funded publication		X						X			
14. Stipulates time period for reporting interests				X		X				X	
15. Income from consulting or speaking related to treatments	X		X	X		X	X	X	X		

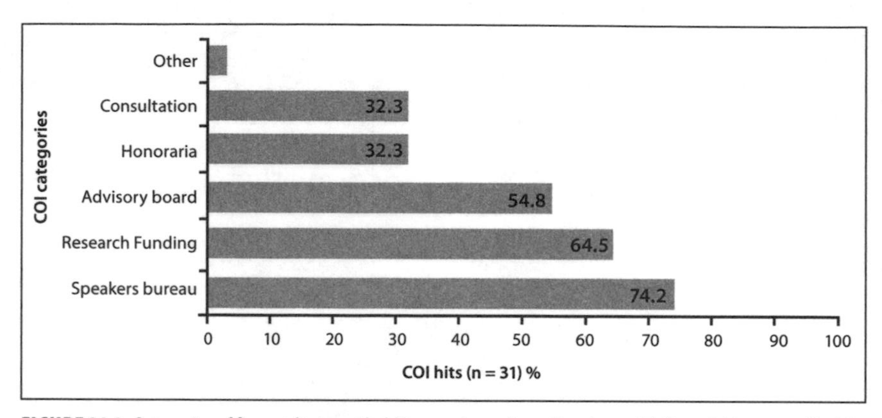

FIGURE 21.2: Categories of financial interest held by panel members: Members with financial interests (N=31).

ANALYSIS OF COI BY CPG CHARACTERISTICS

Panel members on government-produced guidelines were significantly less likely to have FCOI than panel members on non-government-produced guidelines (5.7 percent vs. 26.5 percent, $p < .001$). U.S.-based organizations were significantly less likely to have FCOI than panel members of guidelines produced by organizations from other countries (7.1 percent vs. 28.4 percent, $p < .01$).

DISCUSSION

In this study of guidelines for the treatment of major depressive disorder, we found a significant proportion of guideline development groups with neither an explicit policy regarding FCOI nor a formal declaration of guideline development members' FCOI. Also, the wide range of disclosure requirements found in this study is consistent with Norris et al.'s concern that there are significant gaps in many disclosure policies of organizations that produce guidelines.[31]

In terms of commercial associations, we found that 18 percent of individual panelists had industry financial ties, 57 percent of the guideline development groups had at least one person with an FCOI, and nearly half (43 percent) of the guidelines had a chair with an FCOI. Additionally, it is problematic that almost half of the guidelines had panel members who participated on industry speakers' bureaus and that this type of industry relationship was the most prevalent. This kind of financial association is recognized to have a biasing effect and is typically prohibited.[32,33] A speakers' bureau usually refers to an arrangement between a commercial entity and an individual whereby the individual is hired to give a presentation supporting the company's product.[34] Concerns about these types of commercial interests of guideline panel members led the IOM to make the following recommendation: "Members of the [guideline development group] should divest themselves of financial investments they or their family members have in, and not participate in marketing activities or advisory boards of, entities whose interests could be affected by CPG recommendations."[35] Moreover, an increasing number of medical schools have prohibited speakers' bureau participation on the part of their faculty.[36]

We found that panelists for guidelines produced by U.S.-based organizations were significantly less likely to have FCOI than panelists for guidelines produced by organizations from other countries. Whether or not this finding reflects increased adherence to recent recommendations (e.g., by G-I-N, the IOM) for developing trustworthy guidelines will require further testing on a larger sample of CPGs. Also, one non-U.S. guideline (CANMAT) had 12/14 panel members with FCOI which could account for non-U.S. guidelines appearing more conflicted than U.S. guidelines.

Congruent with Neuman et al.'s study, we found that compared to government-sponsored panels, FCOI were far more common among panel members of non-government organizations.[37] CPGs produced by non-government organizations, and particularly medical specialty societies, have come under scrutiny for methodological weaknesses.[38] Other guidelines produced by specialty organizations in cardiology,[39] in hyperlipidemia,[40,41] and in schizophrenia and bipolar disorder,[42] have been found to have conflicted guideline development groups. These conditions affect large numbers of people and are primarily-treated with pharmaceuticals, making the stakes very high in terms of the impact and cost of industry influenced recommendations. For example, 25.2 million people were taking statins from 2005 to 2010. It has been estimated that following the 2013 ACC/AHA cholesterol guidelines would more than double this number and lead to 87 percent of men over 60 being prescribed statins. (The chair and panel members of this CPG were found to have industry financial ties to the manufacturers of statins[43]). While these industry relationships are sometimes dismissed by leaders within medicine,[44] there is evidence that funding influences decision making.[45–48]

A 2012 study of guideline development groups reported a decrease in the percentage of panelists with FCOI in recent studies when compared with percentages found in older studies.[49] Our finding that less than 20 percent of total panelists had FCOI is perhaps a sign of heightened awareness of the problem of undue industry influence. This finding challenges the oft-cited argument that it is not possible to find expert panel members free of commercial ties. Also, congruent with recent research,[50] we found that although the proportion of panel members with FCOI may be decreasing, the majority of CPGs were developed by panels with one or more members with industry financial ties. Indeed, some guidelines had panels where all or a majority of members had FCOI, and many of the chairs had industry financial ties.

The relationship between FCOI of panel members and guideline recommendations is complex, and it is not known when or under what conditions these conflicts will have a biasing effect. However, because of the far-reaching influence of CPGs, and because of concerns about credibility and trustworthiness, AHRQ, NICE, and G-I-N have issued stricter standards regarding the prohibition, management, and disclosure of industry financial ties (see also, Guyatt et al. 2012[51]). When there are FCOI on the part of panel members, it has been suggested that end users of guidelines should take commercial relationships into account when evaluating the trustworthiness of a guideline.[52–54] Attention to the presence of conflicts of interest among guideline developers may foster a more critical examination of the quality of the guideline. In the present study, both the omission of COI policies in the guidelines and the presence of commercial ties among a majority of guideline development groups and chairs stand in contrast to the recent standards developed by IOM and G-I-N and have the potential to compromise the quality and trustworthiness of those guidelines.

LIMITATIONS

The included sample was limited to those available in English (including non-English CPGs whose organization provided an English translation) and accessible via the search strategy. Also, this study only included a review of guidelines for depression. Thus, the present study's findings may not be generalizable to other disease categories. However, a 2013 study of forty-five CPGs produced by fourteen different specialty societies found that industry financial ties were common.[55] It should also be noted that there are still challenges to identifying FCOI, particularly the lack of public databases for non-U.S. authors. Also, in the U.S., pharmaceutical companies are required to report all payments to physicians to Open Payments, a searchable government database of financial relationships.[56] However, there may be some attempts to prevent the discovery of these payments; in one case, a pharmaceutical company misspelled the name of their own product 953 times in the database, which represented almost one-third of all reports for that product.[57] For those guidelines that did not have financial disclosures of panel members, our methodology for determining the FCOIs might have omitted some. Therefore, for those cases, we can only be confident that our findings are conservative. In addition, disclosures of small payments (<$100) for conflicts such as meals and travel were not included in this analysis.

Finally, the present study focused exclusively on financial conflicts of interest and it is recognized that intellectual COIs (e.g., career advancement), adherence to a particular theoretical model, or guild interests may also be sources of bias.[58] We share the belief that FCOI, unlike other sources of bias, are almost always unidirectional by being favorable toward commercial interests.[59]

CONCLUSIONS

In this cross-sectional study of guidelines for the treatment of depression, the level of specificity of policies and the extent of financial conflicts of interest of guideline development group members varied widely. Our findings suggest that disclosure of FCOI in these guidelines has not kept pace with the current standards for transparency of expert panels and medical journals. Although there may be a recent trend of increased transparency and independence from industry, producers of depression guidelines should take more aggressive steps to try to appoint panel members free of industry ties and prohibit participation on drug companies' speakers' bureaus. In light of the public health burden of depression worldwide, and given the fact that the "funding effect" cannot be explained by standard risk of bias assessments,[60,61] it is critically important that end users of guidelines can be confident that recommendations are not influenced by commercial interests. The fact that only a minority of panel members had industry financial ties shows that it is possible to find independent experts. Indeed, "a guideline development panel that is free from conflicts of interest provides the best safeguard against bias."[62]

ACKNOWLEDGMENTS

The authors would like to thank Darren Freeman-Coppadge, PharmD (University of Massachusetts, Boston), for his assistance with data collection. The authors also thank Amy E. LeVertu of the Hirsh Health Sciences Library at Tufts University for her expertise, assistance, and support with the systematic review.

FUNDING

This work was supported by the Agency for Healthcare Research and Quality under Grant R03HS022940–01A1.

CHAPTER 22

ROUNDUP LITIGATION DISCOVERY DOCUMENTS: IMPLICATIONS FOR PUBLIC HEALTH AND JOURNAL ETHICS*

SHELDON KRIMSKY
Tufts University, Department of Urban and Environmental Policy & Planning
CAREY GILLAM
US Right to Know, Oakland, CA, USA

Abstract

This paper reviews the court-released discovery documents obtained from litigation against Monsanto over its herbicide Roundup and through Freedom of Information Act requests (requests to regulatory agencies and public universities in the United States). We sought evidence of corporate malfeasance and undisclosed conflicts of interest with respect to issues of scientific integrity. The findings include evidence of ghostwriting, interference in journal publication, and undue influence of a federal regulatory agency.

Keywords

Roundup · Monsanto · Ghost writing · Glyphosate · IARC · EPA

J Public Health Pol https://doi.org/10.1057/s41271-018-0134-z

Introduction

Lead, viny[1] chloride, pharmaceuticals,[3,4] asbestos,[5] and tobacco litigation[6,7] cases have resulted in 'discovery documents.' These documents, originally internally held by parties to a lawsuit, have become public in court records from cases filed in the United States (US). Such documents have revealed important information about the actions taken by corporate defendants to withhold, distort, invalidate, ghost-write, or fabricate scientific studies of their products. Among the revelations in the cases are ghost-written articles, withholding of critical public health information, hiring contract research companies to invalidate toxicology studies, funding of nonprofit research centers to create critical reviews of published papers that had cast doubt on the safety of their products, and funding of university faculty to support their agendas. It is now possible for public health scholars to search for and in these documents on open databases.[8,9]

In 2015, the International Agency for the Research on Cancer (IARC), the specialized cancer research arm of the World Health Organization, determined that the chemical glyphosate, the active ingredient in many popular herbicides, is a "probable" human carcinogen. IARC said it found "limited" evidence of cancer links in studies of human exposures, mostly agricultural-related, that had been published since 2001. But IARC said studies in laboratory animals showed "sufficient" evidence that glyphosate can cause cancer. Research also showed "strong" evidence that glyphosate caused DNA and chromosomal damage in human cells, according to IARC.[10]

Subsequent to the finding, several hundred people who believed they had been injured by the herbicide Roundup filed lawsuits against the Monsanto Corporation, its manufacturer.[11] Roundup is one of the most widely used glyphosate-based products. As of November 2017, roughly 3500 plaintiffs had cases ongoing against Monsanto. In each case, the plaintiff alleged that she or he, or their loved ones, developed non-Hodgkin lymphoma due to Roundup exposure. Moreover, the plaintiffs alleged that Monsanto had long covered up the risks of the glyphosate- based herbicide. More than 270 of the cases have been consolidated in multidistrict litigation (MDL) to be overseen by one judge in a federal court in the U.S., the District Court in San Francisco.[12] Many other lawsuits are proceeding in state courts. As part of the litigation, Monsanto has turned over millions of pages of its internal records to plaintiffs' attorneys, and many of those documents have been made public through the court docket.[13] Both these disclosed discovery documents and hundreds of other court documents have been placed in the public domain.[14]

A number of journalists have brought these documents to the attention of the public. Among them are Stéphane Foucart and Stéphane Horel, who received the Varenne Award for their series published in *Le Monde*.[15] One of us, CG, has established a publicly accessible digital repository[16] of the Monsanto litigation documents and has written newspaper and magazine articles describing them.[17] CG has also obtained thousands of pages of documents from the United States (US) Environmental Protection Agency (EPA) through Freedom of Information Act (FOIA) requests. CG has accessed internal email communications emanating from public universities that reveal evidence of the manipulation of science. Much of this has been laid out in a chapter titled "Spinning the Science" in her book *Whitewash*.[18] Many of these FOIA documents are being made available to the public through a database established by the UCSF (University of California San Francisco) Industry Documents Library.[19]

In this Viewpoint, we review the documents gathered through discovery as well as those obtained through Freedom of Information Act requests to regulatory agencies and public universities. We reviewed all these documents for evidence of unethical practices and undisclosed conflicts of interest with respect to issues of scientific integrity.

Ghost writing

Among the Roundup litigation discovery documents are multiple email exchanges authored by Monsanto employees that discuss, as an ostensibly normal business practice, 'ghostwriting' papers that, when published, appear to be authored by independent academic scientists or consultants with academic credentials. In some communications, Monsanto employees themselves used the term "ghostwrite," while in others, they simply describe the strategy and how it can be or has been employed.

A noteworthy example pertains to a paper published in 2000 by Williams, Kroes, and Munro.[20] In a February 2015 email, Monsanto scientist William Heydens discussed with colleagues various papers the company wanted to see published to counter what the company expected IARC to find with respect to glyphosate. (The company internally predicted a "possible" or "probable carcinogenicity classification by IARC.") Heydens wrote:

> A less expensive/more palatable approach might be to involve experts only for the areas of contention, epidemiology and possibly MOA [mode of action] (depending on what comes out of the IARC meeting), and we ghostwrite the Exposure Tox & Genetox sections. An option would be to add Greim and Kier or Kirkland to have their names on the publication, but we would be keeping the cost down by us doing the writing and they would just edit & sign their names so to speak. Recall that is how we handled Williams, Kroes & Munro, 2000.[21]

The Williams et al. paper has been cited hundreds of times and was among those referenced by the Environmental Protection Agency (EPA) in its finding, reported in 2016, that glyphosate was "not likely" carcinogenic.[22] Another example of Monsanto's surreptitious involvement in the science can be found in a memo dated August 4, 2015. Summarizing his "glyphosate activities," Monsanto scientist David Saltmiras, who at that time was a toxicology manager, stated that he "ghost wrote cancer review paper Greim et al. (2015)."[23] That paper too, was among those cited by the EPA in its 2016 glyphosate determination.[24]

A review of glyphosate, published along with four sub-papers in *Critical Reviews in Toxicology* (*CRT*) in September 2016 provides another example.

Monsanto disclosed that it hired Intertek Scientific & Regulatory Consultancy, part of Intertek Group Plc, to develop the review, entitled "An Independent Review of the Carcinogenic Potential of Glyphosate."[25] The review concluded that IARC's classification of glyphosate as a probable human carcinogen was inaccurate and that glyphosate was "unlikely to pose a carcinogenic risk to humans." The internal emails obtained through discovery show that a key goal of the publication of the papers was to influence the European Chemicals Agency (ECHA): "These papers will also be useful for ECHA which is a European Agency that is reviewing the safety of glyphosate. We would very much like to share our manuscripts with them to aid in their deliberations."[26]

The 'declaration of interests' in the special issue of *Critical Reviews in Toxicology (intended for disclosure of any potential conflict of interest)* stated that the authors were "not directly contacted by the Monsanto Company," and that "Neither any Monsanto company employees nor any attorneys reviewed any of the Expert Panel's manuscripts prior to submission to the journal." However, the documents obtained through discovery indicate those statements were not true. The documents demonstrate Monsanto was engaged in organizing, reviewing, and editing the drafts, even arguing with one of the authors and overruling him about language in the manuscript. In one exchange regarding a paper being prepared for publication, Monsanto scientist William Heydens wrote to Intertek: "Here are my suggested edits to the Draft Combined Manuscript. . . . I think I caught all the differences and made the changes in the Combined Manuscript as part of my editing."[27] In a separate email, Heydens wrote to Intertek that he had reviewed the entire draft and indicated "what I think should stay, what can go."[28] The documents also reveal Heydens' direct correspondence by email with at least one of the authors about the papers. Documents also demonstrate that at least one of the authors was under direct contract with Monsanto during the drafting and publication of the paper,[29] a fact not disclosed in the declaration of interest in *CRT* involving that author.[30]

In another email exchange, Heydens stated he had written an introduction to a paper and then proceeded to discuss "who should be the ultimate author" and that he had written a second paragraph in another paper, on neither of which he was listed as an author.[31]

Influencing the retraction of a scientific peer reviewed paper

In 2012, G.-E. Seralini et al. published in the journal, *Food & Chemical Toxicology,* the results of a two-year rat feeding study that found harmful impacts for animals exposed to Monsanto's glyphosate-based Roundup and to genetically modified corn, with and without Roundup application. The paper drew international attention in the media. This provoked a storm of criticisms from industry and academic scientists demanding the journal retract the article. Internal Monsanto documents show that Monsanto officials directed and organized the call for a retraction,[32] while stating internally that it should not appear as though Monsanto was behind the actions.[33]

Litigation discovery documents reveal one internal Monsanto email that stated: "He [editor-in-chief] directly told us [Monsanto] to give him something to work with or else his hands are tied and we will have to deal with the consequences."[34] Also a Monsanto-funded academic spoke directly to the *FCT* Editor-in-Chief and advocated retraction of the Seralini study. He wrote: "Failure of *JFCT* to retract the paper will force the community to be critical of the journal as well as the paper."[35,36] And a Monsanto employee described how he "leveraged his relationship" with the *Food & Chemical Toxicology* Editor-in-Chief and became the "single point of contact between Monsanto and the journal" while he organized a letter campaign to the journal to advocate retraction of the paper.[37]

The journal (*JFCT*) published the criticisms and the authors' responses and ultimately withdrew the article, but not until after this journal appointed a former employee of Monsanto to its editorial board. The *Journal of Environmental Science Europe* promptly republished the paper.[38] That former employee, a scientist named Richard Goodman, was then at the

University of Nebraska and receiving funding from Monsanto and other chemical industry interests to maintain a food allergy database. Email communications obtained through Freedom of Information requests show that around the time Goodman was signing on to the *FCT* journal's editorial board and criticizing the Seralini study, he was also expressing concern to his chemical industry funders about protecting his income stream as a "soft-money professor."[39] In addition, documents reveal that the journal's editor-in-chief, A. Wallace Hayes, entered into a consulting agreement with Monsanto in 2012 for a fee of $400 an hour.[40] Neither Goodman nor Hayes disclosed their financial ties to Monsanto when the Seralini paper was retracted in 2013. In retracting the study, Hayes stated that he found "no evidence of fraud or intentional misrepresentation of the data" and that "the results were not incorrect." There was no misconduct.[41] The paper, he said, was retracted because its results were inconclusive. *Being inconclusive* is not a reason for retraction recognized by the international Committee on Publication Ethics.[42]

Undue influence of a federal agency

The emails among discovery documents and Freedom of Information Act documents obtained from the EPA reveal that Monsanto worked very closely with at least three EPA officials to derail a review of glyphosate by the Agency for Toxic Substances and Disease Registry (ATSDR) that was underway in 2015.[43] The ATSDR announced in February 2015 that it planned to publish a toxicological profile of glyphosate by October of that year. But by October, ATSDR had placed the review 'on hold,' and no such review has yet been published. The documents reveal this was the result of a collaborative effort between Monsanto and a group of high-ranking EPA officials. A series of emails detail how Monsanto sought assistance from EPA officials in persuading ATSDR to drop or delay the review, putting forth the argument that the ATSDR review was unnecessarily "duplicative." It should take a 'back seat' to the EPA review also underway at that time.[44]

But internal documents show that Monsanto's concern was not that the review was a waste of government resources, but that it would find carcinogenicity concerns with glyphosate just as IARC had.

Documents show that Monsanto viewed ATSDR as "very conservative" [meaning too precautionary] and was too "IARC-like."[45] In a text message sent on June 21, 2015, Monsanto scientist Eric Sachs wrote to a former EPA toxicologist asking for contacts at ATSDR: "We're trying to do everything we can to keep from having a domestic IARC occur w this group may need your help."[46] Plaintiffs attorneys filed the text messages in the Federal Court docket and they became part of the court record. The full body of documents revealing the interactions of EPA officials and Monsanto executives is now publicly available.[47]

The litigation discovery emails also reveal that Monsanto used its relationship with EPA regulators to influence the agency to abort convening a Scientific Advisory Panel on glyphosate health risks. Federal regulatory agency personnel are permitted to interact with stakeholders, but they are not, by law, allowed to exhibit preferential treatment or play an advocacy role. The emails suggest that EPA did provide preferential treatment and advocacy for the Monsanto position.

Preparing presentations for "independent" scientists

The documents additionally reveal that Monsanto officials developed presentations for academic scientists to deliver at seminars or in other public fora. In one example from 2012, Monsanto scientist David Saltmiras told colleagues he was arranging for a European scientist to present in a seminar related to glyphosate and that he, Saltmiras, would "likely prepare his presentation and send to him to change/adapt as he sees fit."[48] Scientists who present their findings at scientific meetings are generally expected to disclose any conflicts of interest, as well as any collaborators. The documents show that in multiple instances involving multiple professors, Monsanto scientists prepared presentations for academic scientists. Nondisclosure of these relationships with Monsanto violates the accepted norms of acknowledging help from a commercial stakeholder, as well as failure to acknowledge collaborators.

Conclusion

When vital public health reports are published in refereed journals, there is a heightened expectation that they meet professional standards of scientific integrity. Those standards include full disclosure of conflicts of interest and sources of funding, plus authenticity of authorship. The Roundup litigation disclosure documents and FOIA documents show that these standards were egregiously violated, not by accident but by plan. Journals are the gatekeepers of reliable evidence and credible knowledge. They must set the highest standards of scientific integrity. Journal editors must never manifest a bias to some individual or organization. When a journal learns that an article has been ghost written or that there were undisclosed conflicts of interest, it has an obligation to act appropriately and inform readers. Our study has shown that two journals, *Critical Reviews of Toxicology* and *Food and Chemical Toxicology* did not measure up to these standards. An editor of a journal overseeing submitted papers on a health study of a product cannot be disinterested when he is under contract with the company that manufactures that product. Public regulatory bodies as the guardians of public health cannot allow their scientists to serve one special interest group and still achieve the public trust. The Roundup discovery documents signal serious flaws in the ethics of scientific publication and regulatory processes that must be addressed. The concerns raised in this paper have been discussed in a minority staff report of the congressional Committee on Science, Space & Technology.[49]

Sheldon Krimsky is Lenore Stern Professor of Humanities and Social Sciences in the Department of Urban & Environmental Policy & Planning in the School of Arts & Sciences and Adjunct Professor in Public Health and

Community Medicine in the School of Medicine at the Tufts University. He received his B.S. and M.S. in physics from Brooklyn College, CUNY and Purdue University, respectively, and an M.A. and Ph.D. in philosophy at the Boston University. He is the author of fourteen books including *Science in the Private Interest.*

Carey Gillam is a veteran journalist, researcher, and writer. She is former senior correspondent for Reuters' international news service, member of the Society of Environmental Journalists, and author of *Whitewash: The Story of a Weed Killer, Cancer and the Corruption of Science.* She is a Research Director for the U.S. Right to Know, a consumer group whose mission is "Pursuing Truth and Transparency in America's Food System."

AFTERWORD

I began this book by highlighting the importance of a free and independent knowledge sector as a cornerstone of a democratic society. Absent such a sector, the right to vote and a free press, the two other cornerstones, would be severely compromised. The press depends on academic science when reporting on public health and environmental issues. And the electorate cannot actualize informed choice at the voting booths when they are exposed to information tainted by ideology or controlled by interests other than the unencumbered pursuit of truth.

The research studies and commentaries contained in this volume represent efforts to expose the effects of "sector blending," that is, when the for-profit sector forms a partnership with higher education. As the individual most responsible for the consumer movement in the United States, Ralph Nader understands the danger to a democracy when corporations take over academic science. He wrote:

> Academic science, with its custom of open exchange, its gift relationships, its willingness to provide expert testimony that speaks truth to power, its serendipitous curiosity, and its non-proprietary legacy to the next generation of student scientists, differs significantly from corporate science, which is ridden with trade secrets, profit-determined selection of research, and awesome political power to get its way, whether by domination or servility to its payers.[1]

Given the blending of these sectors, is there a way forward? The funding from the private sector to universities under the model set by the 1980 Bayh–Dole legislation will undoubtedly continue since both institutions benefit. The principles of scientific freedom and integrity, as a system of values, must stand well above the short-term profit interests that would rationalize the distortion of truth for commercial ends. What does the future hold?

Since the 1990s journals, professional societies, and government agencies have become more attentive to financial conflicts of interest. As an example, universities must report and manage significant financial conflicts of interest of federal grant awardees, extended by many to all grantees, that could be perceived to affect the objectivity of their research. Congress also passed the Physician Payment Sunshine Act (PPSA) requiring pharmaceutical companies and medical device manufacturers to disclose gift payments to physicians, which are recorded on a publicly accessible database. The most prestigious medical and science journals have strong conflict-of-interest disclosure policies and proscribe ghost writing and honorific authorships. But there remain many refereed journals that have not adopted strong policies or do not implement the ones they have.

Universities must not accept grants or contracts that limit the autonomy of the investigators. Contract language that gives the sponsor publication rights or authority to edit results should be rejected. Universities should consider ghost authorship or honorific authorship of their faculty as a form of plagiarism and a violation of scientific integrity.

Federal agencies must strictly follow government guidelines that prohibit members of advisory committees from holding a substantial financial conflict of interest, unless regulations call for stakeholder representation on a panel. All financial conflicts of interest of advisory committee members should be publicly acknowledged and listed on any reports of the committee.

Government employees for federal agencies who are responsible for public and environmental health should not be permitted to hold financial interests, including equity, contracts, or consultancy, in for-profit companies that engage in commerce related to the work of the agency.

Companies seeking a license for a new drug, medical device, pesticide, or chemical compound who submit research findings to regulatory agencies should disclose the funding source and whether the company played a role in designing the studies. Research papers submitted to a regulatory body as evidence should be open to public view and not held as proprietary business information.

Post publication, scientists who have failed to disclose personal conflict-of-interest information about any of the authors or misled journal editors should be considered in violation of scientific integrity and subject to sanctions. No company should serve as both manufacturer and sole evaluator of their products. Former editor of the *New England Journal of Medicine* wrote: "Drug companies should no longer be permitted to control clinical testing of their own drugs."[2] She proposed an independent Institute for Prescription Drug Trials to oversee the testing of new drugs. A similar firewall between manufacturer and evaluator should apply to new chemicals.[3] These proposals are essential for addressing the consequences of financial conflicts of interest in science and can mitigate the problems with the erosion of sector boundaries between business and academia.

NOTES

INTRODUCTION

1. Sheldon Krimsky, *Science in the Private Interest* (Lanham, MD: Rowman & Littlefield, 2003); David Michaels, *Doubt Is Their Product* (Oxford: Oxford University Press, 2008); Daniel S. Greenberg, *Science, Money & Politics* (Chicago: University of Chicago Press, 2001).
2. Sheldon Krimsky and David Baltimore, "The Ties that Bind or Benefit," *Nature* 283 (1980): 130–131.
3. Committee on the Institutional Means for Assessment of Risks to Public Health, Commission on the Life Sciences, National Research Council, *Risk Assessment in the Federal Government: Managing the Process*, Washington, DC: National Academy Press, 1983.
4. National Academies of Science, Engineering and Medicine (NASEM), Committee on Genetically Engineered Crops, Genetically Engineered Crops: Experiences and Prospects. Washington, DC. National Academies Press, 2016. http://www.nap.edu/catalog/23395/. genetically-engineered-cropsexperiences-and-prospects.
5. Barry Marshal, Helicobacter Connection, *ChemMedChem* 1:783–802 (2006).
6. Thilo Grüning, Anna B. Gilmore and Martin McKee. Tobacco industry influence on science and scientists in Germany. *American Journal of Public Health* 96(1): 20–32 (January 2006).
7. Jocelyn Kaiser, Rigorous replication effort succeeds for just two of five cancer papers, *Science* 359 (6380) (January 18, 2017).
8. Helga Nowotney, "Controversies in Science: Remarks on the Different Modes of Production of Knowledge and their Use," *Zeitschrift für Sociology* 4 (1975): 37.
9. Garland Allen, "The Role of Experts in Scientific Controversy," in: *Scientific Controversies*, H. T. Engelhardt and A, Caplan, eds. (Cambridge, UK: Cambridge University Press, 1987).
10. U.S. Congress, Office of Technology Assessment (OTA). *New Developments in Biotechnology 4. U.S. Investment in Biotechnology* (Washington, DC: USGPO, July 1988), 113–115.
11. J. E. Bekelman, Y. Li, and C. P. Gross, "Scope and Impact of Financial Conflicts of Interest on Biomedical Research: A Systematic Review," *JAMA* 289 (January 22/29, 2003): 454–465, p. 463.
12. Sheila Kaplan, "Dr. Brenda Fitzgerald, C.D.C. Director, Resigns over Tobacco and other Investments," *New York Times*, January 31, 2018.

13. Arnold S. Relman, "Dealing with Conflicts of Interest," *New England Journal of Medicine* 1985; 313 (Sept. 19, 1985): 749–751.
14. Proceedings of an International Symposium: Universities in the Twenty-first Century, October 23–25, 1985, University of New Hampshire, Durham, NH.
15. Sheldon Krimsky, "The New Corporate Identity of the American University," *Alternatives* 14, no. 2 (May/June 1987): 20–29.
16. Sheldon Krimsky, "Science, Society and the Expanding Boundaries of Moral Discourse," in: *Science, Politics and Social Practice. Boston Studies in the Philosophy of Science* (Dordrecht: Kluwer, 1995).
17. Sheldon Krimsky. *Biotechnics and Society: The Rise of Industrial Genetics* (New York: Prager, 1991).
18. B. C. Pilecki, W. J. Clegg and D. McKay, "The Influence of Corporate and Political Interests on Models of Illness in the Evolution of the DSM," *European Psychiatry* 26 (2011): 194–200.
19. Lisa Cosgrove, Harold J. Bursztajn, and Sheldon Krimsky, "Developing Unbiased Diagnostic and Treatment Guidelines in Psychiatry," *New England Journal of Medicine* 360, no. 19 (May 7, 2009): 2035.
20. Stephen Ehrhardt, Lawrence J. Appel, and Curtis L. Meinert, "Trends in National Institutes of Health Funding for Clinical Trials Registered in ClinicalTrials.gov" *JAMA* 314, no. 23 (Dec. 15, 2015): 2566–7.
21. B. Lo and M. J. Field, *Conflict of Interest in Medical Research, Education and Practice* (Washington, DC: National Academies Press, 2009).
22. Institute of Medicine, *Clinical Practice Guidelines We Can Trust* (Washington, DC: Institute of Medicine, 2011).
23. Public Health Service, Department of Health and Human Services, Objectivity in Research, *Federal Register* 60, no. 132 (July 11, 1995): 35810–19.
24. National Academies of Sciences, Engineering and Medicine (NASEM), Committee on Genetically Engineered Crops, *Genetically Engineered Crops: Experiences and Prospects.* (Washington, DC. National Academies Press. 2016). http://www.nap.edu/catalog/23395/genetically-engineered-cropsexperiences-and-prospects.
25. Committee on Genetically Engineered Crops, National Academies of Sciences, Engineering and Medicine. Letter to the Editor, "National Academies Report on Genetically Engineered Crops Guarded Against Bias," *Chronicle of Higher Education* June 12, 2017.

CHAPTER 1

1. Charles Weiner, personal communication to Sheldon Krimsky, 1984.
2. Office of Technology Assessment, *Commercial Biotechnology: An International Analysis* (Washington, DC: U.S. Government Printing Office, 1984).
3. Subcommittee on Investigations and Oversight, Committee on Science and Technology, U.S. House of Representatives, 1981. *Hearing on Commercialization of Academic Biomedical Research*, Washington, DC: U.S. Government Printing Office, June 8–9.

4. Subcommittee on Investigations and Oversight, Committee on Science and Technology, U.S. House of Representatives, July 16–17, 1982. *Hearings on University/Industry Cooperation in Biotechnology* (Washington, DC: U.S. Government Printing Office, 1982); and Subcommittee, *Hearing on Commercialization* (note 3).
5. *California Rural Legal Assistance vs. Board of Trustees of the University of California*, 1984. (California Judicial Court).
6. Office of Technology Assessment, *Commercial Biotechnology.*
7. Office of Technology Assessment, *Commercial Biotechnology*, 417.
8. J. Walsh, "Universities: Industry Links Raise Conflict of Interest Issue," *Science* 164 (1969), 411–2.
9. Ibid., 412.
10. J. Cone and J. Robinson, "DBCP-UC Research." *Synapse* (University of California at San Francisco) 22 (1977): 4.
11. L. Orr, ed. "Corporate Money and Co-opted Scholars," *Business and Society Review* 37 (1980–1981): 4–11; 5.
12. M. Liebert (publisher), "Letter to Readers," *Genetic Engineering News* 3 no. 6 (1983): 4.
13. Barbara Culliton, "The Academic-Industrial Complex." *Science* 216 (1982): 960.
14. Biogen, N.V., 1983, *Company Prospectus.*
15. Centocor, Inc., 1982, *Company Prospectus.*

CHAPTER 2

1. U.S. Congress, Office of Technology Assessment, *New Developments in Biotechnology, Vol. 4: U.S. Investment in Biotechnology* (Washington, DC: GPO, 1988).
2. Mark Crawford, "Biotech Market Changing Rapidly," *Science* 231 (1986): 12–14.
3. Henry Etzkowitz, "Entrepreneurial Scientists and Entrepreneurial Universities in American Academic Science," *Minerva* 21 (1983): 198–233.
4. Charles Weiner, "Universities, Professors and Patents: A Continuing Controversy," *Technology Review* 35 (1986): 33–43.
5. David Blumenthal, Sherrie Epstein, and James Maxwell, "Commercializing University Research," *New England Journal of Medicine* 341 (1986): 1621–26.
6. Martin Kenney, *Biotechnology: The University-Industrial Complex* (New Haven, CT: Yale University Press, 1986).
7. David Dickson, *The New Politics of Science*, (New York: Pantheon, 1984).
8. Henry Etzkowitz, *A Research University in Flux: University/Industry Interactions in Two Departments. A report to the National Science Foundation,* 83–GB-0040, June 15, 1984.
9. Henry Etzkowitz, "Entrepreneurial Science in the Academy: A Case of the Transformation of Norms," *Social Problems* 36 (1989): 14–29.

10. David Blumenthal, Michael Gluck, Karen Seashore Louis, and David Wise, "Industrial support of university research in biotechnology," *Science* 231 (1986): 242–46.
11. David Blumenthal, Michael Gluck, Karen Seashore Louis, Michael Stoto, and David Wise, "University-Industry Research Relationships in Biotechnology: Implications for the University," *Science* 232 (1986): 1361–66.
12. U.S. Congress, Office of Technology Assessment, *Commercial Biotechnology: An International Analysis* (Washington, DC: GPO, 1984).
13. Office of Technology Assessment, *New Developments*, 68.
14. Blumenthal, et al., "University-Industry Research Relationships," 1364.
15. Crawford, "Biotech Market," 198–233.
16. Aubrey Milunsky and George J. Annas, eds. *Genetics and the Law,* Vol. 3 (New York: Plenum, 1985), 67.

CHAPTER 3

1. Henry Etzkowitz, "Entrepreneurial Science in the Academy: A Case of the Transformation of Norms," *Social Problems* 36, no. 1 (1989): 14–29.
2. U.S. Congress, House Committee on Government Operations, "Are Scientific Misconduct and Conflicts of Interest Hazardous to Our Health?" (Washington, DC: U.S. Government Printing Office, 1990).
3. Sheldon Krimsky, "Academic Corporate Ties in Biotechnology," *Science, Technology & Human Values* 16, no. 3 (Summer 1991): 275–287.
4. Blumenthal, et al., "University-Industry Research Relationships," 1361–1366.
5. Judith Swazey, "Protesting the "Animal of Necessity": Limits to Inquiry in Clinical Investigations," *Daedalus: Limits to Scientific Inquiry* 107, no. 2 (Spring 1978): 129–145.
6. Dorothy Nelkin, *Science as Intellectual Property* (New York: MacMillan, 1984), 93.

CHAPTER 4

1. C. L. Emerson, "The Contributions of Industry to Scientific Education," in Allen T. Bounell and Ruth C. Christman, eds, *Industrial Science: Present and Future* (Washington, DC: AAAS, 1952), 113–124.
2. Blumenthal, et al., "University-Industry Research Relationships," 1361–1366.
3. Blumenthal, et al., "Industrial Support."
4. Blumenthal, et al., "University-Industry Research Relationships."
5. Ibid.
6. Sheldon Krimsky, *Biotechnics and Society: The Rise of Industrial Genetics* (New York: Praeger, 1991).
7. Lawrence M. Fisher, "Profits and Ethics Collide in a Study of Genetic Coding," *New York Times*, sect. l, January 30, 1994; 1, 16.
8. Steven Benowitz, "The Road to University Technology Licensing Is Littered with Patents that Languish," *The Scientist* (September 18, 1995): l, 11.

9. David Blumenthal, “Growing Pains for New Academic/Industry Relationships,” *Health Affairs* 13, no. 3 (1994): 176–193.
10. Raymond Spier, “Ethical Aspects of the University-Industry Interface,” *Science and Engineering Ethics* 1(1995): 151–162.
11. Carl Djerassi, “Managing Competing Interests: Chastity vs. Promiscuity,” in: *Ethics, Values, and the Promise of Science: Forum Proceedings* (Research Triangle Park, NC: Sigma Xi, The Scientific Research Society, 1993), 31–45.
12. Association of American Medical Colleges, Ad Hoc Committee on Misconduct and Conflict of Interest in Research, “Guidelines for Dealing with Faculty Conflicts of Commitment and Conflicts of Interest in Research,” *Academic Medicine* 65 (1990): 488–496; 491.
13. Arnold S. Relman, “Dealing with Conflict of Interest,” *New England Journal of Medicine* 310 (1984): 1182–1183.
14. Martin Kenney, *Biotechnology: The University-Industrial Complex* (New Haven: Yale University Press, 1986).
15. M. Therese Southgate, “Conflict of Interest and the Peer Review Process [editorial],” *JAMA* 258 (1987): 1375.
16. R. S. Johnson, “Conflict of Interest Issue Heats Up,” *FASEB* J.3 (1989): 2005–2006.
17. Krimsky and Baltimore, “The Ties that Bind or Benefit.”
18. Wil Lepkowski, “University/Industry Research Still Viewed with Concern,” *Chemical & Engineering News* 65 (June 25, 1984): 7–11.
19. Norton D. Zinder and Jackie Winn, “A Partial Summary of Relationships in the United States,” *Recombinant DNA Technical Bulletin* 7 (1984): 8–19.
20. Barbara Culliton, “Biomedical Research Enters the Marketplace.” *New England Journal of Medicine* 20 (1981): 1195–1201.
21. Roger J. Porter and Thomas E. Malone, eds. *Biomedical Research: Collaboration and Conflict of Interest* (Baltimore: Johns Hopkins Press, 1992).
22. Daniel E. Koshland, “Conflicts of Interest Policy [editorial],” *Science* 249 (1990): 109.
23. Daniel E. Koshland, “Conflicts of Interest Policy [editorial],” *Science* 257 (1992): 595.
24. Daniel E. Koshland, “Simplicity and Complexity in Conflict of Interest [editorial],” *Science* 261 (1993): 11.
25. P. J. Friedman, “Scientific Research Conflict of Interest: Policies and Tests,” *FASEB Journal* 5 (1991): 2001.
26. J. Maddox, “Conflict of Interest Declared,” *Nature* 360 (1992): 205.
27. William W. Parmley, “Conflict of Interest: An Issue for Authors and Reviewers [editorial],” *Journal of the American College of Cardiology* 20 (1992): 1017–1018.
28. Jerome P. Kassirer and Marcia Angell, “Financial Conflicts of Interest in Biomedical Research,” *New England Journal of Medicine* 329 (1993): 570–571.
29. Richard Smith, “Conflict of Interest and the *BMJ*,” *The BMJ* 308 (1994): 4–5.

30. U.S. Congress, Committee on Government Operations, Subcommittee on Human Resources and Intergovernmental Relations, *Federal Response to Misconduct in Science: Are Conflicts of Interest Hazardous to our Health?* (Washington, DC: U.S. Government Printing Office, 1989).
31. U.S. Congress, Committee on Government Operations, Subcommittee on Human Resources and Intergovernmental Relations, *Are Scientific Misconduct and Conflict of Interest Hazardous to our Health?* (Washington, DC: U.S. Government Printing Office, 1990).
32. U.S. Department of Health and Human Services, Public Health Service, "Objectivity in Research," *Federal Register* 60 (132) (July 11,1995): 35812.
33. National Science Foundation, "Investigator Financial Disclosure Policy," *Federal Register* 60 (132) (July 11, 1995): 35820–35823.
34. Southgate, "Conflict of Interest."
35. Friedman, "Scientific Research."
36. D. J. Lawrence, "Questions of Authorship and Financial Conflicts of Interest [editorial]," *Journal of Manipulative and Physiological Therapeutics* 13 (1990): 61–62.
37. Drummond Rennie, Annette Flanagin, and Richard M. Glass, "Conflicts of Interest in the Publication of Science [editorial]," *JAMA* 266 (1991): 266–267.
38. Lawrence C. Parish, Joseph A. Witkowski, and Larry E. Millikan, "Conflict of Interest and Scientific Publications," *International Journal of Dermatology* 30 (1991): 250–251.
39. A. Kohn and C. Putterman, "Problems and Conflicts in Peer Review [review]," *International Journal of Impotence Research* 5 (1993): 133–137.
40. Michael D. Witt and Lawrence O. Gostin, "Conflict of Interest Dilemmas in Biomedical Research," *JAMA* 271 (1994): 547–551.
41. U.S. Department of Health and Human Services, "Objectivity in Research."
42. Ibid.
43. Ibid.
44. National Science Foundation, "Financial Disclosure."
45. National Science Foundation, "Investigator Financial Disclosure Policy," *Federal Register* 59 (123) (June 28, 1994): 33308–33312.
46. U.S. Department of Health and Human Services, Food and Drug Administration, "Financial Disclosure by Clinical Investigators," *Federal Register* 59 (183) (September 22, 1994): 48708–48719.
47. George D. Lundberg and Annette Flanagin, "New Requirements for Authors: Signed Statements of Authorship Responsibility and Financial Disclosure [editorial]," *JAMA* 262 (1989): 2003–2004.
48. International Committee of Medical Journal Writers, "Conflict of Interest," *The Lancet* 341 (1993): 742–743.
49. Relman, "Conflict of Interest."
50. Rennie, et al., "Conflicts of Interest."
51. International Committee of Medical Journal Writers, "Conflict of Interest."

52. Southgate, "Conflict of Interest."
53. Friedman, "Scientific Research."
54. Sheldon Krimsky, James Ennis, and Robert Weissman, "Academic-Corporate Ties in Biotechnology: A Quantitative Study," *Science, Technology, & Human Values* 16, no. 3 (1991): 275–287.
55. Ibid.
56. Genetic Engineering News, *The Genetic Engineering News Guide to BiotechnologyCompanies,* (New York: Mary Ann Liebert, 1994).
57. Institute for Scientific Information, *1992 Science Citation Index Journal Citation Reports: A Bibliometric Analysis of Science Journals in the ISI Database* (Philadelphia: Institute for Scientific Information, 1993).
58. Andrew H. Berks, "Patent Information in Biotechnology," *Trends in Biotechnology* 12(1994): 352–38.
59. U.S. Department of Health and Human Services, "Objectivity in Research."
60. National Science Foundation, "Financial Disclosure."
61. U.S. Department of Health and Human Services, "Objectivity in Research." 25, 812.
62. Koshland, "Conflicts of Interest" (1992).
63. A. S. Relman, "New Information for Authors and Readers [editorial]," *New England Journal of Medicine* 323 (1990): 56.
64. Arnold S. Relman, "Information for Authors," *New England Journal of Medicine* 327 (1992): 721.
65. Arnold S. Relman, "Information for Contributors," *Science* 257 (1995): 601.
66. Arnold S. Relman, "Information for Contributors," *Proceedings of the National Academy of Sciencesof the USA*: 93 (1996): 1.
67. Kenneth J. Rothman, "Conflict of Interest: The New McCarthyism in Science," *JAMA* 269 (1993): 2782–2784.
68. Carl Djerassi, "Basic Research: The Gray Zone [policy forum]," *Science* 261 (1993): 972–973.

CHAPTER 5

1. Dennis F. Thompson, "Understanding Financial Conflicts of Interest: Sounding Board" *New England Journal of Medicine* 329(1993): 573–576.
2. International Committee of Medical Journal Editors, "Conflict of Interest: Writing for Publication," *Lancet*, 341(1993): 742.
3. Molla S. Donaldson and Alexander M. Capron, eds., "Patient Outcomes Research Teams: Managing Conflict of Interest," *National Academy Press*, (1991): 61–62.
4. Richard Horton, "Conflict of Interest in Clinical Research: Opprobrium or Obsession?" *Lancet*, 349 (1997): 1112–1113.
5. "Avoid financial "correctness"" [editorial], *Nature*, 385 (1997): 469.
6. Council on Scientific Affairs and Council on Ethical and Judicial Affairs, "Conflicts of Interest in Medical Center/Industry Research Relationships," *JAMA*, 263(1990): 2790–2793.

7. Barbara J. Culliton, "Biomedical Research Enters the Marketplace," *New England Journal of Medicine* 304(1981): 1195–1201.
8. U.S. Congress, Committee on Government Operations, Subcommittee on Human Resources and Intergovernmental Relations. *Are Scientific Misconduct and Conflict of Interest Hazardous to Our Health? Nineteenth Report* (Washington, DC: U.S. Government Printing Office, 1990).
9. Frank Davidoff, "Where's the Bias?" *Annals of Internal Medicine* 126 (1997): 986–988.
10. James L. Bernat, Michael L. Goldstein, Steven P. Ringel, "Conflicts of Interest in Neurology," *Neurology,* 50 (1998): 327–331.
11. Michael S. Wilkes and Richard L. Kravitz, "Policies, Practices, and Attitudes of North American Medical Journal Editors," *Journal of General Internal Medicine* 10(1995): 443–450.
12. "Objectivity in research," *Federal Register,* 60 (1995): 35810–35819.
13. "Investigator Financial Disclosure Policy," *Federal Register* 60(1995): 35820–35823.
14. Sheldon Krimsky, James Ennis, and Robert Weissman, "Academic Corporate Ties in Biotechnology: A Quantitative Study," *Science, Technology, & Human Values,* 16 (1991): 275–287.
15. David Blumenthal, Eric G. Campbell, Nancyanne Causino, and Karen Seashore Louis, "Participation of Life-Science Faculty in Research Relationships with Industry," *New England Journal of Medicine* 335 (1996): 1734–1739.
16. International Committee of Medical Journal Editors, "Uniform Requirements for Manuscripts Submitted to Biomedical Journals," *JAMA* 277 (1997): 927–934.
17. Sheldon Krimsky, L.S. Rothenberg, P. Stott, and G. Kyle, "Financial Interests of Authors in Scientific Journals: a Pilot Study of 14 Publications," *Science and Engineering Ethics* 2(1996): 395–410.
18. Henry Thomas Stelfox, Chua Grace, Keith O'Rourke, and Allan S. Detsky, "Conflict of Interest in the Debate over Calcium Channel Antagonists," *New England Journal of Medicine* 338(1998): 101–106.

CHAPTER 6

1. Hanbury Brown, *The Wisdom of Science: Its Relevance to Culture and Religion* (Cambridge: Cambridge University Press, 1986).
2. Sheldon Krimsky, "Science, Society, and the Expanding Boundaries of Moral Discourse," in *Science, Politics and Social Practice* (Dordrecht: Kluwer Academic Publishers, 1995), 113–128.
3. Porter and Malone, eds. *Biomedical Research,* 49.
4. Ibid.
5. R. C. Lewontin, "The Cold War and the Transformation of the Academy," in: *The Cold War and the University,* edited by N. Chomsky, et al. (New York: The New Press, 1997), 1–34.
6. Ibid.

7. Richard Florida, "The Role of the University: Leveraging Talent, Not Technology," *Issues in Science and Technology* 15, no. 4 (1999): 67–68.
8. David Goodman, Bernardo Sorj, and John Wilkinson, *From Farming to Biotechnology: A Theory of Agro-Industrial Development* (New York: Basil Blackwell, 1987), 45.
9. Al Gore, *Earth in the Balance: Ecology and the Human Spirit* (New York: Penguin Group, 1992), 138.
10. Jack Doyle, *Altered Harvest: Agriculture, Genetics, and the Fate of the World's Food Supply* (New York: Penguin Books, 1985), 255–63.
11. Jack Ralph Kloppenburg Jr, *First the Seed: The Political Economy of Plant Biotechnology* (Cambridge: Cambridge University Press, 1988), 5.
12. Sheldon Krimsky, *Genetic Alchemy: The Social History of the Recombinant DNA Controversy* (Cambridge, MIT Press, 1982), p. 72.
13. Goodman et al., *From Farming to Biotechnology*, 37.
14. Sheldon Krimsky, *Biotechnics and Society: The Rise of Industrial Genetics* (Westport: Greenwood Press Inc., 1991), 30–33.
15. *Nature* 283 (1980), cover page.
16. Alexander G. Bearn, "The Pharmaceutical Industry and Academe: Partners in Progress," *The American Journal of Medicine* 71, no. 1 (1981): 81.
17. "Biotechnology Back in the Limelight," *Nature* 283 (1980), 119.
18. Krimsky, *Biotechnics and Society*, 21–30.
19. Beth Savan, *Science Under Siege: The Myth of Objects in Scientific Research* (Montreal: Canadian Broadcasting Corporation, 1988), 73.
20. Martin Kenney, *Biotechnology: The University-Industrial Complex* (New Haven: Yale University Press, 1988): 35; *see also* Philip J. Hilts, *Scientific Temperaments: Three Lives in Contemporary Science*, (New York: Simon & Schuster, 1982), 185.
21. Frederic Golden, "Shaping Life in the Lab," *TIME*, March 9, 1981.
22. Ibid.
23. Ibid., 50, 52.
24. Ibid., 51.
25. Judith A. Johnson, *Biotechnology Commercialization of Academic Research* (Washington, DC, Library of Congress, 1982).
26. Sheldon Krimsky, James Ennis, and Robert Weissman, "Academics-Corporate Ties in Biotechnology: A Quantitative Study," *Science, Technology, & Human Values* 16 (1991): 275.
27. Kenney, *Biotechnology*, 199.
28. "Biotechnology Back in the Limelight," 119.
29. Krimsky, *Biotechnics and Society*, 66–67.
30. Barbara J. Culliton, "Biomedical Research Enters the Market Place," *New England Journal of Medicine* 304 (1981): 1196, 1197.
31. R. M. Rosenzweig, *The Research Universities and Their Patrons*, (Berkeley: University of California Press, 1982), 15–21.
32. Paul E. Gray, "Advantageous Liaisons," *Issues in Science and Technology* 6, no. 3 (Spring 1990): 40.

33. 15 U.S.C. 0 3701(1994).
34. 35 U.S.C. 00 200–212 (1994).
35. 26 U.S.C. 8 1 (1994).
36. 15 U.S.C. 83701(1994).
37. Porter and Malone, *Biomedical Research*, 216.
38. Barbara Culliton, "NIH, Inc: The CRADA Boom," *Science* 245, no. 4922 (Fall 1989): 1034.
39. Bernadine Healy, "Business Opportunities, and Technology of the Committee on Small Businesses: Hearing on S. Doc. C. No. 103–5," Before the Subcommittee on Regulation House of Rep., 103rd Congress (1993): 70.
40. Ibid.
41. Ibid.
42. Ibid.
43. *Patents and the Constitution: Transgenic Animals Hearings Before the Subcomm. on Courts, Civil Liberties, and the Admin. of Justice, H.R. Comm. on the Judiciary,* 100th Congress (1987).
44. Linda Marsa, *Prescription for Profits* (New York: Scribner, 1997), 202; Azra T. Sayeed, "Consumers Pay Billions for Patent Extensions on Medications," in *The Ownership of Life: When Patents and Values Clash* (Minneapolis: Institute for Agriculture and Trade Policy,1997), 48; see also Jonathan King & Doreen Stabinsky, "Patents on Cells, Genes, and Organisms Undermine the Exchange of Scientific Ideas," *The Chronicle of Higher Education*, (Feb. 5, 1999), B8.
45. Office of Technology Assessment, 101st Congress, "New Developments in Biotechnology," *Patenting Life* no. 5, (Washington, DC: U.S. Government Printing Office, 1989), 7.
46. Ibid.
47. Ibid.
48. 447 U.S. 303 (1980).
49. Ibid., 307.
50. Ibid., 308.
51. Office of Technology Assessment, 97th Congress, *Impacts of Applied Genetics: Micro-Organisms, Plants, and Animals* (Washington, DC: U.S. Government Printing Office, 1981), 240.
52. Sheldon Krimsky, "Patents for Life Forms Sui Generis: Some New Questions for Science, Law and Society," *Recombinant DNA Technical Bulletin* 4, no. 1 (1981): 11.
53. Ibid.
54. Diamond, 447 U.S., 310.
55. Ibid.
56. 35 U.S.C. 101 (1994).
57. See note 54, p. 310.
58. Ibid., 313.
59. Ibid., 307–08.
60. Ibid., 310.

61. Ibid., 313.
62. Ibid.
63. Ibid., 308.
64. Ibid., 313–14.
65. Ibid., 320.
66. *The Code of Codes: Scientific and Social Issues in the Human Genome Project,* edited by Daniel J. Kevles and Leroy Hood (Cambridge: Harvard University Press, 1992), 376.
67. Eliot Marshall, "Companies Rush to Patent DNA," *Science* 275 (1997): 780; see also *To Profit or Not to Profit: The Commercial Transformation of the Non-profit,* edited by Burton A. Weisbrod (Cambridge: Cambridge University Press, 1998): 176–182.
68. Evelyn Fox Keller, "Nature, Nurture and the Human Genome Project," in *The Code of Codes: Scientific and Social Issues in the Human Genome Project,* edited by Daniel J. Kevles and Leroy Hood (Cambridge: Harvard University Press, 1992), 281, 313.
69. Rebecca S. Eisenberg, "Patent Rights in the Human Genome Project," in *Gene Mapping: Using Law and Ethics as Guides* (New York: Oxford University Press, 1992), 226.
70. Ibid., 239.
71. Ibid.
72. Marshall, "Companies Rush," 780–81.
73. Jon F. Merz, Mildred Cho, Madeline Robertson, and Debra Leonard, "Patenting Genetic Tests: Putting Profits Before People," *Gene Watch,* 11, no. 4 (Fall 1998): 1.
74. Ibid.
75. Ibid.
76. Ibid.
77. Marshall, "Companies Rush," 180.
78. Sheldon Krimsky, "The New Corporate Identity of the American University," *Alternatives* 14, no. 2 (May/June 1987): 20.
79. Ibid.
80. Marshall, "Companies Rush," 180.
81. Henry Etzkowitz, "Entrepreneurial Science in the Academy: A Case of the Transformation of Norms," *Social Problems,* 36, no. 1 (February 1989): 14.
82. Krimsky, *Biotechnics and Society,* 28–33.
83. Ibid.
84. Ibid.
85. Kenney, *Biotechnology,* 58.
86. Ibid.
87. Ibid, 67–69.
88. Ibid.
89. Ibid.
90. Ibid.
91. Ibid.

92. Ibid.
93. Krimsky, et al., "Academics-Corporate Ties," 275–87.
94. Ibid.
95. Ibid, 281.
96. Ibid, 282.
97. Ibid, 286.
98. Hilts, *Scientific Temperaments*, 185.
99. David Blumenthal, Nancyanne Causino, Eric Campbell, and Karen Seashore Louis, "Relationship between Academic Institutions and Industry in the Life Sciences-An Industry Survey," *New England Journal of Medicine* 334 (February 1996): 368, 371; see also Blumenthal et al., "University-Industry Research Relationships," 1361.
100. Blumenthal et al., "University-Industry Research Relationships."
101. Ibid.
102. Ibid.
103. Ibid, 1361–66.
104. Blumenthal, Causino, et al., "Relationship between Academic Institutions," 368.
105. Ibid.
106. Ibid.
107. Ibid.
108. Eric G. Campbell, Karen Seashore Louis, and David Blumenthal, "Looking a Gift Horse in the Mouth," *JAMA* 279 no. 13 (April 1998): 995; see also Sheryl Gay Stolberg, "Gifts to Science Researchers Have Strings, Study Finds," *The New York Times*, Apr. 1, 1998, A17.
109. Campbell, et al., "Gift Horse," 995.
110. Ibid.
111. Krimsky and Baltimore, "The Ties that Bind or Benefit," 130.
112. "Biotechnology Back in the Limelight."
113. Zina Moukheiber, "Science for Sale," *Forbes* (May 17, 1999): 136; David Shenk, "Money + Science = Ethics Problems on Campus," *The Nation*, (March 22, 1999): 11; Daniel Zalewski, "Bind: Do Corporate Dollars Strangle Scientific Research?" *Lingua Franca*, 7 (June/July 1997): 51.
114. Marsa, *Prescription for Profits*, 199–222.
115. Ibid, 201.
116. Christopher Anderson, "Scandal Scars Minnesota Medical School," *Science* 262 no. 5141 (1993): 1812.
117. Lawrence C. Soley, *Leasing the Ivory Tower: The Corporate Takeover of Academia* (Boston: South End Press, 1995), 49.
118. Ibid.
119. Porter and Malone, *Biomedical Research*, 27.
120. S. Krimsky, L.S. Rothenberg, P. Stott, and G. Ryle. "Financial Interest of Authors in Scientific Journals: A Pilot Study of 14 Publications," *Science & Engineering Ethics* 2 (1996): 395–410.
121. Ibid.

122. Ibid.
123. Ibid.
124. Ibid.
125. Ibid.
126. Kenneth J. Rothman, "Conflict of Interest: The New McCarthyism in Science," *JAMA* 269, no. 21 (1993): 2782.
127. Ibid.
128. Ibid.
129. *House Report*, no. 101–688 (1990).
130. "Avoid Financial "Correctness,'" *Nature* 385 (1997): 469.
131. See note 120, p. 395. Krimsky, et al., "Financial Interest," 395.
132. Ibid.
133. R. A. Davidson, "Source of Funding and Outcomes of Clinical Trials," *Journal of General Internal Medicine* 1, no. 3 (May/June 1986): 155.
134. Stelfox, et al., "Conflict of Interest," 101.
135. Ibid.
136. Ibid.
137. Ibid.
138. Ibid.
139. Christopher Anderson, "Conflict Concerns Disrupt Panels, Cloud Testimony," *Nature* 355 (1992): 753.
140. Leon R. Kass, *Toward a More Natural Science: Biology and Human Affairs* (New York: The Free Press, 1985), 285.
141. Ibid., 231.
142. Eliot Marshall, "Is Data-Hording Slowing the Assault on Pathogens?" *Science* 275, no. 5301 (1991): 797.
143. Seth Shulman, *Owning the Future* (Boston, Houghton Mifflin Co., 1999), 33–36.
144. Ibid.
145. Ibid.
146. Ibid., 36.
147. Marshall, "Companies Rush."
148. Gerard O'Neill, et al., "Public Handouts Enrich Drug Makers, Scientists," *Boston Globe*, April 6, 1998, Al; Gerald O'Neill, et al., "Tax Dollars Fuel University Spinoffs," *Boston Globe* April 7, 1998, Al, A24.
149. Ibid.
150. Ibid.
151. Jonathan King, "Gene Patents Retard the Protection of Human Health," *Gene Watch* (Oct. 1996): 10.
152. Merz, et al., "Patenting Genetic Tests," 4.
153. Ibid.
154. Gary Taubes, "Scientists Attacked for 'Patenting' Pacific Tribe," *Science* 270 (1995): 1112.
155. Vandana Shiva, *Monocultures of the Mind: Perspectives on Biodiversity and Biotechnology* (London, New Jersey: Zed Books, 1993), 151–84.

156. Ibid.
157. George J. Annas, *Standard of Care the Law of American Bioethics* (New York: Oxford Press, 1993), 167–80; see also Beth Burrows, "Second Thoughts about U.S. Patent #4," *Gene Watch (Council for Responsible Genetics, Cambridge, Mass.)* (October 1996): 4.
158. Moore v. Regents of the Univ. of California, 51 Cal. 3d 120 (Cal. 1990).
159. Ibid., 141–42.
160. Ibid., 142.
161. Ibid., 137, 142.
162. Robert K. Merton, *Social Theory and Social Structure* (New York: Free Press, 1957), 595.
163. Ibid.
164. *Science in Context*, edited by Barry Barnes & David Edge (Cambridge: MIT Press, 1982), 16.

CHAPTER 7

1. Jacob Bronowski, *Science and Human Values* (New York, NY: Harper; 1965), 59.
2. Relman, "Conflicts of Interest," 1182–1183.
3. M. Therese Southgate, "Conflict of Interest and the Peer Review Process," *JAMA* 258 (1987): 1375.
4. Richard M. Glass and Mindy Schneiderman, "A Survey of Journal Conflict of Interest Policies," Presentation at the Third International Congress on Peer Review in Biomedical Publication, Prague, Czech Republic, September 18, 1997.
5. Rothman, "The New McCarthyism," 2782–84.
6. Frank Davidoff, "Where's the Bias?" *Annals of Internal Medicine* 126 (1997): 986–988.
7. Kenneth J. Rothman, "Journal Policies on Conflict of Interest," *Science* 261 (1993): 1661–8.
8. "Avoid Financial "Correctness,"" 469.
9. *Federal Response to Misconduct in Science: Are Conflicts of Interest Hazardous to Our Health?* (Washington, DC: U.S. Government Printing Office; 1989), A report of the 100th Congress, 2nd session, September 29, 1988.
10. *Are Scientific Misconduct and Conflicts of Interest Hazardous to Our Health?* (Washington, DC: U.S. Government Printing Office; 1990), A report of the 101st Congress, 2nd session, September 10, 1990.
11. Alan L. Hillman, Mark V. Pauly, and Joseph J. Kerstein, "How Do Financial Incentives Affect Physicians' Clinical Decisions and the Financial Performance of Health Maintenance Organizations?" *New England Journal of Medicine* 321 (1989): 86–92.
12. Mary-Margaret Chren and C. Seth Landefeld, "Physicians' Behavior and Their Interactions with Drug Companies," *JAMA* 271, no. 9 (1994): 684–689.
13. Richard A. Davidson, "Source of Funding and Outcome of Clinical Trials," *Journal of General Internal Medicine* 1 (1986): 155–158.

14. Stelfox, et al., "Conflict of Interest," 101–108.
15. Richard Smith, "Beyond Conflict of Interest," *BMJ* 317 (1998): 291.
16. Mark Friedberg, Bernard Saffran, Tammy J. Stinson, Wendy Nelson, and Charles L. Bennett, "Evaluation of Conflict of Interest in Economic Analyses of New Drugs Used in Oncology," *JAMA* 282 (1999): 1453–1457.
17. Louise B. Russell, Marthe R. Gold, Joanna E. Siegel, Norman Daniels, and Milton C. Weinstein, "The Role of Cost-Effectiveness Analysis in Health and Medicine," *JAMA* 276 (1996): 1172–1177.
18. American Public Health Association, American Water Works Association, Water Environmental Federation, *Standard Methods for the Examination of Water and Waste Water, 19th edition.* (Washington, DC: American Public Health Association, 1996).
19. "Clinical Resources, Practice Guidelines," American Psychiatric Association, APA Online, accessed September 17, 1999, http://www.psych.org.
20. William L. Simonich, "Cost-Effectiveness Analysis," *New England Journal of Medicine* 332 (1995): 124.
21. Anne M. PausJenssen and Allan S. Detsky, "Guidelines for Measuring the Costs and Consequences of Adopting New Pharmaceutical Products: Are They on Track?" *Medical Decision Making* 18 (1998): S19–S22.
22. Jerome P. Kassirer and Marcia Angell, "The Journal's Policy on Cost-Effectiveness Analyses," *New England Journal of Medicine* 331 (1994): 669–670.

CHAPTER 8:

1. Dennis F. Thompson, "Understanding Financial Conflicts of Interest," *New England Journal of Medicine* 329 (1993): 573–576.
2. International Committee of Medical Journal Editors, "Conflicts of Interest," *The Lancet* 341(1993): 742–743.
3. International Committee of Medical Journal Editors, "Uniform Requirements for Manuscripts Submitted to Biomedical Journals," *JAMA* 277 (1997): 927–934.
4. Richard Smith, ed., "Beyond conflict of interest: editorial," *British Medical Journal,* 317 (1998): 291.
5. Frank Davidoff, "Where's the Bias?," *Annals of Internal Medicine* 126 (1997): 986–988.
6. Stelfox, et al., "Conflict of Interest," 101–106.
7. Howard Frumkin, "Review of *Living Downstream,*" *New England Journal of Medicine* 338 (1998): 268.
8. "Omission of Financial Disclosure Information: Corrections," *JAMA* 281 (1999): 1174.
9. Michael Day, "Salt and Vitriol: News," *New Scientist* 159 (August 22, 1998): 4.
10. Ralph T. King Jr., "Did Ties to Alzheimer's Test Maker Sway NIH Report?" *Wall Street Journal* (December 1, 1998): B1.
11. Ralph T. King Jr., "Keyhole Heart Surgery Arrived with Fanfare, But Was It Premature?" *Wall Street Journal* (May 5, 1999): A1.

12. Terence Monmaney, "Medical Journal's Article Raises Conflict Concerns," *Los Angeles Times,* September 28, 1999, A1.
13. Krimsky, et al., "Financial Interests," 395–410.
14. *1996 Science Citation Index—Journal Citation Reports: A Bibliometric Analysis of Science Journals in the ISI Database* (Philadelphia, PA: Institute for Scientific Information, 1997), 10.
15. Ibid, 11–12.
16. Krimsky, et al., "Financial Interests."
17. Blumenthal, et al., "Academic Institutions and Industry," 368–373.
18. Kurt Eichenwald and Gina Kolata, "Hidden Interest: When Physicians Double as Entrepreneurs," *New York Times,* November 30, 1999, A1, C16.
19. Sheryl G. Stolberg, "Biomedicine is Receiving New Scrutiny as Scientists Become Entrepreneurs," *New York Times,* February 20, 2000, A20.
20. Dennis Cauchon, "FDA Advisers Tied to Industry," *USA TODAY,* September 25, 2000, 1.

CHAPTER 9

1. Sheila Slaughter and Larry L. Leslie, *Academic Capitalism: Politics, Policies and the Entrepreneurial University* (Baltimore: Johns Hopkins Univ. Press, 1997); Daniel G. Greenberg, *Science, Money and Politics* (Chicago: University of Chicago Press, 2001); Sheldon Krimsky, *Science in the Private Interest* (Lanham, MD: Rowman & Littlefield, 2003); Derek Bok, *Universities in the Marketplace* (Princeton: Princeton University Press, 2003); Jennifer Washburn, *University Inc.* (New York: Basic Books, 2005); Burton A. Weisbrod (ed), *To Profit or Not To Profit: The Commercial Transformation of the Nonprofit Sector* (Cambridge, UK: Cambridge University Press, 1998); Donald G. Stein (ed), *Buying In or Selling Out? The Commercialization of the American Research University* (New Brunswick, NJ: Rutgers University Press, 2004).
2. *Academe,* "Annual Report on the Economic Status of the Profession, 2004–5," March/April 2005.
3. Science and Engineering Indicators—Academic Research and Development Expenditures. Chapter 6, Academic Research and Development: Financial and Personnel Resources, Support for Graduate Education and Outputs. Washington, DC: National Science Foundation. http://www.nsf.gov/statistics/seind00/c6/c6s1.htm. Accessed December 30, 2005.
4. Stanley Aronowitz, *The Knowledge Factory* (Boston: Beacon Press, 2000).
5. National Science Foundation, *Science and Engineering Indicators—Academic Research and Development Expenditures.* Chapter 6, "Academic Research and Development: Financial and Personnel Resources. Support for Graduate Education and Outputs." Appendix Table 4–4. U.S. inflation-adjusted R&D expenditures, by performing sector and source of funds. 1953–2000 (millions of constant 1996 dollars). (Washington, DC: National Science Foundation, 2000). http://www/nsf.gov/statistics/seind00/frames.htm Accessed December 30, 2005.

6. Sheldon Krimsky, "Reforming Research Ethics in an Age of Multivested Science," *in: Buying in or Selling Out?* Donald G. Stein, ed. (Rutgers, NJ: Rutgers University Press, 2004), 133–152.
7. Ibid., 136.
8. Ronald B. Sandler, "Academic Freedom in the USA." http://www.rbs2.com/afree.htm. Accessed December 30, 2005.
9. Robert Merton, *Social Structure and Social Theory* (Glencoe, IL: The Free Press, 1957), 541.
10. "[Autonomy] is infringed on when scientists are unable to discuss, publish, or circulate their work to other scientists interested in the same or in related problems. It is infringed on when scientists are unable to leave their country or to enter another country to attend a scientific congress because the government in the country from they came or to which they wish to go is concerned about their ideological adequacy." Edward A. Shils, "The autonomy of science," in: *The Sociology of Science*, B. Barber and W. Hirsch (eds). (Wesport, CT: Greenwood Press, 1962).
11. Chris Mooney, *The Republican War on Science* (New York: Basic Books, 2005).
12. Merton, 1957, 543.
13. R.K. Merton, "Science and the Social Order," *Philosophy of Science* 5, no. 3 (July 1938): 328.
14. Goldie Blumestyk, "The Story of Syngenta & Tyrone Hayes at UC Berkeley: The Price of Research," *Chronicle of Higher Education* 50: (Oct. 31, 2003); Jonathan Knight, "Scientists Attack Industrial Influence," *Nature* 426 (December 18/25, 2003): 741.
15. "MIT's Lincoln Laboratory received more than $70 million this year from the Missile Defense Agency for work on missile defense. Postol has accused researchers at the facility of being complicit in fudging a report on a missile defense test in 1998. And he has charged that MIT is covering up the cover up. The dispute's final layer is Postol's claim that MIT retaliated against him by removing him from the institute's Security Studies Program, by increasing his program's overhead costs, and by attempting to move him to offices outside the MIT campus." Charles P. Pierce, "Going Postol," *The Boston Globe Magazine* October 23, 2005, 32.
16. Merton, 1938, 327.
17. Merton, 1938, 328.
18. Norman W. Storer, *The Social System of Science* (New York: Holt, Rinehart and Winston, 1966), 79.
19. Storer, 1966, 102.
20. Storer, 1966, 80.
21. Kenneth J. Rothman, "The New McCarthyism," 2782.
22. Office of Government Ethics, "Interpretation, Exemptions and Waiver Guidance Concerning 18 U.S.C. 208 (Acts Affecting Personal Financial Interest)," *Federal Register* 61, no. 244 (Dec. 18, 1996): 66830–66851.

23. Sang-Hun Choe and Nicholas Wade, "Korean Cloning Scientist Quits Over Report He Faked Research," *New York Times*, December 24, 2005, A1.
24. Karl Popper, *The Logic of Scientific Discovery* (New York: Harper & Row, 1959), 42.
25. L. S. Friedman and E. D. Richter, "Relationship between Conflict of Interest and Research Results," *Journal of General Internal Medicine* 19 (2004): 54.27. R.A.
26. Davidson, "Sources of Funding and Outcomes of Clinical Trials," *Journal of General Internal Medicine* 1 (1986): 155–158.
27. Bekelman et al., "Scope and Impact," 463.
28. John Ziman, *Real Science* (Cambridge, UK: Cambridge University Press, 2000), 74.
29. Ziman, *Real Science*, 174. "The ethical code supporting the norm of disinterestedness cannot stand up to the external pressures to exploit the ever-growing instrumental power of science." (p. 162).
30. Ibid, 174. "The trick is to nullify these individual interests by setting them against one another. In effect, the scientific ethos delineates an agonistic arena, where a hidden melodrama of clashing ego is transformed into a apparently dispassionate intellectual debate. As in a free commercial market, the particular bias of each individual in neutralized in the collective outcome." Ziman, p. 159.
31. Mathias Adam, "Promoting Disinterestedness or Making Use of Bias?: Interests and Moral Obligations in Commercialized Research," Forthcoming in: M. Carrier, D, Howard, J. Kourany (eds), *The Challenge of the Social and the Pressure of Practice: Science and Values Revisited,* University of Pittsburgh Press (since published in 2008).
32. Craig Calhoun, "Is the University in Crisis?" *Society* (May/June 2006), 9–18.

CHAPTER 10

1. Bekelman et al., "Scope and Impact," 454–465.
2. David O. Antonuccio, William G. Danton, Garland Y. DeNelsky, Roger P. Greenberg, and James S. Gordon, "Raising Questions about Antidepressants," *Psychotherapy Psychosomatics*; 68 (1999): 3–14.
3. Miles Little, "Research Ethics and Conflicts of Interest," *Journal of Medical Ethics* 25 (1999): 259–262.
4. Rose Gutfield, "Panel Urges FDA to Act on Adviser Bias," *Wall Street Journal* (Dec 9, 1992), B6.
5. George Monbiot, "Guard Dogs of Perception: Corporate Takeover of Science," *Science and Engineering Ethics* 9 (2003): 49–57.
6. Sheldon Krimsky, "The Redemption of Federal Advisory Committees" in *Science in the Private Interest* (Lanham: Rowman & Littlefield, 2003).
7. Sara Schroter, Julie Morris, Samena Chaudhry, and Helen Barratt, "Does the Type of Competing Interest Statement Affect Readers' Perceptions of the Credibility of Research? Randomised Trial," *BMJ* 328 (2004): 742–743.

8. Sheryl Gay Stolberg, "Study Says Clinical Guides Often Hide Ties of Doctors," *New York Times*, February 6, 2002, A17.
9. Giovanni A. Fava, "Conflict of Interest in Psychopharmacology: Can Dr. Jekyll Still Control Mr. Hyde?" *Psychotherapy Psychosomatics* 73 (2004): 1–4.
10. Karen Eriksen and Victoria E. Kress: *Beyond the* DSM *Story: Ethical Quandaries, Challenges, and Best Practices* (Thousand Oaks: Sage, 2005), x.
11. Thomas Bodenheimer, "Uneasy Alliance: Clinical Investigators and the Pharmaceutical Industry," *New England Journal of Medicine* 342 (2000): 1539–1544.
12. Marcia Angell, "Is Academic Medicine for Sale?" *New England Journal of Medicine* 342 (2000): 1516–1518.
13. B. Goldstein, "APA-pharmaceutical Relations," *Psychiatry News* 38 (2003): 35.
14. James H. Scully, "Advertising Revenue Helps APA Meet its Objectives," *Psychiatry News* 39 (2004): 4.
15. American Psychiatric Association, *Diagnostic and Statistical Manual of Mental Disorders:* DSM-IV, *Edition 4, Text Revision*, (Washington: American Psychiatric Association, 2000).
16. Rachel T. Hare-Mustin and Jeanne Marecek, "Abnormal Psychology and Clinical Psychology: The Politics of Madness," in *Critical Psychology: An Introduction* (Thousand Oaks: Sage, 1997), 104–120.
17. "Mental health practitioners and trainees," *Mental Health* (Rockville, MD: USDHHS Substance Abuse and Mental Health Services Administration, Center for Mental Health Services, United States, 2002), accessed February 16, 2005, http://www.mentalhealth.samhsa.gov/publications/allpubs/SMA04-3938/Chapter21.asp.
18. Lisa Cosgrove and Bethany Riddle, "Constructions of Femininity and Experiences of Menstrual Distress," *Women Health* 38 (2003): 37–58.
19. Krimsky, et al., "Financial Interests," 395–410.
20. Teddy D. Warner and John P. Gluck, "What Do We Really Know About Conflicts of Interest in Biomedical Research?" *Psychopharmacology* 171 (2003): 36–46.
21. Elizabeth A. Boyd and Lisa A. Bero, "Assessing Faculty Financial Relationships with Industry: A Case Study," *JAMA* 284 (2000): 2209–2214.
22. Marcia Angell, *The Truth about Drug Companies*, (New York: Random House, 2004), 142.
23. Sheldon Krimsky and L. S. Rothenberg, "Conflict of Interest Policies in Science and Medical Journals: Editorial Practices and Author Disclosures," *Science and Engineering Ethics* 7 (2001): 205–218.
24. "Leading Therapy Classes by Global Pharmaceutical Sales, 2004," accessed March 24, 2005, www.imshealth.com/ims/portal/front/articleC/0,2777,6599_7123402471234109,00.html.
25. "Leading Products by Global Pharmaceutical Sales, 2004," accessed March 24, 2005, www.imshealth.com/ims/portal/front/articleC/0,2777,6599_71234024_71234109,00.html.

26. "Global Market Forecast," accessed March 24, 2005, www.woodmac.com /pdf/pharmaquantpluspressrelease.pdf.
27. "Editorial: Experts and the Drug Industry," *New York Times*, March 4, 2005.
28. Mildred K. Cho, Ryo Shohara, Anna Schissel, and Drummond Rennie, "Policies on Faculty Conflict of Interest at U.S. Universities," *JAMA* 284 (2000): 2203–2208.
29. Drummond Rennie, "Fair Conduct and Fair Reporting of Clinical Trials," *JAMA*; 282 (1999): 1766–1768.
30. David Korn, "Conflicts of Interest in Biomedical Research," *JAMA* 284 (2000): 2234–2237.
31. Vladan Starcevic, "Opportunistic 'Rediscovery' of Mental Disorders by the Pharmaceutical Industry," *Psychotherapy Psychosomatics* 71 (2002): 305–310.
32. Giovanni A. Fava: "Conflict of Interest and Special Interest Groups: The Making of a Counter Culture," *Psychotherapy Psychosomatics* 70 (2001): 1–5.

CHAPTER 11

1. Marcia Angell and Jerome P. Kassirer, "Editorials and Conflicts of Interest," Editorial *New England Journal of Medicine* 335 (1996): 1055; Catherine D. DeAngelis, Phil B. Fontanarosa, and Annette Flanagin, "Reporting Financial Conflicts of Interest and Relationships between Investigators and Research Sponsors," Editorial *JAMA* 286 (2001): 89.
2. Dennis F. Thompson, "Understanding Financial Conflicts of Interest," *New England Journal of Medicine* 329 (1993): 573.
3. Diana Crane, *Invisible Colleges: Diffusion of Knowledge in Scientific Communities* (Chicago: University of Chicago Press, 1972), 35.
4. John Herman Randall, *The Making of the Modern Mind* (Boston: Houghton Mifflin, 1976).
5. Henry Margenau, *Ethics and Science* (Princeton, NJ: D. Van Nostrand, 1964).
6. Jacob Bronowski, *Science and Human Values* (New York: Harper & Row, 1965), p. 70.
7. Merton, *Social Theory and Social Structure.*
8. Andrew Stark, *Conflict of Interest in American Public Life* (Cambridge, MA: Harvard University Press, 2000).
9. Ibid., 123.
10. 18 U.S.C. §208(a)-(b).
11. 5 C.F.R. §2640.103(a).
12. Dennis Cauchon, "FDA Advisors Tied to Industry," *USA Today*, 25 September 2000, IA, 10A.
13. *Objectivity in Research*, 42 C.F.R. §50.601–7 (1995).
14. The norms of science cited by Robert Merton are communalism (or communism), disinterestedness, organized skepticism, and universalism. See: Merton, *Social Theory and Social Structure*, 552—61.

15. One editor reported removing a previously published paper from all citation indexes after he learned the author failed to report a conflicting interest. Griffith Edwards, "Addiction's Decision to Withdraw a Published Paper from Citation on the Grounds of Undisclosed Conflict of Interest" (2002) *Addiction 97*, 756.
16. Jennifer Washburn, "Informed Consent," *Washington Post*, December 20, 2001, W16.
17. Ibid, W23.
18. *Moore v. The Regents of the University of California*, 793 P.2d 479 (Cal. 1990).
19. John Ziman, *Real Science* (Cambridge: Cambridge University Press, 2000), p. 162.
20. Ibid., 174.
21. Ibid., 171.
22. Ibid., 174.
23. Richard C. Lewontin, Steven Rose, and Leon J. Kamin, *Not in Our Genes* (New York: Pantheon Books, 1984), 107.
24. "Avoid Financial 'Correctness,'" 469.
25. "Declaration of Financial Interests," Editorial, *Nature* 412 (2000): 751.
26. Ibid.
27. Stelfox, et al., "Conflict of Interest," 101–108.
28. Ibid., 101.
29. Lise L. Kjaergaard and Bodil Als-Nielsen, "Association Between Competing Interests and Authors' Conclusions: Epidemiological Study of Randomized Clinical Trials Published in BMI" *BMJ* 325 (2002): 249; John Yaphe, et al., "The Association Between Funding by Commercialized Interests and Study Outcome in Randomized Controlled Drug Trials" *Family Practice* 18 (2001).
30. Mark Friedberg, et al., "Evaluation of Conflict of Interest in Economic Analyses of New Drugs Used in Oncology," *JAMA* 282 (1999): 1453; Sheldon Krimsky, "Conflict of Interest and Cost-Effectiveness Analysis," *JAMA* 282 (1999): 1474.
31. Christina Turner and George J. Spilich, "Research into Smoking or Nicotine and Human Cognitive Performance: Does the Source of Funding Make a Difference?" *Addiction* 92 (1997): 1423.
32. Marcia Angell, Testimony (Plenary Presentation at the Conference on Human Subject Protection and Financial Conflicts of Interest, Department of Health and Human Services, National Institutes of Health, Bethesda, MD, 15–16 August 2000), 36.
33. Bekelman et al., "Scope and Impact," 454.
34. Ibid., 463.
35. Deborah E. Barnes and Lisa A. Bero, "Why Review Articles on the Health Effects of Passive Smoking Reach Different Conclusions," *JAMA* 279 (1998): 1566; Deborah E. Barnes and Lisa A. Bero, "Industry-Funded Research and Conflict of Interest: An Analysis of Research Sponsored by

the Tobacco Industry Through the Center for Indoor Air Research" *J. Health Policy* 21 (1996): 516.
36. Sheldon Rampton and John Stauber, *Trust Us, We're Experts* (New York: Putnam, 2001); Marvin S. Legator, "Industry Pressures on Scientific Investigators," *International Journal of Occupational and Environmental Health* 4 (1998): 133.
37. Marion Nestle, *Food Politics* (Berkeley: University of California Press, 2003), 118.
38. "Safeguards at Risk: John Graham and Corporate America's Back Door to the Bush White House" (Washington, DC: Public Citizen, 2001).
39. Brian C. Martinson, Melissa S. Anderson, and Raymond DeVries, "Scientists Behaving Badly," *Nature* 435 (9 June 2005): 737–78.
40. Ziman, *Real Science*, 175.
41. Sheldon Krimsky, *Science in the Private Interest: Has the Lure of Profits Corrupted Biomedical Research?* (Lanham, MD: Rowman & Littlefield, 2003), 227.
42. Lisa A. Bero, "Accepting Commercial Sponsorship: Disclosure Helps—But Is Not a Panacea," *BMJ* 319 (1999): 653.

CHAPTER 12

1. Sheldon Krimsky, "The Transformation of the American University, " *Alternatives* 14 (May/June 1987): 20–29; Sheldon Krimsky, "University Entrepreneurship and Public Purpose," in *Biotechnology: Professional Issues and Social Concerns*, D. Deforest, et al., eds. (Washington, DC: American Association for the Advancement of Science, 1988); Sheldon Krimsky, "Science and Wall Street," chapter 4 in *Biotechnics and Society* (New York: Praeger, 1991), 64–65.
2. National Council of University Research Administrators (NCURA) and the Industrial Research Institute, Guiding Principles for University-Industry Endeavors, April 2006, http://www7.nationalacademies.org/guirr/Guiding_Principles.pdf.
3. Sen. John Cornyn (R-TX) and Sen Joseph Lieberman (D-CT) introduced the *Federal Research Public Access Act of 2006*. S. 2695, which requires federal agencies that fund over $100 million in annual external research to make electronic manuscripts of journal articles stemming from their research publicly available via the Internet within six months of publication.
4. David B. Resnik, "Industry-Sponsored Research: Secrecy Versus Corporate Responsibility," *Business and Society Review* 99 (1999): 32.
5. Krimsky, *Biotechnics and Society,* 38.
6. Drummond Rennie, "Thyroid Storm," *Journal of the American Medical Association* 227 (April 16, 1997): 1238–43.
7. Henry Miller, "Score 2 for Academic Freedom," *Chronicle of Higher Education* 54 (October 2, 2007): 107.
8. Resnik, "Industry-Sponsored Research," 31.

9. Stephanie Saul, "Merck Wrote Drug Studies for Doctors," *New York Times*, April 10, 2008.
10. Edward S. Herman and Robert J. Rutman, "University of Pennsylvania's CB Warfare Controversy," *BioScience* 17, no. 8 (August 1967): 526–9.
11. Charles Weiner, unpublished research notes and personal communications.
12. Paul Thacker, "Vote Postponed on Tobacco Research Ban," *Inside Higher Education* (January 19, 2007).
13. T. Zeltner, D. A. Kessler, A. Martiny, F. Randera, *Tobacco Company Strategies to Undermine Tobacco Control Activities of the World Health Organization,* World Health Organization Report of the Committee of Experts on Tobacco Industry Documents, The International Agency for Research on Cancer (IARC), July 2000.
14. Ibid.
15. David Grimm, "Is Tobacco Research Turning over a New Leaf?" *Science* 307 (January 7, 2005): 37.
16. Turner and Spilich, "Research into Smoking or Nicotine," 1423–6.
17. Gardiner Harris, "Cigarette Company Paid for Lung Cancer Study," *New York Times*, March 26, 2008.
18. Joan Wallach Scott, "Academic Freedom and Rejection of Research Funds from Tobacco Corporations," American Association of University Professors, Report of Committee A. 2002–3, *Academe* 89, no. 5 (September-October 2003). http://www.aaup.org/AAU comm/rep/A/2002–03Comm-A-report.htm.
19. Grimm, "Is Tobacco Research Turning?" 36.
20. Donna Euben (staff council), American Association of University Professors, "Legal Issues in Academic Research," paper presented at the 13th Annual Conference on Legal Issues in Higher Education, October 6, 2003. http://www.aaup.org/AAUP/protect/legal/topics/researchissues.htm.
21. Alan Charles Kors and Harvev A. Silvergate, *The Shadow University: The Betrayal of Liberty on America's Campuses* (New York: The Free Press, 1998), 140.
22. Scott Glassman, "Pioneer Sparks Debate at Delaware," *Daily Pennsylvanian*, September 23, 1994.
23. The American Legacy Foundation states in its instructions under the Small Innovative Grants Program, "Legacy will not award a grant to any applicant that is in current receipt of any grant monies or in-kind contribution from any tobacco manufacturer distributor, or other tobacco-related entity," http://www.americanlegacy.org/1710.aspx.
24. Resolution of the University of California's Academic Senate, in Grimm, "Is Tobacco Research Turning?" 37.
25. Susan Wright, ed., *Preventing a Biological Arms Race* (Cambridge, MA: The MIT Press, 1991).
26. John W Servos, "The Industrial Relations of Science: Chemical Engineering At MIT: 1900–1939," *ISIS* 71, no. 4 (December 1980): 547.

27. See http://workgroups.Clemson.edu.
28. Colin MacIlwain, "Carbon Sequestration Gains Support," *Nature* 407 (October 26, 2000): 932.
29. Sylvia Wright, "Chevron Fuels $25 M Alternative Energy Endeavor," *Dateline UC Davis* September 22, 2006.
30. Andrew C. Revkin, "Exxon-led Group Is Giving a Climate Grant to Stanford," *New York Times*, November 21, 2002.
31. Sheldon Krimsky, *Science in the Private Interest* (Lanham, MD: Rowman & Littlefield, 2003), 36.
32. Master Agreement between BP Technology Ventures Inc. and The Regents of the University of California, November 9, 2007, http://www.stopbp-berkeley.org/docs/FINAL_Execution_11-19.pdf.
33. Rick DelVecchio, "UC Faculty to Join Talks on Big BP Biofuels Deal," *San Francisco Chronicle*, March 31, 2007.
34. See http://www.stopbp-berkeley.org/corporate.html.
35. See http://EHP.niehs.nih.gov/cfi.pdf.
36. Scott, "Academic Freedom."

CHAPTER 13

1. Niteesh K. Choudhry, Henry Thomas Stelfox and Allan S. Detsky, "Relationships between Authors of Clinical Practice Guidelines and the Pharmaceutical Industry," *JAMA* 287 (2002): 612–617.
2. Stephen J. Genuis, "The Proliferation of Clinical Practice Guidelines: Professional Development or Medicine-by-Numbers?" *Journal of the American Board of Family Practice* 18 (2005): 419–425.
3. Roberto Grilli, Nicola Magrini, Angelo Penna, Giorgio Mura, and Alessandro Liberati, "Practice Guidelines Developed by Specialty Societies: The Need for Critical Appraisal," *The Lancet* 355 (2000): 103–106.
4. Bodil Als-Nielsen, Wendong Chen, Christian Gluud, and Lise L. Kjaergard, "Association of Funding and Conclusions in Randomized Drug Trials," *JAMA* 290 (2003): 921–928.
5. Giovanni A. Fava, "Financial Conflicts of Interest in Psychiatry," *World Psychiatry* 6, no. 1 (2007): 9–24.
6. Benedict Carey and Gardiner Harris, "Psychiatric Association Faces Senate Scrutiny over Drug Industry Ties," *New York Times*, July 12, 2008, A13.
7. Alison Bass, *Side Effects* (Chapel Hill: Algonquin Books, 2008).
8. www.forensic-psych.com (accessed August 21, 2008).
9. A. Herelin, "What is the Impact of Financial Conflicts of Interest on the Development of Psychiatry?" *World Psychiatry* 6, no. 3 (2007): 6–37.
10. Giovanni A. Fava, "The Intellectual Crisis of Psychiatric Research," *Psychotherapy and Psychosomatics* 7, no. 5 (2006): 202–208.
11. Vladan Starcevic, "Opportunistic 'Rediscovery' of Mental Disorders by the Pharmaceutical Industry," *Psychotherapy and Psychosomatics* 71 (2001): 305–310.

12. Eduard Vieta, "Psychiatry: From Interest in Conflicts to Conflicts of Interest," *World Psychiatry* 6 (2007): 27–29.
13. Harold J. Bursztajn , Richard I. Feinbloom, Robert M. Hamm, and Archie Brodsky, *Medical Choices, Medical Chances: How Patients, Families, and Physicians Can Cope with Uncertainty* (New York: Delacorte Press/ Seymour Lawrence, 1981; New York: Routledge, 1990).
14. Lisa Cosgrove, Sheldon Krimsky, Manisha Vijayraghavan, and Lisa Schneider, "Financial Ties between DSM-IV Panel Members and the Pharmaceutical Industry," *Psychotherapy and Psychosomatics* 75 (2006): 154–160.
15. American Psychiatric Association, "APA names DSM-V Task Force Members: Leading Experts to Revise Handbook for Diagnosing Mental Disorders," (press release), Washington, DC, APA, July 23, 2007.
16. American Psychiatric Association (APA), "Meet the Work Groups." www.psych.org/MainMenu/Research/DSMIV/DSMV/ WorkGroups.aspx (accessed June 20, 2008).
17. National Institute of Mental Health, "The Numbers Count: Mental Disorders in America." http://www.nimh.nih.gov/health/publications/the-numbers-count-mental-disorders-in-america/index.shtml (accessed August 20, 2008).
18. Cannen Moreno, Gonzalo Laje, Carlos Blanco, Huiping Jiang, Andrew Schmidt, and Mark Olfson, "National Trends in the Outpatient Diagnosis and Treatment of Bipolar Disorder in Youth," *Archives of General Psychiatry* 64 (2007): 1032–1039.
19. IMS Health (now Quntiles IMS), "2007 Top Therapeutic Classes of Drugs by U.S. Sales," www.imshealth.com (accessed August 21, 2008).
20. Lisa Cosgrove and Harold J. Bursztajn, "Towards Credible Conflict of Interest Policies in Psychiatry," *Psychiatric Times* 26 (2009): 40–41.
21. Stelfox, et al., "Conflict of Interest," 101–106.
22. Sheldon Krimsky, *Science in the Private Interest* (Lanham, MD: Rowman & Littlefield, 2003).
23. Marcia Angell, "The Truth about Drug Companies: How They Deceive Us and What to Do about It" (New York: Random House, 2004).
24. Jerry Avorn, "Dangerous Deception—Hiding the Evidence of Adverse Drug Effects," *New England Journal of Medicine* 355 (2006): 2169–2171.
25. An-Wen Chan, Asbjørn Hróbjartsson, Mette T. Haahr , Peter C. Gotzsche, and Douglas G. Altman, "Empirical Evidence for Selective Reporting of Outcomes in Randomized Trials," *JAMA* 291 (2004): 2457–2465.
26. Richard Balon, "By Whom and How Is the Quality of Research Data Collection Assured and Checked?" *Psychotherapy and Psychosomatics* 74 (2005): 331–335.
27. Robert B. Cialdini, *Influence: The Psychology of Persuasion* (New York: Quill William Morrow, 1993).
28. Dana Katz , Jon F. Mertz, and Arthur L. Caplan, "All Gifts Large and Small: Toward an Understanding of the Ethics of Pharmaceutical Industry Gift Giving," *American Journal of Bioethics* 3 (2003): 39–46.

29. Ashley Wazanza, "Physicians and the Pharmaceutical Industry: Is a Gift Ever Just a Gift?" *JAMA* 283 (2000): 373–380.
30. Asaf Bitton, Mark D. Neuman, Joaquin Barnoya, and Stanton A. Glantz, "The p53 Tumour Suppressor Gene and the Tobacco Industry: Research Debate, and Conflict of Interest," *Lancet* 365 (2005): 531–340.
31. Lisa Cosgrove and Harold J. Bursztajn, "Undoing Undue Industry Influence: Lessons from Psychiatry as Psychopharmacology," *J. Org. Ethics* 3 (2006): 131–133.

CHAPTER 14

1. Krimsky and Baltimore, "The Ties that Bind or Benefit."
2. Relman, "Dealing with Conflicts of Interest," 1182–3.
3. "Avoid Financial "Correctness,'" 469.
4. Philip Campbell, "Declaration of Financial Interests," *Nature* 412 (Aug. 23, 2001): 751.
5. Sheldon Krimsky, "Publication Bias, Data Ownership, and the Funding Effect in Science: Threats to the Integrity of Biomedical Research," in: W. Wagner and R. Stenzor, eds., *Rescuing Science from Politics: Regulation and the Distortion of Scientific Research* (New York: Cambridge University Press; 2006), 61–85.
6. Sheldon Krimsky and Erin Sweet, "An Analysis of Toxicology and Medical Journals Conflict-of-Interest Policies," *Accountability in Research* 16 (2009): 235–53.
7. Public Health Service, Department of Health and Human Services, "Objectivity in Research," *Federal Register* 60, no. 132 (July 11, 1995): 35810–9.
8. Daniel R. Levinson, Inspector General, Department of Health and Human Services, National Institutes of Health, "Conflicts of Interest in Extramural Research (OEI-03–00460)," Bethesda, MD: NIH, 2008.
9. Meredith Wadman, "Money in Biomedicine: The Senator's Sleuth," *Nature* 461 (Sept. 17, 2009): 330–4.
10. Gardiner Harris and Benedict Carey, "Researchers Fail to Reveal Full Drug Pay," *New York Times,* June 8, 2008.
11. Senator Charles Grassley [homepage on the Internet]. Available from: http://grassley.senate.gov.
12. Letter from Elias A. Zerhouni, Director, National Institutes of Health, to Senator Charles E. Grassley, June 20, 2008. http://finance.senate.gov/search/?q=Zerhouni+t0+ Grassley&accesud=l Accessed December 7, 2010.
13. U.S. Senate Committee on Finance, "News release: Grassley Says NIH Conflicts Policy Would Be a Step in the Right Direction," [cited 2010 Aug 18]. Available from: http://finance.senate.gov/newsroom/ranking/release/?id=d8ad1da7–259c-41e7–9d0c-885795dcda29.
14. Committee on Finance, U.S. Senate. Hearing on the Confirmation of Governor Kathleen Sebelius, "Finance Committee Questions for the Record," April 2, 2009. http://finance.senate.gov/imo/media/doc/040209QFRs%20for%20SubmissionKS.pdf Accessed December 7, 2010.

15. Daniel R. Levinson, Inspector General, Department of Health and Human Services, National Institutes of Health, "Conflicts of Interest in Extramural Research," (OEI-03-06-00460). Bethesda, MD: NIH; 2008.
16. Daniel R. Levinson, Inspector General, Department of Health and Human Services, National Institutes of Health, "How Grantees Manage Financial Conflicts of Interest in Research Funded by the National Institutes of Health," (OEI-03-00700). Bethesda, MD: NIH, 2009.
17. Department of Health and Human Services, "Responsibility of Applicants for Promoting Objectivity in Research for which Public Health Funding Is Sought and Responsible Prospective Contractors," *Federal Register* 75 (May 21, 2010): 28687–712.
18. Sheldon Krimsky, "When Sponsored Research Fails the Admissions Test: A Normative Framework," in: J. L. Turk , ed., *Universities At Risk: How Politics, Special Interests, and Corporatization Threaten Academic Integrity* (Toronto: James Lorimer & Co., 2008), 70–94.
19. Ben Raines, "BP Buys Up Gulf Scientists for Legal Defense, Roiling Academic Community," [cited Aug 22, 2010]. Press Register. Available from: http://blog.al.com/live/2010/07/bp_buys_up_gulf_scientists_for.html.
20. Sheldon Krimsky, "Combating the Funding Effect in Science: What's Beyond Transparency?" *Stanford Law & Policy Review* 21, no. 1 (2010): 101–23.

CHAPTER 15

1. Dennis F. Thompson, "Understanding Financial Conflicts of Interest," *New England Journal of Medicine* 329 (1993): 573.
2. University and Small Business Patent Procedures Act, 35 U.S.C. §§ 200–212 (2006). (For regulations promulgated under this Act, see 37 C.F.R. § 401 (2010).)
3. See, e.g., Sheldon Krimsky, *Genetic Alchemy: The Social History of the Recombinant DNA* Controversy (Cambridge, MA: The MIT Press, 1982).
4. Mark H. Cooper, "Commercialization of the University and Problem Choice by Academic Biological Scientists," *Science, Technology, & Human Values* 34 (2009): 629, 647.
5. U.S. Constitution, Art. I, § 9 ("No title of nobility shall be granted by the United States: and no person holding any office of profit or trust under them, shall, without the consent of the Congress, accept of any present, emolument, office, or title, of any kind whatever, from any king, prince, or foreign state.").
6. U.S. Constitution, Art. I, § 6 ("No Senator or Representative shall, during the time for which he was elected, be appointed to any civil office under the authority of the United States, which shall have been created, or the emoluments whereof shall have been increased during such time: and no person holding any office under the United States, shall be a member of either House during his continuance in office.").

7. Kathleen Clark, "Regulating the Conflict of Interest of Government Officials," in *Conflict of Interest in the Professions,* Michael Davis & Andrew Stark, eds. (Oxford, UK, Oxford University Press, 2001).
8. The Federal Advisory Committee Act, 5 U.S.C. app. 2 §§ 1–16 (2006).
9. For an analysis of waivers by FDA, see Dennis Cauchon, "FDA Advisors Tied to Industry," *USA Today,* Sept. 25, 2000, 1A. For a discussion of conflict of interest among members of federal advisory committees, see Sheldon Krimsky, *Science in the Private Interest* (Lanham, MD: Rowman & Littlefield, 2003), 91–106.
10. Ibid.
11. Objectivity in Research, 59 Fed. Reg. 33, 243 (proposed June 28, 1994) (to be codified at 42 C.F.R. pt. 50 and 45 C.F.R. pt. 94).
12. 42 C.F.R. § 50.601 (2010).
13. See, e.g., Stacey Burling, "Jefferson to Pay $2.6 Million to Settle Research-Fraud Case," *Philadelphia Inquirer,* May 20, 2000, A1.
14. Merton, *Social Theory and Social Structure*, 560.
15. Ibid. at 547, 560.
16. Sheldon Krimsky, "Autonomy, Disinterest, and Entrepreneurial Science," *Society,* May-June 2006, 22, 27 (noting that a scientist's financial interest is usually not in the public record).
17. The *New England Journal of Medicine* introduced its COI policy in 1984, and the Journal of the American Medical Association followed with its policy in 1985. Relman, "Dealing with Conflict of Interest," 1182; Elizabeth Knoll & George D. Lundberg, "New Instructions for *JAMA* Authors," *JAMA* 254 (1985): 97.
18. See Lisa Cosgrove, Harold J. Bursztajn, and Sheldon Krimsky, "Developing Unbiased Diagnostic and Treatment Guidelines in Psychiatry," *New England Journal of Medicine* 360 (2009): 2035–36.
19. See Lisa Cosgrove, et al., "Financial Ties Between DSM-IV Panel Members and the Pharmaceutical Industry," *Psychotherapy & Psychosomatics* 75 (2006): 154–155 (arguing that contributors to the DSM should reveal their financial conflicts of interest in light of the effect those conflicts of interest can have on their research).
20. Sheldon Krimsky and L.S. Rothenberg, "Conflict of Interest Policies in Science and Medical Journals: Editorial Practices and Author Disclosures," *Science and Engineering Ethics* 7 (2001): 205.
21. Sheldon Krimsky and Erin Sweet, "An Analysis of Toxicology and Medical Journal Conflict-of-Interest Policies," *Accountability in Research* 16 (2009): 235.
22. See id. at 252.
23. *Journal of Internal Medicine*, Journal Information, http://www.wiley.com / bw/submit.asp?ref=0954–6820&site=1 (last visited Jan. 26, 2010).
24. *Archives of Internal Medicine*, Author Instructions, http://archinte .ama-assn.org/misc/ifora.dtl#ConflictofInterest (last visited Jan. 26, 2010).
25. See, e.g., Sheldon Krimsky, "The Funding Effect in Science and Its Implications for the Judiciary," *Journal of Law & Policy* 13 (2005): 43, 45–46.

26. See Richard Edwards and Raj Bhopal, "The Covert Influence of the Tobacco Industry on Research and Publication: A Call to Arms," *Journal of Epidemiology & Community Health* 53 (1999): 261.
27. See David B. Resnik, "Perspective: Disclosing Hidden Sources of Funding," *Academic Medicine* 84(2009):1226–27; Suzaynn F. Schick and Stanton A. Glantz, "Old Ways, New Means: Tobacco Industry Funding of Academic and Private Sector Scientists Since the Master Settlement Agreement," *Tobacco Control* 16 (2006): 157, 161.
28. See, e.g., Community of Experts on Tobacco Industry Documents, World Health Org., Tobacco Company Strategies to Undermine Tobacco Control Activities at the World Health Organization, at iii, (July 1, 2000), available at http://www.escholarship.org/uc/item/83m9c2wt.
29. Turner and Spilich, "Research into Smoking or Nicotine," 1423.
30. See, e.g., Community of Experts on Tobacco Industry Documents, World Health Org., supra note 28, at 197 ("The tobacco companies planned an ambitious series of studies, literature reviews and scientific conferences, to be conducted largely by front organizations or consultants, to demonstrate the weaknesses of the IARC [International Agency for Research on Cancer] study and of epidemiology, to challenge ETS [environmental tobacco smoke] toxicity and to offer alternatives to smoking restrictions . . . .").
31. Derek Yach and Stella Aguinaga Bialous, "Junking Science to Promote Tobacco," *American Journal of Public Health* 91 (2001): 1745, 1747.
32. Bekelman et al., "Scope and Impact," 454.
33. Ibid. at 463.
34. Ibid.
35. Lisa J. Kjaergard and Bodil Als-Nielsen, "Association between Competing Interests and Authors' Conclusions: Epidemiological Study of Randomized Clinical Trials Published in the British Medical Journal," *BMJ* 325 (2002): 249. ("Authors' conclusions . . . significantly favoured experimental interventions if financial competing interests were declared. Other competing interests were not significantly associated with authors' conclusions.")
36. Stelfox et al., "Conflict of Interest," 101. ("Our results demonstrate a strong association between authors' published positions on the safety of calcium-channel antagonists and their financial relationships with pharmaceutical manufacturers.")
37. Benjamin Djulbegovic et al., "The Uncertainty Principle and Industry-Sponsored Research," *The Lancet* 356 (2000): 635. ("Studies funded by non-profit organizations maintained equipoise favouring new therapies over standard ones (47% vs. 53%; p=0.608) to a greater extent than randomized trials supported solely or in part by profit-making organisations [sic] (74% vs. 26% p=0.004).")
38. David J. Rothman et al., "Professional Medical Associations and Their Relationships with Industry," *JAMA* 301 (2009): 1367.

39. Bodil Als-Nielsen et al., "Association of Funding and Conclusions in Randomized Drug Trials: A Reflection of Treatment Effect or Adverse Events?" *JAMA* 290 (2003): 921.
40. Kjaergardet & Als-Nielsen, supra note 35, at 249.
41. Turner & Spilich, supra note 29, at 1426.
42. Mark Friedberg et al., "Evaluation of Conflict of Interest in Economic Analyses of New Drugs Used in Oncology," *JAMA* 282 (1999): 1453.
43. John Yaphe et al., "The Association Between Funding by Commercial Interests and Study Outcome in Randomized Controlled Drug Trials," *Family Practice* 18 (2001): 565.
44. Merton, supra note 14, at 558.
45. Sheldon Krimsky, *Science in the Private Interest: Has the Lure of Profits Corrupted Biomedical Research?* (Lanham, MD: Rowman & Littlefield, 2003), p. 77.
46. Sheldon Krimsky, "Reforming Research Ethics in an Age of Multivested Science," in *Buying in or Selling Out?* Donald G. Stein ed. (New Brunswick, NJ: Rutgers University Press, 2004), p. 133.
47. John Ziman, *Real Science: What It Is, and What It Means* (Cambridge, UK: Cambridge University Press, 2000), p. 174.
48. See John Ziman, "No Conflict," *New Scientist* Oct. 4, 2003, at 34 (This sentence "The production of objective knowledge thus depends less on genuine personal 'disinterestedness' than on the effective operation of other norms, especially the norms of communalism, universalism and skepticism] refers to the supposed philosophical objectivity of scientific knowledge. I do not believe this has much changed in the transition to 'post academic' science. As I explain in my book, the conventional notion that it is entirely independent of human thought and action is epistemological codswallop. Throughout the book, however, I make it clear that the decline of disinterestedness in science gravely compromises its social objectivity–its hard-won reputation for a reasonable degree of impartiality, political neutrality and fairness. That's the key point.")
49. See, e.g., Benny Haerlin and Doug Parr, "How to Restore Public Trust in Science," *Nature* 400 (1999): 499.
50. See Arnold S. Relman, "Separating Continuing Medical Education From Pharmaceutical Marketing," *JAMA* 285 (2001): 2009.
51. S. 301, 111th Cong. (2009).
52. Press Release, Sen. Chuck Grassley, "Disclosure of Drug Company Payments to Doctors," available at http://grassley.senate.gov/about/Disclosure-of-Drug-CompanyPayments-to-Doctors.cfm.
53. Liz Kowalczyk, "State Bans Drug Firm Gifts to Doctors," *Boston Globe* March 12, 2009.
54. S. 301, 111th Cong. (2009).
55. 2007–2008 Vermont Attorney General, "Pharmaceutical Marketing Disclosures Report 4, available at http://www.atg.state.vt.us/assets/files/2009%20Pharam%20Report.pdf.

56. Pam Belluck, "Child Psychiatrist to Curtail Industry-Financed Activities," *New York Times* Dec. 30, 2008, at A16.
57. Ben Comer, "Survey Says Most Want Disclosure of Gifts to Docs," *International Communications Research,* June 19, 2008, http://www.icrsurvey.com/Study.aspx?f=Community_Catalyst_Prescription_Project_061908 .htm.
58. Kevin P. Weinfurt et al., "Disclosure of Financial Relationships to Participants in Clinical Research," *New England Journal of Medicine* 361 (2009): 916, 917.
59. 105 Mass. Code Regs. 970.009 (2010).
60. 105 Mass Code Regs. 970.007 (2010).
61. See Duff Wilson, "Harvard Medical School in Ethics Quandary," *New York Times* Mar. 2, 2009, at B1.
62. See American Bar Association, "Model Code of Judicial Conduct," (2008), available at http://www.abanet.org/cpr/mcjc/canon_4.html.
63. David Luban, "Law's Blindfold," in *Conflict of Interest in the* Professions , Michael Davis & Andrew Stark eds., (Oxford, UK: Oxford University Press, 2001), 26.
64. Andrew Stark, *Conflicts of Interest in American Public Life* (Cambridge, MA: Harvard University Press, 2003).
65. Christine Crofts and Sheldon Krimsky, "Emergence of a Scientific and Commercial Research and Development Infrastructure for Human Gene Therapy," *Human Gene Therapy* 16 (2005): 169, 173.
66. Ibid.
67. See Donald B. Kohn and Fabio Candotti, "Gene Therapy Fulfilling Its Promise," *New England Journalof Medicine* 360(2009): 521; see also Marina Cavazzana-Calvo et al., "Gene Therapy of Human Severe Combined Immunodeficiency (SCID)-x1 Disease," *Science* 288 (2000): 669.
68. Krimsky, supra note 46, at 133.
69. Jeffrey L. Fox, "Gene-Therapy Death Prompts Broad Civil Lawsuit," *Nature Biotechnology* 18 (2000): 1136.
70. Draft Interim Guidance, Advisory Committee to Office of Human Research Protections, Department of Health & Human Services, "Financial Relationships in Clinical Research: Issues for Institutions, Clinical Investigators, and IRBs to Consider when Dealing with Issues of Financial Interests and Human Subject Protection," Jan. 10, 2001. Available at http://www.hhs.gov/ohrp/nhrpac/mtg12–00/finguid.htm.
71. "Financial Relationships and Interests in Research Involving Human Subjects Protection," 69 Fed. Reg. 26,393 (proposed May 5, 2004) (to be codified at 21 C.F.R. pts. 50, 56 and 45 C.F.R. pt. 46).
72. Institute of Medicine, *Conflict of Interest in Medical Research, Education, and Practice* 5. Bernard Lo & Marilyn J. Field, eds. (Washington, DC: National Academy Press, 2009).
73. Laurence J. Hirsch, "Conflicts of Interest, Authorship, and Disclosures in Industry-Related Scientific Publications: The Tort Bar and Editorial Oversight in Medical Journals," *Mayo Clinic Proc.* 84(2009):811–21, at p. 811. ("Conflicts

of interest are widespread and represent a state of affairs, not a behavior or misconduct. They should be managed, rather than vainly attempting their elimination."); Andrew P. White et al., "Physician-Industry Relationships Can Be Ethically Established, and Conflicts of Interest Can Be Ethically Managed, *Spine* 32(2007):53–57, at p. 53. ("Many conflicts of interest are inevitable, and management is the optimal strategy to eliminate bias.")

74. Rothman et al., supra note 38, at 1368.
75. Ibid. at 1370.
76. Ibid.
77. Ibid. at 1372.
78. Gardiner Harris, "Institute of Medicine Calls for Doctors to Stop Taking Gifts from Drug Makers," *New York Times* Apr. 29, 2009, at A17.
79. Robert Steinbrook, "Controlling Conflict of Interest–Proposals from the Institute of Medicine," *New England. Journal of Medicine* 360 (2009): 2160–63, at p. 2160 (emphasis added).
80. Ibid.
81. Marcia Angell and Jerome P. Kassirer, "Editorials and Conflict of Interest," *New England Journal of Medicine* 335 (1996): 1055–56.
82. Tufts E-News, "Relaxing the Rules," June 19, 2002. http://enews.tufts.edu/stories/1095/2002/06/19/RelaxingTheRules.
83. Jeffrey M. Drazen and Gregory D. Curfman, "Financial Associations of Authors," *New England Journal of Medicine* 346 (2002): 1901.
84. Astrid James et al., "Commentary: The Lancet's Policy on Conflicts of Interest," *The Lancet* 363 (2004): 2–3.
85. Drummond Rennie, "Thyroid Storm," *JAMA* 277 (1997): 1238–43.
86. Hugo Ector et al., "A Statement on Ethics from the HEART Group," *Journal of Cardiovascular Pharmacology* 51 (2008):521.
87. Gardiner Harris, "F.D.A. Says Bayer Failed to Reveal Drug Risk Study," *New York Times*, Sept. 30, 2006, at A1.

CHAPTER 17

1. Charles S. Peirce, "The Fixation of Belief," *Popular Science Monthly* 12, no. 11(1877): 1–15. Accessed October 1, 2010. http://www.peirce.org/writings/p.107.html.
2. Arjen Van Dalen, "Structural Bias in Cross-National Perspective: How Political Systems and Journalism Cultures Influence Government Dominance in the News," *The International Journal of Press/Politics* 17, no. 1(2012): 32–55.
3. Brian Martin, *The Bias of Science* (Canberra, Australia: Society for Social Responsibility of Science, 1979).
4. David B. Resnik, *The Ethics of Science* (London, England: Routledge,1998).
5. DHHS (Department of Health and Human Services, Public Health Service), "Objectivity in Research," *Federal Register* s60: 35810–823 (1995).
6. N.G. Levinsky, "Nonfinancial Conflicts of Interest in Research," *The New England Journal of Medicine* 347 (2002): 759–61.

7. Jeremy Sugarman, "Human Stem Cell Ethics: Beyond the Embryo," *Cell Stem Cell* 2, no. 7 (2008): 529–33.
8. Sugarman, "Human Stem Cell Ethics," 2008.
9. S. Krimsky, "Publication Bias, Data Ownership, and the Funding Effect in Science: Threats to the Integrity of Biomedical Research," in: *Rescuing Science from Politics: Regulation and the Distortion of Scientific Research,* edited by W. Wagner and R. Steinzor (New York: Cambridge University Press, 2006), 61–85.
10. S. Krimsky, "Combating the Funding Effect in Science: What's Beyond Transparency?" *Stanford Law & Policy Review* 21(2010): 101–23.
11. S. Krimsky, "The Funding Effect in Science and its Implications for the Judiciary," *Journal of Law and Policy* 23, no.1 (2005): 43–68.
12. B. Als-Nielsen, W. Chen, C. Gluud, and L.L. Kjaergard, "Association of Funding and Conclusions in Randomized Drug Trials: A Reflection of Treatment Effect or Adverse Events" *JAMA* 290, no. 7 (Aug. 20, 2003): 921–8.
13. Als-Nielsen et al. "Association of Funding," 2003.
14. Als-Nielsen et al. "Association of Funding," 2003.
15 J.R. Yaphe, B. Edman, B. Knishkowy, J. Herman, "The Association Between Funding by Commercial Interests and Study Outcome in Randomized Controlled Drug Trials," *Family Practice* 18, no. 12 (2001): 565–68.
16. P.A. Rochon, J. H. Gurwitz, R. W. Simms, P.R. Fortin, D.T. Felson, K.L. Minaker, and T.C. Chalmers, "A Study of Manufacture-supported Trials of Nonsteroidal Anti-inflammatory Drugs in the Treatment of Arthritis," *Archives of Internal Medicine* 154, no. 2 (1994): 157–63.
17. Rochon et al., "A Study of Manufactured-supported Trials," 1994.
18. Marcia Angell, *The Truth about the Drug Companies* (New York, NY: Random House, 2004).
19. B. Djulbegovic, M. Lacevic, A. Cantor, K. K. Fields, C. L. Bennett, J. R. Adams, N. M. Kuderer, and G. H. Lyman, "The Uncertainty Principle and Industry-sponsored Research," *The Lancet* 356, no. 9230 (2000): 635–8.
20. Djulbegovic et al., "The Uncertainty Principle and Industry-Sponsored Research," 2000.
21. L.S. Friedman and E. D. Richter. 2004. "Relationship between Conflict of Interest and Research Results," *Journal of General Internal Medicine* 19, no.1 (2004): 51–56.
22. Friedman et al., "Relationship between Conflict of Interest and Research Results," 2004.
23. Friedman et al., "Relationship between Conflict of Interest and Research Results," 2004.
24. Friedman et al., Relationship between Conflict of Interest and Research Results," 2004.
25. T. J. Clifford, N. J. Barrowman, and D. Moher, "Funding Source, Trial Outcome and Reporting Quality: Are they Related? Results of a Pilot Study." *BMC Health Services Research* 2, no. 18 (2002): 1–6.

26. Clifford et al., "Funding Source, Trial Outcome and Reporting Quality," 2002.
27. J. E. Bekelman, J. E., Yan Li, and C. P. Gross, "Scope and Impact of Financial Conflicts of Interest in Biomedical Research: A Systematic Review," *Journal of American Medical Association* 289 (2003): 454–65.
28. J. A. Johnson and S. J. Coons, "Evaluation in Published Pharmacoeconomic Studies," *Journal of Pharmacy Practice* 8, no.8 (1995): 156–66, at p. 165.
29. M. Friedberg, B. Saffran, T. J. Stinson, W. Nelson and C. L. Bennett, "Evaluation of Conflict of Interest in Economic Analyses of New Drugs used in Oncology," *Journal of the American Medical Association* 282, no. 15 (1999): 1453–7.
30. C. M. Bell, D. R. Urbach, J. G. Ray, A. Bayoumi, A. B. Rosen, D. Greenberg, and P. J. Neumann, "Bias in Published Cost Effectiveness Studies: Systematic Review," *British Medical Journal* 332, no.7543 (2006): 699–70.
31. L. Garattini, D. Rolova, and G. Casasdei, "Modeling in Pharamacoeconomic Studies: Funding Sources and Outcomes," *International Journal of Technology Assessment Health Care* 26, no. 3 (2010): 330–33.
32. Garattini et al., "Modeling in Pharamacoeconomic Studies," 2010.
33. Garattini et al., "Modeling in Pharamacoeconomic Studies," 2010.
34. Sheldon Krimsky, "Conflict of Interest and Cost Effectiveness Analysis," *The Journal of American Medical Association* 282, no. 10 (1999): 474–75.
35. K. S. Knox, J. R. Adams, B. Djulbegovic, T. J. Stinson, C. Tomor, and C. L. Bennett, "Reporting and Dissemination of Industry versus Non-profit Sponsored Economic Analyses of Six Novel Drugs used in Oncology," *Annals of Oncology* 11, no. 12 (2000): 1591–95.
36. Yaphe et al., "The Association Between Funding by Commercial Interests," 2001.
37. Yaphe et al., "The Association Between Funding by Commercial Interests," 2001.
38. Yaphe et al., "The Association Between Funding by Commercial Interests," 2001.
39. Rochon et al., "A Study of Manufactured-Supported Trials," 1994.
40. Rochon et al., "A Study of Manufactured-Supported Trials," 1994.
41. Djulbegovic et al., "The Uncertainty Principle and Industry-sponsored Research," 2000.
42. Friedberg et al., "Evaluation of Conflict of Interest in Economic Analyses," 1999.
43. Sheldon Krimsky, "Conflict of Interest and Cost Effectiveness Analysis," *The Journal of American Medical Association* 282, no. 10 (1999): 474–75.
44. J. A. Sacristan, J. Soto, and I. Galende, "Evaluation of Pharmacoeconomic Studies: Utilization of a Checklist," *The Annals of Pharmacotherapy* 27 (1988): 1126–33.
45. T. Jefferson, R. Smith, Y. Yee, M. Drummond, M. Pratt, and R. Gale, "Evaluating the BMJ Guidelines for Economic Submissions." *Journal of the American Medical Association* 280, no. 3 (1998): 275–77.

46. Christina Turner and George J. Spilich, "Research into Smoking or Nicotine and Human Cognitive Performance: Does the Source of Funding make a Difference?" *Addiction* 92, no. 11 (1997): 1423–426.
47. J. C. Bailar, "How to Distort the Scientific Record without Actually Lying: Truth and the Arts of Science." *European Journal of Oncology* 11, no. 4 (2006): 217–24.
48. D. E. Barnes and L. A. Bero, "Why Review Articles on the Health Effects of Passive Smoking Reach Different Conclusions," *Journal of American Medical Association* 279, no. 19 (1998): 1566–570.
49. Barnes et al., "Why Review Articles on the Health Effects of Passive Smoking," 1998.
50. Barnes et al., "Why Review Articles on the Health Effects of Passive Smoking," 1998.
51. Barnes et al., "Why Review Articles on the Health Effects of Passive Smoking," 1998.
52. F. S. vom Saal and W. V. Welshons, "Large Effects from Small Exposures," *Environmental Research* 100, no. 1 (2006): 50–76, at p. 62.
53. vom Saal et al., "Large Effects from Small Exposures," 2006.
54. vom Saal et al., "Large Effects from Small Exposures," 2006.
55. Karl Popper, *The Logic of Scientific Discovery* (New York, NY: Harper & Row, 1968).
56. vom Saal et al., "Large Effects from Small Exposures," 2006.
57. Peirce, "The Fixation of Belief," 1877.
58. Robert K. Merton, *Social Theory and Social Structure* (New York, NY: The Free Press, 1968).

CHAPTER 18

1. A. C. Tsai, N. Z. Rosenlicht, J. N. Jureidini, P. I. Parry, G. I. Spielmans et al. "Aripiprazole in the Maintenance Treatment of Bipolar Disorder: A Critical Review of the Evidence and its Dissemination into the Scientific Literature," *PLOS Med* 8 (2011): e1000434.
2. L. Cosgrove, H. J. Bursztajn, S. Krimsky, M. Anaya, and J. Walker, "Conflicts of Interest and Disclosure in the American Psychiatric Association's Clinical Practice Guidelines," *Psychotherapy and Psychosomatics* 78 (2009): 228–232.
3. J. R. Lacasse, and J. Leo, "Ghostwriting at Elite Academic Medical Centers in the United States," *PLOS Med* 7 (2010): e1000230.
4. B. Roehr, "Professor Files Complaint of Scientific Misconduct over Allegation of Ghostwriting," *BMJ* 343 (2011): d4458.
5. G. Harris and B. Carey, "Researchers Fail to Reveal Full Drug Pay," *New York Times*, June 8, 2008. Available: http://www.nytimes.com/2008/06/08/us/08conflict.html?pagewanted = all. Accessed January 11, 2012.
6. S. Heres, J. Davis, K. Maino, E. Jetzinger, W. Kissling, et al., "Why Olanzapine Beats Risperidone, Risperidone Beats Quetiapine, and Quetiapine Beats Olanzapine: An Exploratory Analysis of Head-to-Head Comparison

Studies of Second-Generation Antipsychotics," *American Journal of Psychiatry* 163 (2006): 185–194.
7. R. Perlis, C. Perlis, Y. Wu, C. Hwang, M. Joseph, et al., "Industry Sponsorship and Financial Conflict of Interest in the Reporting of Clinical Trials in Psychiatry," *American Journal of Psychiatry* 162 (2005): 1957–1960.
8. M. Angell, *The Illusions of Psychiatry, New York Review of Books* 58 (2010): 20–22.
9. A. Frances, "Opening Pandora's Box: The 19 Worst Suggestions for DSM5," *Psychiatric Times* 27 (2010): 9.
10. World Health Organization, *International Statistical Classification of Disease and Related Health Problems, Tenth Revision (ICD-10)* (Geneva: World Health Organization, 1992).
11. American Psychiatric Association, "APA Names DSM-V Task Force Members: Leading Experts to Revise Handbook for Diagnosing Mental Disorders," Press Release #07–57, July 23, 2007. Available: http://www.dsm5.org/Newsroom/Documents/0757%20APA%20Announces%20DSM%20Task%20Force%20Members.pdf. Accessed November 16, 2011.
12. Bekelman et al., "Scope and Impact," 454–465.
13. Stelfox et al., "Conflict of Interest," 101–106.
14. Kjaergard and Als-Nielsen, "Association Between Competing Interests," 249.
15. S. Krimsky, "Combating the Funding Effect in Science: What's Beyond Transparency?" *Stanford Law Policy Rev* XXI (2010): 101–123.
16. National Research Council, "Conflict of Interest in Medical Research, Education, and Practice [consensus report]," B. Lo, M. J. Field, eds. (Washington, DC: National Academies Press, 2009). Available: http://www.iom.edu/Reports/ 2009/Conflict-of-Interest-in-Medical-Research-Education-and-Practice.aspx. Accessed January 11, 2012.
17. A. Matheson, "Corporate Science and the Husbandry of Scientific and Medical Knowledge by the Pharmaceutical Industry," *Bio Societies* 3 (2008): 355–382.
18. L. Cosgrove, H. J. Bursztajn, and S. Krimsky, "Developing Unbiased Diagnostic and Treatment Guidelines in Psychiatry," *New England Journal of Medicine* 360 (2009): 2035–2036.
19. B. C. Pilecki, J. W. Clegg, and D. McKay, "The influence of corporate and political interests on models of illness in the evolution of the DSM," *European Psychiatry* 26(2011): 194–200.
20. A. W. Jørgensen, J. Hilden, P. C. Gøtzsche, "Cochrane reviews compared with industry supported meta-analyses and other meta-analyses of the same drugs: systematic review," *BMJ* 331(2006): 782–785.
21. M. Bhandari, J. W. Busse, D. Jackowski, V. M. Montori, H. Schuenemann, et al., "Association between industry funding and statistically significant pro-industry findings in medical and surgical randomized trials," *Canadian Medical Association Journal* 170(2004): 477–480.
22. L. I. Lesser, C. B. Ebbeling, M. Goozner, D. Wypij, and D. S. Ludwig, "Relationship between funding source and conclusion among nutrition-related

scientific articles," *PLOS Medicine* 4(2007): e5. doi:10.1371/ journal.pmed.0040005.
23. D. M. Cain, G. Lowenstein, D. A. Moore, "The dirt on coming clean: Perverse effects of disclosing conflicts of interest," *Journal of Legal Studies* 31(2005): 1–25.
24. L. Cosgrove, S. Krimsky, M. Vijayaraghavan, and L. Schneider, "Financial ties between DSM-IV panel members and the pharmaceutical industry. *Psychotherapy and Psychosomatics* 75(2006): 154–160.
25. J. Lexchin, O. O'Donovan, "Prohibiting or 'managing' conflict of interest?: A review of policies and procedures in three European drug regulation agencies," *Social Science & Medicine* 70(2010): 643–647.
26. D. Katz, A. L. Caplan, J. F. Merz, "All gifts large and small: Toward an understanding of the ethics of pharmaceutical industry gift-giving," *American Journal of Bioethics* 3(2003): 39–46.
27. C. Mather, "The pipeline and the porcupine: Alternate metaphors of the physician-industry relationship," *Social Science & Medicine* 60(2005): 1323–1334.
28. M. Mauss, "The gift: Forms and functions of exchange in archaic societies" (I. Cunnison, Trans.) (New York: W.W. Norton Co., 1967), 130.
29. T. B. Mendelson, M. Meltzer, E. G. Campbell, A. L. Caplan, and J. N. Kirkpatrick, "Conflicts of interest in cardiovascular clinical practice guidelines," *Archives of Internal Medicine* 171(2011): 577–584. doi:10.1001/archinternmed.2011.96.
30. Institute of Medicine (2011) Clinical practice guidelines we can trust: Report brief. Washington (DC): National Academies Press, Available: http://www.iom.edu/,/media/Files/Report% 20Files/2011/Clinical-Practice-Guidelines-We-Can-Trust/Clinical%20Practice%20Guidelines%202011%20Report%20Brief.pdf. Accessed 30 March 2011.
31. American Psychiatric Association (2010) DSM-5 Development: Frequently asked questions. Available: http://www.dsm5.org/about/pages/faq.aspx. Accessed 30 September 2011.

CHAPTER 19

1. Cosgrove et al., "Financial Ties," 154–160.
2. L. Cosgrove and S. Krimsky, "A Comparison of DSM-IV and DSM-5 Panel Members' Financial Associations with Industry: A Pernicious Problem Persists," *PLOS Medicine* 9(2012): e1001190.
3. American Psychiatric Association, "Policy on Conflicts of Interest Principles and Guidelines: with Special Interest for Clinical Practice and Research" (Arlington, VA: APA, 2010).
4. D. F. Thompson, "The Challenge of Conflicts of Interest in Medicine," *Zeitschrift Fur Evidenz, Fortbildung Und Qualitat Im Gesundheitswesen* 103(2009): 136–140.
5. American Psychiatric Association, *Diagnostic and Statistical Manual of Mental Disorders, ed 5* (Arlington, VA, American Psychiatric Association, 2013).

6. A. Lundh, S. Sismondo, J. Lexchin, O.A, Busuioc, and L. Bero, "Industry Sponsorship and Research Outcome," *Cochrane Database of Systematic Reviews* 12(2012): MR000033.
7. Institute of Medicine (IOM), "Clinical Practice Guidelines We Can Trust: Standards for Developing Trustworthy Clinical Practice Guidelines (CPGs)" (Washington, DC, Institute of Medicine, 2011). http://www.iom.edu/Reports/2011/Clinical-Practice-guidelines-We-Can-Trust/Standards.aspx (accessed April 7, 2013).
8. The *PLOS* Medicine Editors, "The Paradox of Mental Health: Over-treatment and Under-recognition," *PLOS Medicine* 10 (2012): e1001456.
9. L. Cosgrove, S. Krimsky, and H. J. Bursztajn, "Developing Unbiased Diagnostic and Treatment Guidelines in Psychiatry," *New England Journal of Medicine* 360 (2009): 2035–2036.
10. S. Krimsky, and L. S. Rothenberg, "Financial Interests and its Disclosure in Scientific Publications," *JAMA* 280 (1998): 225–226.
11. R. H. Perlis, C. S. Perlis, Y. Wu, C. Hwang, M. Joseph, and A. A. Nierenberg, "Industry Sponsorship and Financial Conflict of Interest in the Reporting of Clinical Trials in Psychiatry," *American Journal of Psychiatry* 162 (2005): 1957–1960.
12. A. Frances, "DSM 5 is Guide Not Bible—Simply Ignore Its Ten Worst Changes," *Psychiatric Times* On DSM-5 Blog [Internet]. http://www.psychiatrictimes.com/blog/frances/content/article/10168/2117994 (accessed April 2, 2013).
13. British Psychological Society, "DSM-5: The Future of Psychiatric Diagnosis [Internet]," (Leicester,UK: British Psychological Society, 2012). http://apps.bps.org.uk/publicationfiles/consultation-responses/DSM-5%202012%20-%20BPS%20response.pdf (accessed April 7, 2013).
14. D. Brauser, "DSM-5 Field Trials Generate Mixed Results," [Internet]. Medscape. 2012. http://www.medscape.com/viewarticle/763519 (accessed April 2, 2013).
15. G. M. Callaghan, C. Chacon, J. Botts, and S. Laraway, "An Empirical Evaluation of the Diagnostic Criteria for Premenstrual Dysphoric Disorder: Problems with Sex Specificity and Validity," *Women & Therapy* 32 (2008): 1–21.
16. L. Batstra, and A. Frances, "Diagnostic Inflation: Causes and a Suggested Cure, *Journal of Nervous And Mental Disease* 200 (2012): 474–479.
17. R. L. Spitzer, J. B. W. Williams, and J. Endicott, "Standards for DSM-5 Reliability," *American Journal of Psychiatry* 169 (2012): 537.
18. J. W. Li and J. C. Vederas, "Drug Discovery and Natural Products: End of an Era or an Endless Frontier?" *Science* 325 (2009): 161–165.
19. H. Gupta, S. Kumar, S. K. Roy, and R. S. Gaud, "Patent Protection Strategies," *Journal of Pharmacy and Bioallied Sciences* 2 (2010): 2–7.
20. P. S. Appelbaum, and A. Gold, "Psychiatrists' Relationships with Industry: The Principal-Agent Problem," *Harvard Review of Psychiatry* 18 (2010): 255–265.

21. M. Angell, "The Illusions of Psychiatry," *New York Review of Books* 58 (2011): 20–22.
22. G. A. Fava, "The Intellectual Crisis of Psychiatry," *Psychotherapy and Psychosomatics* 75 (2006): 202–208.
23. A. Kahn, R. M. Leventhal, S. R. Khan, and W. A. Brown, "Severity of Depression and Response to Antidepressants and Placebo: An Analysis of the Food and Drug Administration Database," *Journal of Clinical Psychopharmacology* 22 (2002): 40–45.
24. S. Pettypiece, "Lilly Profit Beats Analyst Estimates as Cymbalta Sales Climb," [Internet]. *Bloomberg News* Jan. 29, 2013. http://www.bloomberg.com/news/2013-01-29/lilly-profitbeats-analyst-estimates-as-cymbalta-salesclimb.html (accessed April 4, 2013).
25. K. Rickels, "Should Benzodiazepines Be Replaced by Antidepressants in the Treatment of Anxiety Disorders?: Fact or Fiction?" *Psychotherapy and Psychosomatics* 82 (2013): 351–352.
26. R. Balon, "Benzodiazepines Revisited," *Psychotherapy and Psychosomatics* 82 (2013): 353–354.
27. G. A. Fava, "Financial Conflicts of Interest in Psychiatry," *World Psychiatry* 6 (2007): 19–24.

CHAPTER 20

1. Institute of Medicine: Committee on Standards for Developing Trustworthy Clinical Practice Guidelines, "Clinical Practice Guidelines We Can Trust" (Washington, DC: The National Academies Press, 2011). Available: http://www.nationalacademies.org/hmd/Reports/2011/Clinical-Practice-Guidelines-We-CanTrust.aspx.
2. Cosgrove et al., "Financial Ties," 154–160. Available: http://www.ncbi.nlm.nih.gov/pubmed/16636630. doi: 10.1159/000091772 PMID: 16636630.
3. See, for example, *Gelsinger v. University of Pennsylvania*, in: Sheldon Krimsky, *Science in the Private Interest* (Lanham, MD: Roman & Littlefield. 2003).
4. David Michaels, "Manufactured Uncertainty," *Annals of the New York Academy of Sciences* 1076 (2006):149–162. doi: 10.1196/annals.1371.058 PMID: 17119200.
5. Benjamin Capps, "Can a Good Tree Bring Forth Evil Fruit?: The Funding of Medical Research by Industry," *British Medical Bulletin* 118 (2016): 5–15. doi: 10.1093/bmb/ldw014 PMID: 27151955.
6. S. Krimsky, "Combatting the Funding Effect in Science: What's Beyond Transparency?" *Stanford Law & Policy Review* 21 (2010): 101–123. Available: https://journals.law.stanford.edu/stanford-law-policy-review/print/volume-21/issue-1–academic-integrity/combatting-funding-effect-science-whats-beyond.
7. A. Lundh, S. Sismondo, J. Lexchin, O.A. Busuioc, and L. Bero, "Industry Sponsorship and Research Outcome," *Cochrane Database of Systematic*

Reviews, 2012. [cited 2012 Dec 12]. Available from: https://www.ncbi.nlm.nih.gov/pubmed/23235689.

8. Stelfox et al., "Conflict of Interest," 101–106. doi: 10.1056/NEJM199801083380206 PMID: 9420342.
9. F. vom Saal, and C. Hughes, "An Extensive New Literature Concerning Low-Dose Effects of Bisphenol A Shows the Need for a New Risk Assessment," *Environmental Health Perspectives* 113, no. 8 (August 2005): 926–933 at 929. Available: https://www.ncbi.nlm.nih.gov/pmc/articles/PMC1280330/doi:10.1289/ehp.7713 PMID: 16079060.
10. L. A. Bero, "Tobacco Industry Manipulation of Research," *Public Health Reports* 120 (Mar–Apr 2005): 200– 208. Available: https://www.ncbi.nlm.nih.gov/pmc/articles/PMC1497700/pdf/15842123.pdf.
11. J. Lopez, S. Lopez, J. Means, R. Mohan, A. Soni, J. Milton, A. P. Tufaro, J. W. May, and A. Dorafshar, "Financial Conflicts of Interest: An Association Between Funding and Findings in Plastic Surgery," *Plastic and Reconstructive Surgery* 136, no. 5 (2015): 690e–697e. Available: https://www.ncbi.nlm.nih.gov/pubmed/26505726.
12. L. E. Van Nierop, M. Röösli, M. Egger, and A. Huss, "Sources of Funding in Experimental Studies of Mobile Phone Use on Health: Update of Systematic Review," *Journal of C. R. Physique* 11 (2010): 622–627. Available: http://www.sciencedirect.com/science/article/pii/S1631070510001465.
13. M. Bes-Rastrollo, M. B. Schulze, M. Ruiz-Canela, and M. A. Martinez-Gonzalez, "Financial Conflicts of Interest and Reporting Bias Regarding the Association Between Sugar-Sweetened Beverages and Weight Gain: A Systematic Review of Systematic Reviews," *PLOS Medicine* 10, no. 12 (Dec. 2013): 1–9.
14. J. Diels, M. Cunha, C. Manaia, B. Sabugosa-Madeira, and M. Silva, "Association of Financial or Professional Conflict of Interest to Research Outcomes on Health Risks or Nutritional Assessment Studies of Genetically Modified Products," *Food Policy* 36, no. 2 (April 2011): 197–203. Available: http://www.sciencedirect.com/ science/article/pii / S0306919210001302.
15. T. Guillemaud, E. Lombaert, and D. Bourguet, "Conflicts of Interest in GM Bt Crop Efficacy and Durability Studies," *PLOS One* 11, no. 12 (Dec. 2016). Available: http://journals.plos.org/plosone/article?id=10.1371/journal.pone.0167777.
16. R. M. Nazarro, U.S. General Accounting Office, "Federal Advisory Committee Act: Issues Related to the Independence and Balance of Advisory Committees," Testimony before the Subcommittee on Information Policy, Census, and National Archives, Committee on Oversight and Government Reform, U.S. House of Representatives. GAO-08–611T, April 2, 2008. Available: http://www.gao.gov/assets/120/119486.pdf.
17. The National Academy of Sciences. IRS filing. Form 990. 2014.
18. P. Boffey, *The Brain Bank of America* (New York: McGraw-Hill; 1975).

19. M. Parascandola, "A Turning Point for Conflicts of Interest: The Controversy over the National Academy of Sciences' First Conflicts of Interest Disclosure Policy," *Journal of Oncology* 25, no. 24 (2007): 3774–3779. Available: http:// jco.ascopubs.org/content/25/24/3774.full.
20. The National Research Council was created in 1916, the National Academy of Engineering in 1964, and the Institute of Medicine in 1970. In 2015 the Institute of Medicine changed its name to the National Academy of Medicine. Since that time, the overarching organization of the individual academies is referred to as the National Academies of Sciences, Engineering and Medicine (NASEM), changed from the National Academy of Sciences.
21. Federal Advisory Committee Act of 1997. 5 U.S.C. app. §15 (b) Available: https://www.gpo.gov/fdsys/ pkg/USCODE-2010–title5/html/USCODE-2010–title5–app-federalad.htm.
22. The National Academies. Policy on Committee Composition and Balance and Conflicts of Interest for Committees used in the Development of Reports. May 12, 2003. Available: http://www.nationalacademies.org/coi/bi-coi_form-0.pdf.
23. National Academy of Sciences, National Academy of Engineering, and Institute of Medicine: Committee on Science, Engineering, and Public Policy. On Being a Scientist: A Guide to Responsible Conduct in Research. 3rd edition. Washington, DC. National Academies Press. 2009. Available: http://www.nap.edu/catalog/12192/on-being-a-scientist-a-guide-to-responsible-conduct-in.
24. U.S. General Accounting Office, "Report to Congressional Requesters: The National Academy of Sciences and the Federal Advisory Committee Act," GAO/RCED-99–17. November 1998. Available: http://www.gao.gov/assets/230/226646.pdf.
25. A. McCook, "Conflicts of Interest at Federal Agencies," *The Scientist*, July 24, 2006. Available: http://www.the-scientist.com/?articles.view/articleNo/24174/title/Conflicts-of-interest-at-Federal-agencies/. Accessed August 22, 2016.
26. M. Goozner, "Ensuring Independence and Objectivity at the National Academies. Center for Science in the Public Interest," July 1 2006. Available: https://cspinet.org/new/pdf/nasreport.pdf. Accessed 2016 August 23.
27. M. Peterson, "Biotech Expert's New Job Casts a Shadow on Report," *The New York Times* Aug 16; 1999. Available: http://www.nytimes.com/1999/08/16/us/biotech-expert-s-new-job-casts-a-shadow-on-report.html. Accessed August 23, 2016.
28. M. Wadman, "GM Advisory Panel is Slanted, Say Critics," *Nature* 399, no. 6731 (May 6, 1999): 7. Available: http://www.nature.com/nature/journal/v399/n6731/full/399007a0.html.
29. S. Strom, "National Biotechnology Panel Faces New Conflict of Interest Questions," *New York Times*, December 27, 2016.

30. National Academies of Science, Engineering and Medicine (NASEM). Committee on Genetically Engineered Crops, "Genetically Engineered Crops: Experiences and Prospects (Washington, DC. National Academies Press. 2016). Available: http://www.nap.edu/catalog/23395/genetically-engineered-cropsexperiences-and-prospects.
31. Cosgrove et al., "Conflicts of Interest and Disclosure," 228–232. Available: http://www.ncbi.nlm.nih.gov/pubmed/19401623.
32. National Institutes of Health, "HHS Tightens Financial Conflict of Interest Rules for Researchers," August 23, 2011. Available: https://www.nih.gov/news-events/news-releases/hhs-tightens-financial-conflictinterest-rules-researchers.
33. Institute of Medicine, "Conflict of Interest in Medical Research, Education, and Practice" (Washington, DC: National Academies Press, 2009). Available: http://www.nap.edu/catalog/12598/conflict-of-interest-in-medical-research-education-and-practice.
34. J. Dana and G. A. Loewenstein, "Social Science Perspective on Gifts to Physicians from Industry," *JAMA* 290, no. 2 (Jul 9, 2003): 252–5. Available: http://med.stanford.edu/coi/journal%20articles/Loewenstein_A_Social_Science_Perspective_on_Gifts.pdf. Accessed October 19, 2016.
35. National Academies, "Policy and Procedures on Committee Composition and Balance and Conflicts of Interest for Committees Used in the Development of Reports," Website National Academies of Sciences, Engineering and Medicine. Available: http://www8.nationalacademies.org/cp/information.aspx?key=Conflict_of_Interest. Accessed July 3, 2016.
36. American Academy of Neurology, "Policy on Conflicts of Interest." Available: http://tools.aan.com/apps/disclosures/index.cfm?event=committee:intro; Deutsches Arzteblatt International. Conflict of Interest Statement. Available: http://www.aerzteblatt.de/int/for-authors/instructions. Accessed November 3, 2016.
37. International Committee of Medical Journal Editors (ICMJE), "Form for Disclosure of Potential Conflicts of Interest." Website. Available: http://icmje.org/conflicts-of-interest/.
38. American Psychiatric Association, "Financial Statement, Disclosure of Affiliations and Conflict of Interests Policy. Website, " March 20, 2016. Available: https://www.psychiatry.org/File%20Library/About-APA/OrganizationDocuments-Policies/apa-discolsure-of-interests-policy.pdf.
39. R. Steinbrook, "Controlling Conflict of Interest—Proposals from the Institute of Medicine," *N Engl J Med* 360 (Mar 21, 2009): 2160–2163.
40. Institute of Medicine, "Conflict of Interest in Medical Research, Education, and Practice," (Washington, DC: National Academies Press, 2009). Available: http://www.nap.edu/catalog/12598/conflict-of-interest-in-medical-research-education-and-practice.
41. Association of American Medical Colleges, American Association of Universities, "Protecting Subjects, Preserving Trust, Promoting Progress II: Principles and Recommendations for Oversight of an Institution's

Financial Interests in Human Subjects Research," Task Force on Financial Conflicts of Interests in Clinical Research, 2002 October. Available: http://ccnmtl.columbia.edu/projects/rcr/rcr_conflicts/misc/Ref/AAMC_2002CoIReport.pdf. Accessed Aug 22, 2016.

42. American Association of Universities Task Force on Research Accountability, "Report and Recommendations on Individual and Institutional Conflicts of Interest," 2001. Available: http://www.aau.edu/workarea/downloadasset.aspx?id=6358. Accessed 2016 Aug 23.
43. National Academy of Sciences, "Report of the Treasurer of the National Academy of Sciences for Year Ended December 2014," 2015. Available: http://www.nap.edu/read/21779/chapter/4. Accessed 2016 Aug 2.
44. National Academies of Sciences, Engineering and Medicine: Public Interface in Life Sciences Roundtable, "Public Engagement on Genetically Modified Organisms: When Science and Citizens Connect, A Workshop Summary." (Washington, DC: National Academies Press, 2015). Available: http://www.nap.edu/read/21750/chapter/1. Accessed Aug 22, 2016.
45. National Academies Roundtable Agenda. Available: http://nas-sites.org/publicinterfaces/files/2014/07/ PILS-02–GMO-Interface-agenda10.pdf.
46. National Academy of Sciences: Committee on a National Strategy for Biotechnology in Agriculture, National Research Council, *Agricultural Biotechnology: Strategies for National Competitiveness* (Washington, DC: National Academies Press, 1987). Available: http://www.nap.edu/read/1005/chapter/ 1. Accessed Aug 22, 2016.
47. National Academy of Sciences: Committee on Genetically Modified Pest-Protected Plants, "Genetically Modified Pest-Protected Plants: Science and Regulation." (Washington, DC: National Academies Press, 2000). Available: http://www.nap.edu/read/9795/chapter/1.
48. National Academy of Sciences, "Impact of Biotechnology on Farm-Level Economics and Sustainability Impact of Genetically Engineered Crops on Farm Sustainability in the United States." (Washington, DC: National Academies Press, 2010). Available: http://www.nap.edu/download/12804.
49. Association of Governing Boards of Universities and Colleges, "Board Policy on Conflicts of Interest," Jan 17, 2007. Available: http://agb.org/statements/2007/agb-statement-on-board-accountability.
50. National Academies of Science, Engineering and Medicine (NASEM), "Background Information and Confidential Conflict of Interest Disclosure Form," March 2016. Available: http://www8.nationalacademies.org/cp/information.aspx?key=Conflict_of_Interest Accessed July 3, 2016.
51. A. Iles, M. Anderson, M. Antoniou, P. Bereano, L. Bunin, L. Carroll et al, "Letter to Kara Laney, National Academies of Sciences, Engineering and Medicine: Genetically Engineered Crops: Past Experience and Future Prospects," Public Access File. Aug. 4, 2014.
52. Wenonah Hauter, "Letter to Kara Laney, National Academies of Sciences, Engineering and Medicine: Genetically Engineered Crops: Past Experience

and Future Prospects." Edited version (with financial COI documentation removed) available in Public Access File. Aug 4, 2014. Unedited version available: http://www.foodandwaterwatch.org/sites/default/files/hauter.pdf.
53. Proceedings of the National Academy of Sciences of the United States of America, "Conflict of Interest Policy." Website. Available: http://www.pnas.org/site/authors/coi.xhtml.
54. Harvest Plus. Pro-Vitamin A maize flyer. Available: http://www.harvest-plus.org/sites./default/files/ HarvestPlus_Maize_Strategy.pdf. Accessed Aug 22, 2016
55. Consortium of International Agricultural Research Centers. Media Release. "CGIAR Generation Challenge Program. New Genomic Resources for Maize Breeding." February 29, 2012. Available: http://www. generationcp.org/communications/media/press-releases/new-genomic-resources-for-maize-breeding. August 22, 2016.
56. W. B. Suwarno, K. V. Pixley, N. Palacios-Rojas, S. M. Kaeppler, and R. Babu, "Genome-Wide Association Analysis Reveals New Targets for Carotenoid Biofortification in Maize," *Theoretical and Applied Genetics* 128, no. 5 (2015): 851–864. Available: http://www.ncbi.nlm.nih.gov/pmc/articles/PMC4544543/.
57. C. B. Kandianis, R. Stevens, W. Liu, N. Palacios, K. Montgomery, K. Pixley et al., "Genetic Architecture Controlling Variation in Grain Carotenoid Composition and Concentrations in Two Maize Populations," *Theoretical and Applied Genetics* 126, no. 11 (2013): 2879–95.
58. Whistler Center, Purdue University. Annual reports 2015, 2014, 2013, 2012, 2011. Available: https:// www.whistlercenter.purdue.edu/.
59. C. N. Stewart, University of Tennessee, C.V. Available: http://plantsciences.utk.edu/pdf/stewart%20_cv_%202015_%20midyear_public_long.pdf. Accessed Aug. 24, 2016. At 18–21.
60. R. Dixon. University of North Texas Faculty Information System online database. Available https://facultyinfo.unt.edu/faculty-profile?query=Richard+Dixon&type= name&profile=rad0169. Accessed Aug 24, 2016.
61. Timothy Griffin, C.V. Dated January 19, 2016. Available at https://nutrition.tufts.edu/sites/default/files/ profiles-cv/Timothy%20Griffin%20CV.pdf.
62. Wolfe's Neck Farm. Press Release. "Wolfe's Neck Farm Secures Major Grant from Stonyfield to launch an Organic Dairy Farmer Training and Research Program" Undated. Available: http://wolfesneckfarm.org/organic-dairy-training-program-press/
63. L. Dube, "The Latest in Stonyfield's GMO Labeling Support," July 25, 2014. Available: http://www.stonyfield.com/blog/stonyfield-supports-gmo-labels/
64. Danone Ecosystem Fund. [Newsletter]. Issue 12 (July 2015): 3. Available at http://ecosysteme.danone.com/nl/2015–07/en/#/2.
65. Dannon. [Press Release]. Dannon Announces Breakthrough Sweeping Commitment for Sustainable Agriculture, More Natural Ingredients and

Greater Transparency," April 27 2016. Available: http://www.dannon.com/the-dannon-pledge-on-sustainable-agriculture-naturality-and-transparency/onfile.
66. Food & Water Watch. Under the Influence: The National Research Council and GMOs, 2016 May. Available: https://www.foodandwaterwatch.org/insight/under-influence-national-research-council-andgmos.
67. F. Gould, "Public Release Event for Genetically Engineered Crops: Experiences and Prospects. National Academies of Science, Engineering and Medicine Keck Center, 500 5th St NW, Washington, DC. May 17, 2016. Available: https://nas-sites.org/ge-crops/2016/04/27/report-release/.
68. D. Resnik, "Institutional Conflicts of Interest in Science," *Science and Engineering Ethics*, 2015.
69. S. Krimsky, "The Ethical and Legal Foundations of Scientific 'Conflict of Interest,'" In: T. Lemmens and D. Waring, eds. *Law and Ethics in Biomedical Research: Regulation, Conflict of Interest, and Liability* (Toronto: University of Toronto Press; 2006), 63–81. Available: http://emerald.tufts.edu/~skrimsky/PDF/Law%20and%20Ethics.PDF.
70. R. Steinbrook, 2009. "Controlling Conflict of Interest–Proposals from the Institute of Medicine," *New England Journal of Medicine* 360, no. 21 (2009): 2160–63.

CHAPTER 21

1. 1. T. M. Shaneyfelt and R. M. Centor, "Reassessment of Clinical Practice Guidelines: Go Gently into that Good Night," *Journal of the American Medical Association* 301, no. 8 (2009): 868–69.
2. S. J. Genuis, "The Proliferation of Clinical Practice Guidelines: Professional Development of Medicine-by-Numbers?" *Journal of the American Board of Family Medicine* 18, no. 5 (2005): 419–425.
3. B. Lo, and M. J. Field, *Conflict of Interest in Medical Research, Education, and Practice* (Washington, DC: National Academies Press, 2009).
4. *Institute of Medicine, Clinical Practice Guidelines We Can Trust* (Washington, DC: Institute of Medicine, 2011).
5. P.M Tereskerz, "Research Accountability and Financial Conflicts of Interest in Industry-Sponsored Clinical Research: A Review," *Accountability in Research* 10 (2003): 137–58.
6. IOM, 2011.
7. H. Bastian, "Nondisclosure of Financial Interest in Clinical Practice Guideline Development: An Intractable Problem?" *Plos Medicine* 13 (2016): (5): e1002030.
8. *National Institute for Health and Care Excellence, Developing NICE Guidelines: The Manual* (London, England: National Institute for Health and Care Excellence, 2014).
9. Guidelines International Network, Resources/International Guideline Library. Guidelines International Network, 2015. http://www.g-i-n.net/library/international-guidelines-library (accessed October 10, 2016).

10. Agency for Healthcare Research and Quality, Inclusion Criteria. Agency for Healthcare Research and Quality, 2015.
11. K. C. Elliott, "Scientific Judgment and the Limits of Conflict-of-Interest Policies," *Accountability in Research* 15 no. 1 (2008): 1–29.
12. S. Krimsky, "Combating the Funding Effect in Science: What's Beyond Transparency?" *Stanford Law & Policy Review* 21, no. 1 (2010): 81–103.
13. S. Krimsky, "Do Financial Conflicts of Interest Bias Research?: An Inquiry into the 'Funding Effect' Hypothesis," *Science, Technology & Human Values* 38, no. 4 (2013): 566–87.
14. D. B. Resnik and K. C. Elliott, "Taking Financial Relationships into Account when Assessing Research," *Accountability in Researc*h 20, no. 3 (2013): 184–205.
15. Krimsky, "Do Financial Conflicts of Interest Bias Research?" 2013.
16. Tereskerz, "Research Accountability and Financial Conflicts of Interest," 2003.
17. A. Lundh, S. Sismondo, J. Lexchin, O. A. Busuioc, and L. Bero, "Industry Sponsorship and Research Outcome," *The Cochrane Database of Systematic Reviews* 12 (2012): MR000033.
18. L. Cosgrove, S. Vannoy, B. Mintzes, and A. F. Shaughnessy, "Under the Influence: The Interplay Among Industry, Publishing, and Drug Regulation," *Accountability in Research* 23 no. 5 (2016): 257–79
19. J. Neuman, D. Korenstein, J. S. Ross, and S. Keyhani, "Prevalence of Financial Conflicts of Interest Among Panel Members Producing Clinical Practice Guidelines in Canada and United States: Cross Sectional Study," *BMJ* (Clinical Research Ed.) 343 (2011): d5621–d5621.
20. L. Cosgrove, H. J. Bursztajn, D. R. Erlich, E. E. Wheeler, and A. F. Shaughnessy, "Conflicts of Interest and the Quality of Recommendations in Clinical Guidelines," *Journal of Evaluation in Clinical Practice* 19 no. 4 (2013): 674–81.
21. Trip. 2016. What is trip? Newport, UK: Trip Database Ltd. https://www.tripdatabase.com/about.
22. Institute of Medicine, Committee on Conflict of Interest in Medical Research, Education, and Practice, Conflict of Interest in Medical Research, Education, and Practice, M.J. Field and Lo eds. (Washington, DC: National Academies Press, 2009).
22. Neuman et al., "Prevalence of Financial Conflicts of Interest," 2011.
23. IOM, 209.
24. ICMJE, 2016.
25. *National Institutes of Health, Financial Conflict of Interest* (Bethesda, MD: National Institutes of Health, 2014).
26. ICMJE, 2016.
27. Cosgrove et al. "Conflicts of Interest and Disclosure in the American Psychiatric Association's Clinical Practice Guidelines," 2009.

28. L. Cosgrove, S. Krimsky, M. Vijayaraghavan, and L. Schneider, "Financial Ties Between DSM-IV Panel Members and the Pharmaceutical Industry," *Psychotherapy and Psychosomatics* 75, no. 3 (2006): 154–60.
29. Neuman et al. "Prevalence of Financial Conflicts of Interest," 2011.
30. ICMJE, 2016.
31. S. L. Norris, H. K. Holmer, L. A. Ogden, S. S. Selph, and F. Rongwei, "Conflict of Interest Disclosures for Clinical Practice Guidelines in the National Guideline Clearinghouse," *PLOS ONE* 7, no 11 (2012): 1–8.
32. *American Medical Students Association, Executive Summary, AMSA Scorecard* 2014 (Sterling, VA: American Medical Student Association, 2009).
33. IOM, 2011.
34. R. Steinbrook, 2009. "Controlling Conflict of Interest–Proposals from the Institute of Medicine," *New England Journal of Medicine* 360, no. 21 (2009): 2160–63.
35. IOM, 2011.
36. AMSA, 2014.
37. Neuman et al., "Prevalence of Financial Conflicts of Interest," 2011.
38. Shaneyfelt et al., "Reassessment of Clinical Practice Guidelines," 2009.
39. T. B. Mendelson, M. Meltzer, E. G. Campbell, A. L. Caplan, and J. N. Kirkpatrick, "Conflicts of Interest in Cardiovascular Clinical Practice Guidelines," *Archives of Internal Medicine* 171 no. 6 (2011): 577–84.
40. J. Lenzer, "Why We Can't Trust Clinical Guidelines," *BMJ* (Clinical Research Ed.) 346 (2013): f3830–f3830.
41. Neuman et al., "Prevalence of Financial Conflicts of Interest," 2011.
42. L. Cosgrove, H. J. Bursztajn, S. Krimsky, M. Anaya, and J. Walker, "Conflicts of Interest and Disclosure in the American Psychiatric Association's Clinical Practice Guidelines," *Psychotherapy and Psychosomatics* 78, no. 4 (2009): 228–32.
43. M. J. Pencina, A. M. Navar-Boggan, R. B. D'Agostino, K. Williams, B. Neely, A. D. Sniderman, and E. D. Peterson, "Application of New Cholesterol Guidelines to a Population-Based Sample," *New England Journal of Medicine* 370, no. 15 (2014): 1422–31.
44. L. Rosenbaum, "Reconnecting the Dots – Reinterpreting Industry-Physician Relations," *New England Journal of Medicine* 372 no. 19 (2015): 1860–64.
45. Krimsky, "Combating the Funding Effect in Science," 2010.
46. S. Krimsky, "Do Financial Conflicts of Interest Bias Research?: An Inquiry into the 'Funding Effect' Hypothesis," *Science, Technology & Human Values* 38 no. 4 (2013): 566–87.
47. C. Ornstein, R. G. Jones, and M. Tigas, "Now There's Proof: Docs Who Get Company Cash Tend to Prescribe More Brand-Name Meds," *ProPublica*, 2016.
48. Shaneyfelt et al., "Reassessment of Clinical Practice Guidelines," 2009.
49. Norris et al., "Conflict of Interest Disclosures for Clinical Practice Guidelines," 2012.

50. Norris et al., "Conflict of Interest Disclosures for Clinical Practice Guidelines," 2012.
51. G. H. Guyatt, S. L. Norris, S. Schulman, M. H. Jack Hirsh, E. A. Eckman, M. C. Akl, P. O. Vandvik, J. W. Eikelboom, M. S. McDonagh, S. Z. Lewis, D. D. Gutterman, D. J. Cook, and H. J. Schünemann, "Methodology for the Development of Antithrombotic Therapy and Prevention of Thrombosis Guidelines: Antithrombotic Therapy and Prevention of Thrombosis, 9th ed: American College of Chest Physicians Evidence-Based Clinical Practice Guidelines," *Chest* 141, no. 2 suppl. (2012): 53S–70S.
52. IOM, 2011.
53. J. Lenzer, "Why We Can't Trust Clinical Guidelines," *BMJ* (Clinical Research Ed.) 346 (2013): f3830–f3830.
54. J. Lenzer, J. and S. Brownlee, 2015. "Diverting Attention from Financial Conflicts of Interest," *BMJ* (Clinical Research Ed.) 350 (2015): h3505–h3505.
55. J. B. Bindslev, B. J. Schroll, P. C. Gøtzsche, and A. Lundh. 2013. "Underreporting of Conflicts of Interest in Clinical Practice Guidelines: Cross Sectional Study," *BMC Medical Ethics* 14 (2013): 19.
56. Centers for Medicare & Medicaid Services, Open Payments, 2011.
57. C. Ornstein, M. Tigas, and R. G. Jones, "Why Pharma Payments to Doctors Were So Hard to Parse," *ProPublica*, 2015. https://www.propublica.org /article/why-pharma-payments-to-doctors-were-so-hard-to-parse" (accessed October 10, 2016).
58. F. Godlee, "Conflict of Interest: Forward Not Backward," *BMJ* 350, no. 8012 (2015): h3176–h3176.
59. Shaneyfelt et al., "Reassessment of Clinical Practice Guidelines," 2009.
60. A. Lundh, S. Sismondo, J. Lexchin, O. A. Busuioc, and L. Bero, "Industry Sponsorship and Research Outcome," *The Cochrane Database of Systematic Reviews* 12 (2012): MR000033.
61. A. T. Wang, C. P. McCoy, M. H. Murad, and V. Montori, 2010. "Association Between Industry Affiliations and Position on Cardiovascular Risk with Rosiglitazone: Cross Sectional Systematic Review," *BMJ* 340 (2010): c1344.
62. D. F. Ransohoff, M. Pignone, and H. C. Sox, "How to Decide Whether a Clinical Practice Guideline is Trustworthy," *Journal of the American Medical Association* 309 no. 2 (2013): 139–40.

CHAPTER 22

1. G. Markowitz, D. Rosner, *Deceit and Denial: The Deadly Politics of Industrial Pollution*, (Oakland: University of California Press, 2002).
2. Ibid.
3. U.S. Department of Justice, Office of Public Affairs, "GlaxoSmith Kline to Plead Guilty and Pay $3 Billion to Resolve Fraud Allegations and Failure to Report Safety Data," (July 2, 2012), https ://www.justice.gov/opa/pr/glaxo smith kline-plead-guilty-and-pay-3-billi on-resolve-fraud -allegations -and-failure-report. Accessed 20 Oct 2017.
4. Plos Medicine Editors, "Ghostwriting: The Dirty Little Secret of Medical Publishing That Just Got Bigger," *PLOS Med.* (2009);6(9): e1000156.

5. B. L. Castleman, *Asbestos: Medical and Legal Aspects.* 5th ed, Frederick, MD: Aspen Publishers; (2004).
6. R. Delafontaine, *Historians as Expert Judicial Witnesses in Tobacco Litigation,* Studies in the History of Law and Justice, (New York: Springer, 2015).
7. *N. Oreskes, E. M. Conway, Merchants of Doubt: How a Handful of Scientists Obscured the Truth on Issues from Tobacco Smoke to Global Warming, (New York: Bloomsbury, 2010).*
8. D. Rosner, G. Markowitz G, M. Showkwanyun, ToxicDocs (www.ToxicDocs.org): "From History Buried in Stacks of Paper to Open, Searchable Archives Online," *J Public Health Pol.* (2018) 39:4-11.
9. S. Horel, "Browsing a Corporation's Mind," *J Public Health Pol.* (2018) 39:12-4.
10. International Agency for Research on Cancer. IARC monographs, vol. 112, "Evaluation of Five Organophosphate Insecticides and Herbicides," (March 20, 2015) http://www.iarc.fr/en/media-centre/iarcnews/pdf/Monograph Volume 112.pdf. Accessed 12 Nov 2017.
11. U.S. Right to Know website of Monsanto litigation documents. https://usrtk.org/pesticides/monsanto-glyphosate-cancer-case-key-documents-analysis/. Accessed 17 May 2018.
12. Roundup Products Liability Litigation, U.S. District Court, Northern District of California, MDL. No. 2741. Case No. 16-md-02741-VC. http://www.cand.uscourts.gov/VC/roundupmdl Accessed 20 Oct 2017.
13. Ibid.
14. www.usrtk .org is a publicly accessible Internet site.
15. S. Foucart, S. Horel, "Monsanto papers: les agencies sous l'influence de la firm," *Le Monde* April 10, 2017; May 10, 2017.
16. U.S. Right to Know website of Monsanto litigation documents.
17. https://www.huffington post.com/author/carey-gillam; http://carey-gillam.com/articles.
18. C. Gillam, Ch. 6, "Spinning the Science," In: *Whitewash.* (Washington DC: Island Press, 2017).
19. UCSF Industry Documents Library https ://www.industrydocumentslibrary.ucsf.edu/.
20. G. M. Williams, R. Kroes, I. C. Munro, "Safety Evaluation and Risk Assessment of the Herbicide Roundup and Its Active Ingredient, Glyphosate, for Humans." *Reg Tox Pharm.* (2000) 31:117-65.
21. Roundup Products Liability Litigation, U.S. District Court, Northern District of California, MDL No. 2741. Case No. 16-md-02741-VC Document 187-12 Filed 03/14/17 https ://usrtk.org/wp-content/uploads/2017/03/187series.pdf, p. 203. Accessed 12 Nov 2017.
22. Environmental Protection Agency, "Glyphosate Issue Paper: Evaluation of Carcinogenic Potential," (September 12, 2016) https://www.epa.gov/sites/production/files/2016-09/documents/glyphosate_issue_paper_evaluation_of_carcincogenic_potential.pdf. Accessed 12 Nov 2017.

23. Roundup Products Liability Litigation, document ID MONGLY01723742 https://usrtk.org/wp-content/uploads/2017/08/18-Monsanto-Scientist-Admits -to-Ghost writing-Cancer-Review-Paper.pdf. Accessed 12 Nov 2017.
24. Environmental Protection Agency. "Glyphosate Issue Paper"
25. G. M. Williams, M. Aardema, J. Acquavella, S. C. Berry, D. Brusick, M. M. Burns, J. L. V. de Camargo, D. Garabrant, H. A. Greim, L. D. Kier, D. J. Kirkland, G. Marsh, K. R. Solomon, T. Sorahan, A. Roberts, D. L. Weed, "A Review of the Carcinogenic Potential of Glyphosate by Four Independent Expert Panels and Comparison to the IARC Assessment," *Crit Rev Tox.* (2016) 46:3-20.
26. Email. Ashley Robert, Intertek to Roger McClellan, editor of *Crit Rev Tox,* (July 5, 2016). http://baumh edlun dlaw.com/pdf/monsanto-documents /0C-Challenged-Documents-Chart .pdf. Accessed 18 May 2018.
27. Email. William F. Heydens, Monsanto to Ashley Roberts, Intertek, (January 13, 2016). http://baumhedlundlaw.com/pdf/monsanto-documents/3 -Internal-Emails-Show-Monsanto-Made-Substantial-Contributions-to-Published -Expert-Panel-Manuscript.pdf. Accessed 13 Nov 2017.
28. Email. William F. Heydens, Monsanto to Ashley Roberts, Intertek, (February 9, 2016). http://baumhedlun dlaw.com/pdf/monsanto-documents /1-Monsanto-Executive-William-Heydens-Edits and-Comments-on-Expert -Consultant-Manuscript.pdf. Accessed 13 Nov 2017.
29. Email. William F. Heydens, Monsanto to consultant Larry Kier, (January 9, 2016). http://baumh edlundlaw.com/pdf/monsanto-documents/3-Internal -Emails-Show-Monsanto-Made-Substantial-Contributions-to-Published -Expert-Panel -Manuscript.pdf. Accessed 13 Nov 2017.
30. Consulting agreement between Larry D. Kier and Monsanto Company, Project Amendment Letter, (August 17, 2015). https ://usrtk.org/wp-content /uploads/2017/08/23-Email-Regarding-Monsanto-Paid-Consultant-and -Expert-Panel-of-Independent-Scientists.pdf. Accessed 13 Nov 2017.
31. Email. William F. Heydens, Monsanto to Ashley Roberts, Intertek, (February 4, 2016). http://baumhedlun dlaw.com/pdf/monsanto-documents/4 -Internal-Email-Further-Demonstrating-Heydens-Involvement-Drafting -Expert-Panel-Manuscript.pdf. Accessed 13 Nov 2017.
32. Email. Email string between Monsanto employees (September 26, 2012). http://baumhedlundlaw.com/pdf/monsanto-documents/7-Monsanto-Personnel -Discusses-Plan-Seeking-Retraction-of-Serlani-Glyphosate-Study.pdf. Accessed 13 Nov 2017.
33. Email. Email string between Eric Sachs, Monsanto and other Monsanto employees, (September 28, 2012). http://baumhedlundlaw.com/pdf/monsanto -documents/14-Monsanto-Emails-Confirming-Undis closed-Involvement -in-Successful-Retraction-of-Serlani-Study.pdf Accessed 13 Nov 2017.
34. Email. William Heydens, Monsanto to Eric S. Sachs, Monsanto. (Sept. 26, 2012). http://baumh edlundlaw.com/pdf/monsanto-documents/7-Monsanto -Personnel-Discusses-Plan-Seeking-Retraction-of-Seralani-Glyphosate -Study.pdf. Accessed 13 Nov 2017.

35. Email. Bruce Chassey, University of Illinois, Monsanto Consultant to Wallace Hayes, September 26, 2012.
36. E. Lipton, "Food Industry Enlisted Academics in G.M.O. Lobbying War, Emails Show," *New York Times* September 2, 2015. https ://www.nytimes .com/2015/09/06/us/food-industry-enlisted-academics-in-gmo-lobbying -war-emails-show.html?r=1.
37. Email. Monsanto Email chain. September 26, 2012; Business Goals of David Saltmiras, Monsanto, (August 20, 2013). http://baumhedlundlaw.com /pdf/monsanto-documents/8-Monsanto-Scientist-Admits-to-Leveraging -Relationship-with-Food-and-Chemical-Toxicology-Journal.pdf. Accessed 13 Nov 2017.
38. G. Seralini, E. Clair, R. Mesnage, S. Gress, N. Defarge, M. Malatesta, D. Hennequin, J. Spiroux de Vendomois, Republished Study: "Long-Term Toxicity of a Roundup Herbicide and a Roundup-Tolerant Genetically Modified Maize," *Environ Sci Eur.* (2014) 26(14):1-17.
39. https ://www.usrtk.org/wp-content/uploads/2016/07/GoodmanRG-00012 6-00000 1.pdf. Accessed 13 Nov 2017.
40. http://baumhedlundlaw.com/pdf/monsanto-documents/10-Monsanto -Consulting-Agreement-with-Food-and-Chemical-Toxicology-Editor.pdf. Accessed 13 Nov 2017.
41. Wallace Hayes A. Retraction notice to "Long Term Toxicity of a Roundup Herbicide and a Roundup-Tolerant Genetically Modified Maize," *Food ChemTox.* (2014) 63:244.
42. https ://publicationethics.org/files/retraction%20guidelines.pdf. Accessed 13 Nov 2017.
43. C. Gillam, "Collusion or Coincidence? Records Show EPA Efforts to Slow Herbicide Review Came in Coordination with Monsanto," *Huffington Post,* (August 18, 2017). https://www.huffington post.com/entry/collusion -or-coincidence-records-show-epa-efforts_us_5994dad4e4b056a2b0ef 02f1. Accessed 13 Nov 2017.
44. Ibid.
45. Emails. Daniel Jenkins (Monsanto) to William Heydens (Monsanto), Jennifer Listello, (Monsanto), (June 24, 2015). http://baumhedlundlaw.com/pdf /monsanto-documents/56-Email-Showing-Communications-Between -Monsanto-and-EPA-in-Furtherance-of-Avoiding-Round up-and-Glyphosate -Testing.pdf. Accessed 13 Nov 2017.
46. https ://usrtk.org/wp-content/uploads/2017/08/Text-Messages.pdf. Accessed 13 Nov 2017.
47. Gillam, "Collusion or Coincidence?"
48. Email. David Saltmiras, Monsanto to Monsanto colleagues, (August. 13, 2012) http://baumhedlundlaw.com/pdf/monsanto-documents/11-Email -Confirming-Monsantos-Intention-to-Pay-Wallace-Hayes.pdf. Accessed 13 Nov 2017.
49. U.S. House of Representatives. Minority Staff Report Prepared for Members of the Committee on Science, Space & Technology. "Spinning Science

& Silencing Scientists: A Case Study in How the Chemical Industry Attempts to Influence Science," (February 2018). https://www.baumhedlundlaw.com/pdf/monsanto-documents/Final-minority-report-glyphosate-spinning-science-silencing-scientists.pdf.

AFTERWORD

1. Ralph Nader, foreword, *Science in the Private Interest*, Sheldon Krimsky, (Lanham, MD: Rowman & Littlefield, 2003), xiv.
2. Marcia Angell, *The Truth about the Drug Companies.* (New York: Random House, 2004), 244.
3. Thomas G. Neltner, Heather M. Alger, James T. O'Reilly, Sheldon Krimsky, Lisa A. Bero, Maricel V. Maffini, "Conflicts of Interest in Approvals of Additives to Food Determined to Be Generally Recognized as Safe," *JAMA Internal Medicine* 13(5) (May 2013): 16–17.

INDEX